# THE THEORY OF SEARCH GAMES AND RENDEZVOUS

**INTERNATIONAL SERIES IN**
**OPERATIONS RESEARCH & MANAGEMENT SCIENCE**
**Frederick S. Hillier, Series Editor**
*Stanford University*

Weyant, J./ *ENERGY AND ENVIRONMENTAL POLICY MODELING*
Shanthikumar, J.G. & Sumita, U. / *APPLIED PROBABILITY AND STOCHASTIC PROCESSES*
Liu, B. & Esogbue, A.O. / *DECISION CRITERIA AND OPTIMAL INVENTORY PROCESSES*
Gal, T., Stewart, T.J., Hanne, T. / *MULTICRITERIA DECISION MAKING: Advances in MCDM Models, Algorithms, Theory, and Applications*
Fox, B.L. / *STRATEGIES FOR QUASI-MONTE CARLO*
Hall, R.W. / *HANDBOOK OF TRANSPORTATION SCIENCE*
Grassman, W.K. / *COMPUTATIONAL PROBABILITY*
Pomerol, J-C. & Barba-Romero, S. / *MULTICRITERION DECISION IN MANAGEMENT*
Axsäter, S./ *INVENTORY CONTROL*
Wolkowicz, H., Saigal, R., & Vandenberghe, L. / *HANDBOOK OF SEMI-DEFINITE PROGRAMMING: Theory, Algorithms, and Applications*
Hobbs, B.F. & Meier, P. / *ENERGY DECISIONS AND THE ENVIRONMENT: A Guide to the Use of Multicriteria Methods*
Dar-El, E. / *HUMAN LEARNING: From Learning Curves to Learning Organizations*
Armstrong, J.S. / *PRINCIPLES OF FORECASTING: A Handbook for Researchers and Practitioners*
Balsamo, S., Personé, V., & Onvural, R./ *ANALYSIS OF QUEUEING NETWORKS WITH BLOCKING*
Bouyssou, D. et al. / *EVALUATION AND DECISION MODELS: A Critical Perspective*
Hanne, T. / *INTELLIGENT STRATEGIES FOR META MULTIPLE CRITERIA DECISION MAKING*
Saaty, T. & Vargas, L. / *MODELS, METHODS, CONCEPTS and APPLICATIONS OF THE ANALYTIC HIERARCHY PROCESS*
Chatterjee, K. & Samuelson, W. / *GAME THEORY AND BUSINESS APPLICATIONS*
Hobbs, B. et al. / *THE NEXT GENERATION OF ELECTRIC POWER UNIT COMMITMENT MODELS*
Vanderbei, R.J. / *LINEAR PROGRAMMING: Foundations and Extensions, 2nd Ed.*
Kimms, A. / *MATHEMATICAL PROGRAMMING AND FINANCIAL OBJECTIVES FOR SCHEDULING PROJECTS*
Baptiste, P., Le Pape, C. & Nuijten, W. / *CONSTRAINT-BASED SCHEDULING*
Feinberg, E. & Shwartz, A. / *HANDBOOK OF MARKOV DECISION PROCESSES: Methods and Applications*
Ramík, J. & Vlach, M. / *GENERALIZED CONCAVITY IN FUZZY OPTIMIZATION AND DECISION ANALYSIS*
Song, J. & Yao, D. / *SUPPLY CHAIN STRUCTURES: Coordination, Information and Optimization*
Kozan, E. & Ohuchi, A. / *OPERATIONS RESEARCH/ MANAGEMENT SCIENCE AT WORK*
Bouyssou et al. / *AIDING DECISIONS WITH MULTIPLE CRITERIA: Essays in Honor of Bernard Roy*
Cox, Louis Anthony, Jr. / *RISK ANALYSIS: Foundations, Models and Methods*
Dror, M., L'Ecuyer, P. & Szidarovszky, F. / *MODELING UNCERTAINTY: An Examination of Stochastic Theory, Methods, and Applications*
Dokuchaev, N. / *DYNAMIC PORTFOLIO STRATEGIES: Quantitative Methods and Empirical Rules for Incomplete Information*
Sarker, R., Mohammadian, M. & Yao, X. / *EVOLUTIONARY OPTIMIZATION*
Demeulemeester, R. & Herroelen, W. /*PROJECT SCHEDULING: A Research Handbook*
Gazis, D.C. / *TRAFFIC THEORY*
Zhu, J. /*QUANTITATIVE MODELS FOR PERFORMANCE EVALUATION AND BENCHMARKING*
Ehrgott, M. & Gandibleux, X. / *MULTIPLE CRITERIA OPTIMIZATION: State of the Art Annotated Bibliographical Surveys*
Bienstock, D. / *Potential Function Methods for Approx. Solving Linear Programming Problems*
Matsatsinis, N.F. & Siskos, Y. / *INTELLIGENT SUPPORT SYSTEMS FOR MARKETING DECISIONS*

# THE THEORY OF SEARCH GAMES AND RENDEZVOUS

*by*

**Steve Alpern**
*London School of Economics, UK*

**Shmuel Gal**
*University of Haifa, Israel*

*(in alphabetical order)*

**KLUWER ACADEMIC PUBLISHERS**
**Boston / Dordrecht / London**

**Distributors for North, Central and South America:**
Kluwer Academic Publishers
101 Philip Drive
Assinippi Park
Norwell, Massachusetts 02061 USA
Telephone (781) 871-6600
Fax (781) 871-9045
E-Mail: kluwer@wkap.com

**Distributors for all other countries:**
Kluwer Academic Publishers Group
Post Office Box 322
3300 AH Dordrecht, THE NETHERLANDS
Telephone 31 786 576 000
Fax 31 786 576 254
E-mail: services@wkap.nl

Electronic Services <http://www.wkap.nl>

**Library of Congress Cataloging-in-Publication Data**

A C.I.P. Catalogue record for this book is available from the Library of Congress.

Alpern, Steve and Gal, Shmuel/ THE THEORY OF SEARCH GAMES AND RENDEZVOUS

ISBN 978-1-4419-4908-0 e-ISBN 978-0-306-48212-0



*Printed on acid-free paper.*

Printed in the United States of America.

# Contents

**Appendices**

# Preface

Search Theory is one of the original disciplines within the field of Operations Research. It deals with the problem faced by a Searcher who wishes to minimize the time required to find a hidden object, or "target." The Searcher chooses a path in the "search space" and finds the target when he is sufficiently close to it. Traditionally, the target is assumed to have no motives of its own regarding when it is found; it is simply stationary and hidden according to a known distribution (e.g., oil), or its motion is determined stochastically by known rules (e.g., a fox in a forest).

The problems dealt with in this book assume, on the contrary, that the "target" is an independent player of equal status to the Searcher, who cares about when he is found. We consider two possible motives of the target, and divide the book accordingly. Book I considers the zero-sum game that results when the target (here called the Hider) does not want to be found. Such problems have been called Search Games (with the "zero-sum" qualifier understood). Book II considers the opposite motive of the target, namely, that he wants to be found. In this case the Searcher and the Hider can be thought of as a team of agents (simply called Player I and Player II) with identical aims, and the coordination problem they jointly face is called the Rendezvous Search Problem. This division of the book according to Player II's motives can be summarized by saying that in a Search Game the second player (Hider) wishes to maximize the capture time $T$, while in a Rendezvous Problem the second player (Rendezvouser) wishes to minimize $T$. (In both cases, the first player wishes to minimize $T$.)

Of the two problems dealt with in the book, the area of Search Games (Book I) is the older. These games stem in part from the "The Princess and the Monster" games proposed by Rufus Isaacs (1965) in his well known book on Differential Games. Beginning with the first search game with mobile hider to be solved (that on the circle, by Alpern (1974), Foreman (1974), and Zelinkin (1972)), and the subsequent solutions of search games on networks and regions in space by Gal (1979), the early work on such games culminated in the classic book of Gal (1980). This work has stimulated much subsequent research in the field including applications in computer science, economics, and biology. Much of this research is covered in Book I which contains many new results on Search Games as well as the classical results, presented with simpler exposition and proofs. However there are many open questions, even some of a fairly elementary nature, which are also covered here. For an extensive introduction to the area of Search Games, see Chapter 1.

The Rendezvous Search Problem (Book II) is a more recent area of interest. It asks how quickly two (or maybe more) players can meet together at a single location, when placed in a known search region, without a common labelling of locations. Although posed informally by Alpern as early as 1976, a rigorous formulation for the continuous time version did not appear until Alpern (1995). Beginning with the early subsequent papers of Alpern and Gal (1995) and Anderson and Essegaier (1995) on rendezvous on the line, the interest in this problem has expanded to encompass many variations, including multiple player rendezvous and different forms studied by V. Baston, A. Beck, S. Fekete, S. Gal, J. V. Howard, W. S. Lim, L. Thomas, and others. Particular interest has been paid to some discrete time rendezvous models, which have a separate history going back to the original papers of Crawford and Haller (1990) on coordination games in the economics literature, and Anderson and Weber (1990) in a search theory context. Much of this work is surveyed in the paper of Alpern (2002a). An extensive introduction to the field of Rendezvous Search can be found in Chapter 10.

Although both authors have worked in the two fields of Search Games and Rendezvous Search Theory, the division of this book into two parts reflects the emphasis of their work. As such, Book I (Search Games) was mainly written by Shmuel Gal, and results in this part which are not otherwise ascribed are due to him. Similarly, Book II (Rendezvous Search) was mainly written by Steve Alpern, with unascribed results there due to him. Of course both authors take joint responsibility for this book as a whole.

We would like to put the work of this book into its historical context with respect to earlier survey articles and books on search. Search Theory is usually considered to have begun with the work of Koopman and his colleagues on "Search and Screening" (1946). (An updated edition of his book appeared in 1980.) The problem of finding the optimal distribution of effort spent in search is the main subject of the classic work of Stone (1989, 2nd ed.), "Theory of Optimal Search", which was awarded the 1975 Lanchester Prize by the Operations Research Society of America. Much of the early work on search theory surveyed by Dobbie (1968) was concerned with aspects other than optimal search trajectories, and as such is very different from our approach. The later survey of Benkoski, Monticino, and Weisinger (1991) shows how the determination of such trajectories has come to be studied more extensively. Recent books on Search Theory include those of Ahlswede and Wegener (1987), Haley and Stone (1980), Iida (1992), and Chudnovsky and Chudnovsky (1989). The first book to introduce game theoretic aspects of search problems was of course Gal (1980), but these are also considered in Ruckle (1983a) and form the basis of the recent stimulating book of Garnaev (2000). This volume is the first to cover the new field of rendezvous search theory.

# Frequently Used Notations

## Search Games

$\mu$ Lebesgue measure of the search space
$\bar{\mu}$ Minimal length of a tour that covers the search space
$\rho$ Rate of discovery of the searcher
$C(S, H)$ Cost function (the payoff to the hider)
$c(s, h)$ Expected cost
$\hat{C}(S, H) = C(S, H)/|H|$ Normalized cost function
$\hat{c}(s, h)$ Expected normalized cost
$d(Z_1, Z_2)$ Distance between $Z_1$ and $Z_2$
$D$ Diameter of the search space
$E$ Expectation
$H$ A pure hiding strategy
$\mathcal{H}$ The set of all pure hiding strategies
$h$ A mixed hiding strategy
$h_\mu$ The uniform hiding strategy
$O$ Origin (usually, the starting point of the searcher)
$Pr$ $(A)$ Probability of an event $A$
$Q$ Search space
$r, r(Z)$ Discovery (or detection) radius
$S$ A pure search strategy (a search trajectory)
$\mathcal{S}$ The set of all admissible search trajectories
$s$ A mixed search strategy
$T$ Capture time
$t$ Time parameter
$v(H), v(h)$ Value of the hiding strategy ($v(h) = \inf_s c(s, h)$)
$v(S), v(s)$ Value of the search strategy ($v(s) = \sup_h c(s, h)$)
$\bar{V}$ Minimal value obtained by a pure search strategy (the "pure value" $\bar{V} = \inf_S v(S)$)
$v$ Value of the search game ($v = \inf_s v(s) = \sup_h v(h)$)
$w$ Maximal velocity of the hider
$Z$ A point in the search space
$\lfloor . \rfloor$ Integer part

## Rendezvous

| | |
|---|---|
| $\hat{T}$ | Expected rendezvous time |
| $R^a$, $R^s$ | Player-asymmetric, player-symmetric, rendezvous values |
| $G$ | A given group of symmetries of $Q$ |
| $t_{f,g}$ | The maximum rendezvous time for $f$, $g$ |
| $H(Q, p, q)$ | The H-network based on $Q$ |
| $\mathcal{N}_e, \mathcal{N}_o$ | The sets of even and odd nodes of an H-network |
| $(var\ s)(r)$ | Total variation of function $s$ up to time $r$ |
| $Var(s)$ | Total variation of $s$ over its domain |

# Acknowledgment

The authors would like to express their gratitude to Vic Baston who did a tremendous job in looking over the manuscript for this book, correcting mistakes and misprints, and suggesting new examples. We also wish to thank Wei Shi Lim, who contributed similar work on the rendezvous portion of the book. This research was supported in part by grants from EPSRC, London Mathematical Society, STICERD, and NATO.

# Book I

# SEARCH GAMES

# Chapter 1

# Introduction to Search Games

In this book we are mainly concerned with finding an "optimal" search trajectory for detecting a target. In the *search game* part (Book I) we shall usually not assume any knowledge about the probability distribution of the target's location, using instead a minimax approach. The minimax approach can be interpreted in two ways. One could either decide that because of the lack of knowledge about the distribution of the target, the searcher would like to assure himself against the worst possible case (this worst-case analysis is common in computer science and operations research), or, as in many military situations, the target is a hider who wishes to evade the searcher as long as possible. This approach leads us to view the situation as a game between the searcher and the hider. In general, we shall consider search games of the following type. The search takes place in a set $Q$ called the *search space*. We shall distinguish between games in compact search spaces, which are considered in Part I and games in unbounded domains which are considered in Part II. The searcher usually starts moving from a specified point $O$ called the origin and is free to choose any continuous trajectory inside $Q$, subject to a maximal velocity constraint. As to the hider, in some of the problems it will be assumed that the hider is immobile and can choose only his hiding point, but we shall also consider games with a mobile hider who can choose any continuous trajectory inside $Q$. It will always be assumed that neither the searcher nor the hider has any knowledge about the movement of the other player until their distance apart is less than or equal to the discovery radius $r$, and at this very moment capture occurs.

Each search problem will be presented as a two-person zero-sum game. In order to treat a game mathematically, one must first present the set of strategies available to each of the players. These strategies will be called *pure strategies* in order to distinguish between them and probabilistic choices among them, which will be called *mixed strategies*. We shall denote the set of pure strategies of the searcher by $\mathcal{S}$ and the set of pure strategies of the hider by $\mathcal{H}$. A pure strategy $S \in \mathcal{S}$ is a continuous trajectory inside $Q$ such that $S(t)$ represents the point that is visited by the searcher at time $t$. As to the hider, we have to distinguish between two cases: If the hider is immobile, then he can choose only his hiding point $H \in Q$. On the other hand, if he is mobile, then his strategy $H$ is a continuous trajectory $H(t)$ so that, for any $t \geq 0$, $H(t)$ is the point occupied by the hider at time $t$. The next step in describing the search game is to present a cost function (the game-theoretic payoff to the maximizing hider) $c(S, H)$, where $S$ is a pure search strategy and $H$ is a pure hiding strategy. The cost $c(S, H)$ has to represent the loss of the searcher (or the effort spent in searching) if the searcher uses strategy $S$ and the hider uses strategy $H$. Since the game is assumed to be zero-sum, $c(S, H)$ also represents the gain of the hider, so that the players have opposite goals: the searcher wishes to make the cost as small as possible, while the hider wishes to make it large. A natural choice for the cost function is the time spent until the hider is captured (the *capture time*). For the case of a bounded search space $Q$, this choice presents no problems. But if $Q$ is unbounded and if no restrictions are imposed on the hider, then he can make the capture time as large as desired by choosing points that are very far from the origin. We overcome that difficulty either by imposing a restriction on the expected distance of the hiding point from the origin or by normalizing the cost function. The details concerning the choice of a cost function for unbounded search spaces are presented in Chapter 6. Given the available pure strategies and the cost function $c(S, H)$, the value $v(S)$ guaranteed by a pure search strategy $S$ is defined as the maximal cost that could be paid by the searcher if he uses the strategy $S$; thus,

$$v(S) = \sup_{H \in \mathcal{H}} c(S, H). \tag{1.1}$$

We define the *minimax value of the game* $\bar{V}$ as

$$\bar{V} = \inf_{S \in \mathcal{S}} v(S). \tag{1.2}$$

Then for any $\varepsilon > 0$, the searcher can find a pure strategy $S$, which guarantees that the loss will not exceed $(1 + \varepsilon)\bar{V}$. A pure strategy $S_\varepsilon$ that satisfies

$$v(S_\varepsilon) < (1 + \varepsilon)\bar{V}$$

will be called an *ε-minimax search trajectory*. If there exists a pure strategy $\bar{S}$ which satisfies

$$v(\bar{S}) = \bar{V}, \tag{1.3}$$

then $\bar{S}$ will be called a *minimax search trajectory*.

The value $\bar{V}$ represents the minimal capture time that can be guaranteed by the searcher if he uses a fixed trajectory, but in all the interesting search games the searcher can do better on the average if he uses random choices out of his pure strategies. These choices are called *mixed strategies* (see Appendix A).

If the players use mixed strategies, then the capture time is a random variable, so that each player cannot guarantee a fixed cost but only an expected cost. Obviously, any pure strategy can be looked on as a mixed strategy with degenerate probability distribution concentrated at that particular pure strategy, so that the pure strategies are included in the set of mixed strategies. A mixed strategy of the searcher will be denoted by $s$ and a mixed strategy of the hider will be denoted by $h$. The expected cost of using the mixed strategies $s$ and $h$ will be denoted by $c(s, h)$. We will use the notation $v(.)$ to denote the expected cost guaranteed by a player (either the searcher or the hider) if he uses a specific strategy. Thus, $v(s)$ is the maximal expected cost of using a search strategy $s$:

$$v(s) = \sup_h c(s, h) = \sup_{H \in \mathcal{H}} c(s, H). \tag{1.4}$$

$v(s)$ will be called the *value of strategy* $s$. Similarly, the minimal expected cost $v(h)$ of using a hiding strategy $h$

$$v(h) = \inf_s c(s, h) = \inf_{S \in \mathcal{S}} c(S, h) \tag{1.5}$$

will be called the *value of strategy* $h$. It is obvious that for any $s$ and $h$, $v(s) \geq v(h)$, because $v(s) \geq c(s, h) \geq v(h)$.

If there exists a real number $v$ that satisfies

$$v = \inf_s v(s) = \sup_h v(h), \tag{1.6}$$

then we say that the game has a value $v$. In this case, for any $\varepsilon > 0$, there exist a search strategy $s_\varepsilon$ and a hiding strategy $h_\varepsilon$ that satisfy

$$v(s_\varepsilon) < (1 + \varepsilon)v \quad \text{and} \quad v(h_\varepsilon) > (1 - \varepsilon)v. \tag{1.7}$$

Such strategies will be called $\varepsilon$-optimal strategies. In the case that there exists $\bar{s}$ (resp. $\bar{h}$) such that $v(\bar{s}) = v$ (resp. $v(\bar{h}) = v$), then $\bar{s}$ (resp. $\bar{h}$) is called an *optimal strategy*.

The reader will have noticed that to avoid more cumbersome notation we have made a rather versatile use of the letter $v$. Its meaning will depend on the context: without an argument, it denotes the value, as defined in (1.6). When its argument is a search strategy, it is defined by equation (1.4); when a hiding strategy, by (1.5).

In general, if the sets of pure strategies of both players are infinite, then the game need not have a value (for details, see Luce and Raiffa, 1957, Appendix 7). However, in Appendix A we shall show that any search game of the type already described has a value and an optimal search strategy. (The hider need not have an optimal strategy and in some games he has only $\varepsilon$-optimal strategies.)

Keeping the previous framework in mind, we present a general description of the material covered in Book I (Parts I and II). This book contains many new results on search games, as well as the classical results presented with simpler exposition and proofs. It also contains many open problems, even some of elementary nature, which hopefully will stimulate further research.

In Part I we consider search games in compact spaces within the framework presented in Chapter 2. In Chapter 3 we analyze search games with an immobile hider in networks and in multidimensional regions. Among the topics considered we investigate the performance of the following natural search strategy: Find a minimal closed curve $L$ that covers all the search space $Q$. ($L$ is called a *Chinese postman tour*.) Then, encircle $L$ with probability 1/2 for each direction. This *random Chinese postman tour* is indeed an optimal search strategy for Eulerian networks and for trees, or if $Q$ is a two-dimensional region. An intriguing problem is to characterize the family of graphs for which the optimality property of the random Chinese postman holds. The solution, found recently by Gal (2000), is presented. The difficulties associated with solving search games for networks outside of the above family are presented. We also present a dynamic programming algorithm for numerically finding an optimal search trajectory against a known hiding strategy. This algorithm can sometimes help us to solve search games that are difficult to handle analytically.

Problems with a mobile hider are usually more difficult. (This statement, however, is not always true. For example, the solution of the search game on three arcs is easy for a mobile hider but very difficult for an immobile hider.) Search games with a mobile hider are analyzed in Chapter 4. We first present the solutions for the search game on the circle and on $k$ unit arcs connecting two points. We also present several new results on networks that can be relatively quickly searched, including the figure eight network. The $k$ arcs game can serve as a useful introduction for the Princess and Monster game in two (or more) dimensional regions, analyzed later in this chapter.

In Chapter 5 we consider four types of search games in compact spaces, which do not fall into the framework of Chapters 3 and 4. We present in detail new results for searching in a maze (i.e., a network with an unknown structure) and "high–low" search in which the searcher gains a directional information in each observation. Then we survey the problems of searching for an infiltrator who would like to reach a sensitive zone and searching in discrete locations.

In Part II we consider search games in unbounded domains. The general framework of such problems is described in Chapter 6. We introduce the normalized cost function (called the *competitive ratio* in Computer Science literature). We show that solving a search game using a normalized cost function is usually equivalent to restricting the absolute moment of the hiding strategy by an upper bound.

In Chapter 7 we develop a general tool for obtaining minimax trajectories for problems involving homogeneous unimodal functionals. We show that the minimax trajectory is a geometric sequence. This enables us to easily find it by minimizing over a single parameter (the generator of this sequence) instead of searching over the whole trajectory space. The results obtained in Chapter 7 are used in Chapters 8 and 9 but the proofs of the theorems are mainly for experts and can be skipped at first reading.

The *linear search problem (LSP)*, i.e., finding a target with a known distribution on the line, has been attracting much attention over several decades. This problem was analyzed as a search game by Beck and Newman (1970) and by Gal (1980). In Chapter 8 we present the above classical results along with several variants. In addition we present a new model of the linear search game when changing the direction of motion requires some time and cannot be done instantaneously (as originally assumed in the LSP). We also present a new dynamic programming algorithm for computing, with any desired accuracy, the optimal search trajectory of the LSP for any known hiding distribution.

In Chapter 9, the last chapter of Book I, we use the tools developed in Chapter 7 to solve several search games. At first we find a minimax trajectory for searching a set of rays. This problem has recently attracted a considerable attention in computer science literature. We then present some new results for the minimax search trajectory on the boundary of a region in the plane. Then we analyze the minimax search trajectory for a point in the plane. We also discuss several classical and new "swimming in the fog" problems in which we have to find a minimax trajectory to reach a shoreline of a known shape, starting from an unknown initial point. We then conclude by presenting an open problem of searching for a submarine with a known initial location.

# Part One

# Search Games in Compact Spaces

# Chapter 2

# General Framework

The search spaces considered in Part I are closed and bounded subsets of a Euclidean space. They are usually either a compact region (i.e., the closure of a connected bounded open set) in a Euclidean space with two or more dimensions, or a network. In this book a *network* will mean a finite connected set of arcs, called *edges*, which can intersect only at their endpoints, called *nodes*. Examples of such networks are a circle, a tree, a set of $k$ arcs connecting two points, etc. Obviously, if a graph is given in the combinatorial form of nodes and edges, then it can be embedded in a three-dimensional Euclidean space $R^3$ in such a way that the edges intersect only at nodes of the network. (Two dimensions are not sufficient for nonplanar networks.) Thus, we shall look upon each Network as a subset of $R^3$. Each arc in the Network has a given length and an associated distance function defined on it.

**Definition 2.1** *The distance $d(x, y)$ between any two points $x$ and $y$ in a network $Q$ is defined as the minimum length among all the paths that connect $x$ and $y$ within $Q$. The diameter $D$ of $Q$ is defined as the maximum distance between two of its points, that is,* $D = \max_{x,y \in Q} d(x, y)$.

We now describe more specifically a search game $\Lambda$ in the space $Q$, with the outline given in Chapter 1. A pure search strategy $S$ is a continuous trajectory inside $Q$ that does not exceed a fixed maximal velocity. The time unit will be chosen so as to normalize this maximal velocity to 1. Such a trajectory $S(t)$ is a continuous mapping $S : [0, \infty) \to Q$ satisfying

$$d(S(t_1), S(t_2)) \le t_2 - t_1 \quad \text{for any } t_2 > t_1 \ge 0.$$

We shall usually assume that the searcher has to start from a fixed point $O$ to be called the *origin* (i.e., $S(0) = O$), but we shall sometimes consider other possibilities such as a chosen or a random starting point. The set of all pure search strategies is denoted by $\mathcal{S}$. A pure hiding strategy $H$ is an arbitrary continuous trajectory inside $Q$ with maximal velocity not exceeding a given maximum hider velocity $w$. In the case $w = 0$, the hider is immobile and $H$ is a single point, while if $w > 0$, then the hider is mobile and $H$ is a trajectory that satisfies

$$d(H(t_1), H(t_2)) \leq w \times (t_2 - t_1) \quad \text{for all } t_2 > t_1 \geq 0.$$

The case of a mobile hider also includes the possibility of $w = \infty$, i.e., a hider, moving along a continuous trajectory, with an unbounded velocity. The set of all pure hiding strategies is denoted by $\mathcal{H}$.

We assume that the searcher and the hider cannot see one another until their distance is less than or equal to the discovery (or detection) radius $r$ and at that very moment capture occurs (and the game terminates). In cases where $Q$ is a network, then (for convenience) $r$ will be taken as zero. (Actually, $r$ can usually be chosen as a small positive number without introducing any significant changes in the results.) If $Q$ is a multidimensional region, then it will be assumed that $r$ is very small in comparison with the magnitude of $Q$. (To be more precise, we will assume that $\gamma r \ll \mu$, where $\mu$ and $\gamma$ are, respectively, the Lebesgue measure, of appropriate dimension, of $Q$ and the boundary of $Q$.) In order to simplify the presentation of the results, we shall generally consider the case in which both the maximal velocity of the searcher and the radius of detection are constants. However, we shall also extend the results to the case where the maximal velocity of the searcher depends on his location and the radius of detection depends on the location of the hider. We will call such a case an *inhomogeneous search space*.

Whenever the search space $Q$ is a Network or a subset of Euclidean space, it is endowed with Lebesgue measure of the appropriate dimension (corresponding to length, area, volume, etc.). To avoid a separate notation for the total measure of $Q$, we make the following simplifying definition.

**Definition 2.2** *The Lebesgue measure of any measurable subset $B$ of $Q$ is denoted by $\mu(B)$. The total measure of $Q$ is denoted by $\mu = \mu(Q)$.*

The set of points of $Q$ which have been "searched" by a trajectory $S$ by time $t$ is denoted by $X_S(t)$. That is,

$$X_S(t) = \{x \in Q : d(S(t'), x) \leq r \quad \text{for some } 0 \leq t' \leq t\}. \tag{2.1}$$

Obviously, the set that is discovered at time 0 does not depend on $S$. Its measure will be denoted by $\mu_0$, i.e., $\mu_0 = \mu(X_S(0))$.

The following notion describes the maximum rate at which new points of $Q$ can be discovered.

**Definition 2.3** *The maximal discovery rate $\rho$ of the searcher is defined as,*

$$\rho = \sup_{S, t > 0} \frac{\mu(X_S(t)) - \mu_0}{t}.$$

Since the maximal velocity of the searcher is 1 it follows that $\rho = 1$ for search in a network. In case that $Q$ is a two-dimensional region, the sweep width is $2r$, so that the maximal area $\rho$ of the strip that can be swept in one unit of time is $2r$. By a similar reasoning, $\rho$ is equal to $\pi r^2$ for three-dimensional regions, and so on.

The *capture time*, which is denoted by $c(S, H)$ (and sometimes by $T$) represents the loss of the searcher (and the gain of the hider). It is formally defined as

$$c(S, H) = \min\{t : d(S(t), H(t)) \leq r\}.$$

If no such $t$ exists, then we say $c(S, H) = \infty$.

A mixed strategy $s$ (resp. $h$) of the searcher (resp. hider) is a probability measure on $\mathcal{S}$ (resp. $\mathcal{H}$). In order to rigorously present such strategies, one has to introduce a substantial amount of measure-theoretic machinery for $\mathcal{S}$ and $\mathcal{H}$. Such a construction is briefly presented in Appendix A. In Gal (1980) full details are presented, including the result that $c(S, H)$ is Borel measurable in both variables, so that we can define the payoff $c$, in the case that the searcher uses $s$ and the hider uses $h$, as the expected value of $c$ with respect to the product measure $s \times h$:

$$c(s, h) = \int C(S, H)\, d(s \times h). \tag{2.2}$$

The fundamental results (see Appendix A) are that any search game as described above has a value $v$, i.e.,

$$\inf_s \sup_h c(s, h) = \sup_h \inf_s c(s, h) \tag{2.3}$$

and that the searcher always has an optimal strategy. Thus, for any such search game, the searcher can always guarantee an expected payoff not exceeding $v$, while the hider can guarantee that the expected payoff exceeds $(1 - \varepsilon)v$.

For the search games presented in this book, we shall generally use constructive methods to find the value and the optimal (or $\varepsilon$-optimal) strategies of the players. In the case of a network, whenever we can obtain a solution of the game, it will be an exact solution. On the other hand, the solutions that we get for the search games in multidimensional regions depend on the fact that the detection radius $r$ is small. In this case, we shall present two strategies $\bar{s}_\varepsilon$ and $\bar{h}_\varepsilon$ and a function $f(r)$ which satisfy (see (1.7))

$$v(\bar{s}_\varepsilon) < (1 + \varepsilon) f(r) \quad \text{and} \quad v(\bar{h}_\varepsilon) > (1 - \varepsilon) f(r), \quad \text{where } \varepsilon \to 0 \text{ as } r \to 0.$$

Thus, $\bar{s}_\varepsilon$ and $\bar{h}_\varepsilon$ are $\varepsilon$-optimal strategies and $v \sim f(r)$ for small $r$.

In calculating the expected capture time of the search games to be considered, we shall often use the following result, which is well known in probability theory (see, e.g., Feller, 1971, p. 150).

**Proposition 2.4** *The expected value $E(T)$ of a nonnegative random variable $T$ satisfies*

$$E(T) = \int_0^\infty Pr(T > t)\, dt. \tag{2.4}$$

Since $Pr(T > t)$ is monotonic nonincreasing in $t$, it follows from (2.4) that for any positive number $\beta$

$$E(T) = \sum_{i=0}^\infty \int_{i\beta}^{(i+1)\beta} Pr(T > t)\, dt \leq \beta \sum_{i=0}^\infty Pr(T > i\beta). \tag{2.5}$$

Similarly,

$$E(T) \geq \beta \sum_{i=0}^{\infty} Pr(T > (i+1)\beta) = \beta \sum_{i=1}^{\infty} Pr(T > i\beta). \tag{2.6}$$

We now present a simple but useful result known as the *scaling lemma*, which will enable us to normalize the arc lengths in some networks and will also be used for search games in unbounded domains. It actually states that changing the unit length in $Q$ affects the search game in a very simple manner.

**Proposition 2.5 (Scaling Lemma)** *Let $\Lambda$ be a search game in a set $Q$ with an origin $O$ and a detection radius $r$. Assume that the value of $\Lambda$ is $v$ and that $s, h$ are optimal ($\varepsilon$-optimal) strategies. Consider a set $\hat{Q}$ with a metric $\hat{d}$, which is obtained from $Q$ by an onto mapping $\Phi : Q \to \hat{Q}$ with the following property for some $\alpha > 0$:*

$$\hat{d}(\Phi(x), \Phi(y)) = \alpha\, d(x, y), \quad \textit{for all } x, y \in Q.$$

*Define a search game $\hat{\Lambda}$ in $\hat{Q}$ with an origin $\hat{O} = \Phi(O)$, a detection radius $\hat{r} = \alpha r$, and the same maximal velocities for the searcher and the hider as in $\Lambda$. Then the value $\hat{v}$ of $\hat{\Lambda}$ satisfies $\hat{v} = \alpha v$ and the optimal ($\varepsilon$-optimal) strategies of $\hat{\Lambda}$ are obtained by applying the mapping $\Phi$ to the trajectories in $Q$ and changing the time scale by a factor of $\alpha$.*

The proof is based on the simple observation that for any pair of trajectories $S$ and $H$, in $Q$, the capture time corresponding to $\Phi(S)$ and $\Phi(H)$, in $\hat{Q}$, would be multiplied by $\alpha$. A formal proof is given in Gal (1980).

An identical argument shows that an analogous result holds for the rendezvous search problems discussed in Book II.

# Chapter 3

# Search for an Immobile Hider

## 3.1 Introduction

In this chapter, we consider search games in compact spaces with an immobile hider. In this case, a pure hiding strategy $H$ is simply a point in the search space $Q$, and a mixed hiding strategy $h$ is a probability measure on $Q$. A pure search strategy $S$ is a continuous trajectory in $Q$, starting at the origin $O$, with maximal velocity not exceeding 1. Since the hider is immobile, it can be assumed that the searcher will always use his maximal velocity because any trajectory that does not use the maximal velocity is dominated by a trajectory that uses the maximal velocity along the same path. A mixed search strategy $s$ is a probability measure on the set $\mathcal{S}$ of these pure strategies.

A hiding strategy that plays an important role in some of the games to be presented is the uniform strategy $h_\mu$, which chooses the hiding point in $Q$ "completely randomly."[1] More precisely:

**Definition 3.1** *The uniform strategy $h_\mu$ is a random choice of the hiding point $H$ such that for all measurable sets $B \in Q$,*

$$Pr(H \in B \mid h_\mu) = \mu(B)/\mu.$$

Recall (see Definition 2.2) that the use of $\mu$ without an argument means that the argument is $Q$. That is, $\mu \equiv \mu(Q)$.

Note that it makes more sense for the hider not to hide within distance $r$ from $O$ using the uniform distribution on the rest of $Q$. However, since $r$ is either 0 or very small with respect to the magnitude of $Q$, we will not use this $\varepsilon$-improvement.

Our next result shows that if the hider chooses his hiding point according to the uniform strategy $h_\mu$, he ensures an expected evasion (capture) time of at least $\mu/2\rho$, where $\rho$ is the searcher's maximal discovery rate, as introduced earlier in Definition 2.3. This result holds not only for search strategies in $\mathcal{S}$ (continuous search paths) but even for the following larger class of generalized search strategies.

[1] Using normalized Lebesgue measure on $Q$.

**Definition 3.2** *A generalized search strategy is defined by the sets $X(t) \subset Q$ that it has "discovered" by time $t$. The sets $X(t)$ are only required to satisfy the conditions*

$$X(t') \subset X(t) \quad \text{for } t' < t, \quad \text{and} \quad \mu(X(t) - X(0)) \leq \rho t.$$

*In particular, every continuous strategy $S \in \mathcal{S}$ defines a generalized strategy by the formula (2.1).*

(Note that we usually restrict the searcher to move along a continuous trajectory and so do not allow generalized search strategies. These strategies will be discussed only in this section in order to introduce the *unrestricted search game*, which will be solved in Theorem 3.7.)

**Theorem 3.3** *If the hider chooses his hiding point according to the uniform strategy $h_\mu$, then he ensures an expected capture time of at least $\mu/2\rho$ against any generalized search strategy and in particular against any trajectory $S \in \mathcal{S}$.*

**Proof.** Let $\mu_0 = \mu(X(0))$ denote the measure of the set of points discovered at time 0. (In the case that $r = 0$, which we shall generally assume for networks, we have $\mu_0 = 0$; for the multidimensional spaces $r$ is very small so that $\mu_0 \ll \mu$.) In any case it follows from the definition of a generalized strategy that

$$\mu(X(t)) = \mu(X(0)) + [\mu(X(t) - X(0))] \leq \mu_0 + \rho t,$$

or simply

$$\mu(X(t)) \leq \rho t, \quad \text{in the case } r = 0 \text{ and } \mu_0 = 0.$$

Consequently, for $r = 0$, the probability that a hider hidden according to the distribution $h_\mu$ has been found by time $t$ is given by

$$Pr(T \leq t) \leq \min\left[\frac{\rho t}{\mu}, 1\right]$$

and hence by (2.4) we have

$$c(S, h_\mu) = E(T) \geq \int_0^\infty \max\left[1 - \frac{\rho t}{\mu}, 0\right] dt$$

$$= \int_0^{\mu/\rho} \left(1 - \frac{\rho t}{\mu}\right) dt = \frac{\mu}{2\rho}.$$

(If we do not assume that $\mu_0 = 0$, the same analysis gives the slightly more complicated estimate $c(S, h_\mu) \geq \left(\mu^2 - 2\mu\mu_0 + \mu_0^2\right)/2\mu\rho \sim \mu/2\rho$.) ∎

The following result is an immediate consequence of the considerations used in the above theorem.

**Corollary 3.4** *If $S$ satisfies $c(S, h_\mu) = \mu/2\rho$, then for all $0 < t < \mu/\rho$ the measure of the points swept by $S$ in the time interval $(0, t]$ is equal to $\rho t$ (i.e., $S$ sweeps without overlapping).*

Since the hiding strategy $h_\mu$ guarantees $\mu/2\rho$ against *any* starting point of the searcher, we also have the following.

**Corollary 3.5** *Let $\Lambda$ be a search game with value $v = \mu/2\rho$ and $\Lambda'$ be the search game obtained from $\Lambda$ by allowing the searcher to choose his starting point. Then the value of $\Lambda'$ is also $\mu/2\rho$.*

The extension of the preceding discussion to search games with more than one searcher is presented in the following result.

**Corollary 3.6** *Consider a search game with one immobile hider and $J$ searchers, with the $j$-th searcher having a maximal velocity $V_j$. Assume that all the searchers cooperate in order to discover the hider (by at least one of them) as soon as possible. Let $\rho_j$ be the Lebesgue measure of a set, which can be swept by the $j$-th searcher in one unit of time, and define the total rate of discovery $\hat{\rho} = \sum_{j=1}^{J} \rho_j$. Then Theorem 3.3 holds for this game, with $\hat{\rho}$ replacing $\rho$.*

**Proof.** Let $X(t)$ denote the set of all points discovered by at least one of the $J$ searchers by time $t$. Then since $X(t)$ is easily seen to be a generalized strategy with respect to the parameter $\hat{\rho}$, Theorem 3.3 applies to this game as claimed. ■

Note that since the uniform strategy $h_\mu$ is always available to the hider, Theorem 3.3 shows that $\mu/2\rho$ is a lower bound for the value of any search game, even if generalized search strategies are allowed. In fact, the following result of the authors shows that if we allow generalized strategies (and mixtures of them), $\mu/2\rho$ is always the value of the resulting "unrestricted game." Note that for all the search spaces $Q$, which we will consider in this book, the measure space $(Q, \mu)$ has the following properties: there are no atoms, and any subset of a measure zero set is measurable. Such a measure space is called a *Lebesgue space*.

**Theorem 3.7** *The value of the unrestricted search game on any Lebesgue space $Q$ is given by $\mu/2\rho$, where $\mu$ denotes the total measure $\mu(Q)$.*

**Proof.** According to the Theorem 3.3, we need only present a generalized search strategy that finds any hiding point in expected time not exceeding $\mu/2\rho$. A simple construction is to find any generalized search strategy $X(t)$ that sweeps without overlapping during the time interval $[0, \mu/\rho]$ and define $\tilde{X}(t)$ as the "reverse" of $X(t)$ (i.e., any point first covered by $X$ at time $t$ is first covered by $\tilde{X}$ at time $t - \mu/\rho$). Then the generalized (mixed) strategy that adopts $X$ and $\tilde{X}$ equiprobably, discovers any $H \in Q$ in expected time $\mu/2\rho$.

For readers who are familiar with measure theory we present a formal proof of the theorem as follows. For any Lebesgue space $(Q, \mu)$ there exists an invertible bi-measurable map $\phi : [0, \mu(Q)] \to Q$, which takes one-dimensional Lebesgue measure $\mu_1$ into the measure $\mu$ (that is, $\mu(\phi([a, b])) = b - a$, for $0 \le a \le b \le \mu(Q)$) (see Halmos, 1950). Define two generalized strategies $X$ and $\tilde{X}$ by $X(t) = \phi([0, \rho t])$ and $\tilde{X}(t) = \phi([\mu(Q) - \rho t, \mu(Q)])$, for $0 \le t \le \mu(Q)/\rho$. For any hiding point $H \in Q$, choose $t = \phi^{-1}(H)$ and observe that $H \in X(t/\rho) \cap \tilde{X}((\mu(Q)/\rho) - t/\rho)$. Consequently, if the searcher adopts $X$ and $\tilde{X}$ equiprobably, then any point $H$ will have been discovered in expected time $\frac{1}{2}(t/\rho + ((\mu(Q)/\rho) - t/\rho)) = \mu(Q)/2\rho$, as claimed. ■

The unrestricted game is similar to a discrete search game in which $Q$ consists of $n$ cells of equal size. We now formulate and solve a more general discrete version of the search game. In the game to be considered, $Q$ consists of $n$ cells of sizes $\mu_1, \ldots, \mu_n$ and the measure of $Q$ is defined as $\mu = \sum \mu_i$. It is assumed that the maximal rate of discovery of the searcher is $\rho$, so that it takes him $\mu_i/\rho$ units of time to look at cell number $i$. It is also assumed that if the hider is located in cell $i$ and if the searcher starts to look in this cell at time $t$, then the hider is discovered at time $t + \mu_i/2\rho$. A pure hiding strategy $H$ is an element of the set $\{1, 2, \ldots, n\}$, while a pure search strategy $s$ is a permutation $(i_1, \ldots, i_n)$ of the numbers $(1, 2, \ldots, n)$. We now show that the result $v = \mu/2\rho$ also holds for this discrete version.

**Proposition 3.8** *The value of the discrete search game is $\mu/2\rho$. An optimal hiding strategy $\bar{h}$ assigns a probability of $\mu_i/\mu$ to each cell, and an optimal search strategy $\bar{s}$ is to choose any permutation $(i_1, \ldots, i_n)$ and to assign a probability $1/2$ to this permutation and a probability $1/2$ to its "reverse" $(i_n, \ldots, i_1)$.*

**Proof.** For any permutation $S = (i_1, \ldots, i_n)$, the strategy $\bar{h}$ satisfies

$$\begin{aligned} c(S, \bar{h}) &= \frac{1}{\rho} \sum_{j=1}^{n} \left[ \left( \sum_{m=1}^{j-1} \mu_{i_m} \right) + \frac{\mu_{i_j}}{2} \right] \frac{\mu_{i_j}}{\mu} \\ &= \frac{1}{\mu\rho} \left[ \sum_{1 \le m < j \le n} \mu_{i_m} \mu_{i_j} + \frac{1}{2} \sum_{j=1}^{n} \mu_{i_j}^2 \right] \\ &= \frac{1}{2\mu\rho} \left( \sum_{i=1}^{n} \mu_i \right)^2 = \frac{\mu}{2\rho}. \end{aligned}$$

For all $H \in \{1, \ldots, n\}$, the mixed strategy $\bar{s}$ satisfies

$$\begin{aligned} c(\bar{s}, H) &= \tfrac{1}{2}(C((i_1, \ldots, i_n), H) + C((i_n, \ldots, i_1), H)) \\ &= \frac{\mu}{2\rho}. \end{aligned}$$

Thus $v(\bar{h}) = v(\bar{s}) = v = \mu/2\rho$. ■

A description of some other discrete search games is given in Chapter 5.

The expression $\mu/2\rho$ can be looked upon as the value that is obtained if the searcher is able to carry out his search with maximal efficiency. The games considered in this book are obviously restricted by the fact that the searcher has to move along a continuous trajectory so that the value does depend on the structure of $Q$. We shall have cases, such as Eulerian networks (Section 3.2), in which the searcher can perform the search with "maximal efficiency" which assures him a value of $\mu/2\rho$. A similar result holds for search in two-dimensional regions with a small detection radius (Section 3.7), and in this case the searcher can keep the expected capture time below $(1 + \varepsilon)\mu/2\rho$. On the other hand, in the case of a non-Eulerian network, we shall prove that the value is greater than $\mu/2\rho$ and that the maximal value is $\mu/\rho$ (Section 3.2). This value is obtained in the case that $Q$ is a tree (Section 3.3). A more general family that contains both the Eulerian networks and the trees as subfamilies is the *weakly Eulerian* networks (see Definition 3.24) for which the optimal search strategy has a simple structure similar to that for Eulerian networks and the trees (Section 3.4). We shall also demonstrate the complications encountered in finding optimal strategies in the case that $Q$ is not weakly Eulerian, even if the network simply consists of three unit arc connecting two points (Section 3.5). *Dynamic programming* is sometimes an effective technique for numerically computing an optimal search trajectory against a given hiding strategy (Section 3.6). For example, this technique can be used to numerically verify the optimality of rather complex strategies for the three-arcs search game.

## 3.2 Search in a Network

In our discussion, "network" will mean any finite connected set of arcs that intersect only at their end points which we call *nodes* of $Q$. Thus, $Q$ can be represented by a set in a three-dimensional[2] Euclidean space with nodes consisting of all points of $Q$ with degree $\neq 2$ plus, possibly, a finite number of points with degree 2. (As usual, the *degree* of a node is defined as the number of arcs incident at that node.) Note that we allow more than one arc to connect the same pair of nodes. The sum of the lengths of the arcs in $Q$ will be denoted by $\mu$ and called either the *total length* or the *measure*.

In studying search trajectories in $Q$, we shall often use the term *closed trajectory*, defined as follows.

**Definition 3.9** *A trajectory $S(t)$ defined for $0 \leq t \leq \tau$ is called "closed" if $S(0) = S(\tau)$. (Note that a closed trajectory may cut itself and may even go through some of the arcs more than once.) If a closed trajectory visits all the points of $Q$, then it is called a tour.*

We now consider a family of networks that lend themselves to a simple solution of the search game. These are the Eulerian networks defined as follows.

**Definition 3.10** *A network $Q$ is called Eulerian if there exists a tour $L$ with length $\mu$, in which case the tour $L$ will be called an Eulerian tour. A trajectory $S(t), 0 \leq t \leq \mu$, which covers all the points of $Q$ in time $\mu$, will be called an Eulerian path. (Such a path need not be closed.)*

It is well known that $Q$ is Eulerian if and only if the degree of every node is even and that it has an Eulerian path starting at $O$ if and only if the only nodes of odd degree are $O$ and another node $A$. In this case every Eulerian path starting at $O$ must end at $A$ (see Harary, 1972).

Since the maximal rate of discovery $\rho$ in networks is 1, it follows from Theorem 3.3 that $\mu/2$ is a lower bound for the value of the search game in any network. We now show that this bound is attained if and only if $Q$ is Eulerian.

**Theorem 3.11** *The value of the search game for an immobile hider on a network $Q$ is equal to $\mu/2$ (half the total length of $Q$) if and only if $Q$ is Eulerian.*

[2] Two dimensions are sufficient for planar networks.

**Proof.** First suppose that $Q$ is Eulerian and that $L$ is an Eulerian tour. Define $\tilde{L}$ as the reverse path given by $\tilde{L}(t) = L(\mu - t)$ and define the mixed strategy $\hat{s}$ to pick $L$ and $\tilde{L}$ equiprobably. For any hiding point $H$ in $Q$, there is at least one $t_0 < \mu$ with $L(t_0) = H$ and hence $\tilde{L}(\mu - t_0) = L(t_0) = H$. Consequently $c(\hat{s}, H) \leq (1/2)(t_0 + (\mu - t_0)) = \mu/2$. On the other hand, it follows from Theorem 3.3 that if the hider uses the uniform strategy $h_\mu$, we have $c(S, h_\mu) \geq \mu/2$ for any pure search strategy $S$. So if $Q$ is Eulerian, the value is half its total length.

Suppose now that $v = \mu/2$ and $Q$ is not Eulerian. By the first assumption, Corollary 3.4 (with $t = \mu$ and $\rho = 1$) says that any optimal strategy $\bar{s}$ must be supported by pure strategies $S$ for which

$$\mu\{S(x) : 0 \leq x \leq \mu\} = \mu, \text{ and hence } \{S(x) : 0 \leq x \leq \mu\} = Q.$$

In other words, $S$ must be an Eulerian path (not tour) starting at $O$. Consequently there is a unique node $A$ of $Q$ with odd degree such that every Eulerian path ends at $A$. We will construct a small modification $h'$ of the uniform hider distribution $h_\mu$ such that for every Eulerian path $S$ we have

$$c(S, h') > c(S, h_\mu) = \mu/2, \tag{3.1}$$

and hence

$$c(\bar{s}, h') > c(\bar{s}, h_\mu) = \mu/2,$$

which contradicts our optimality assumption for $\bar{s}$.

To construct $h'$, let $\alpha > 0$ be the minimum length of the arcs incident at $A$. Define the mixed strategy $h'$ by first using $h_\mu$ and then simply moving any hider $H$ with $\alpha/2 < d(H, A) < \alpha$ to the point $H'$ on the same arc with

$$d(H', A) = d(H, A) - \alpha/2.$$

That is, we move such hiders $H$ a distance $\alpha/2$ closer to $A$. For any Eulerian path $S$ and for $H$ as above, we have $c(S, H') - c(S, H) = \delta \times \alpha/2$, where $\delta = 1$ if $S$ traverses the arc containing $H$ toward $A$ and $\delta = -1$ if this arc is traversed away from $A$. Since any Eulerian path $S$ traverses one more of the arcs incident at $A$ *toward* $A$ than *away from* $A$, we have

$$c(S, h') > c(S, h_\mu),$$

completing the proof by establishing the required inequality (3.1). ■

**Corollary 3.12** *For an Eulerian network, the optimal strategies and the value of the search game remain the same if we remove the usual restriction $S(0) = O$ and instead allow the searcher to choose his starting point.*

The claim of the corollary is an immediate consequence of Corollary 3.5.

**Remark 3.13** *Corollary 3.12 does not hold for non-Eulerian networks because (unlike the Eulerian case) the optimal hiding strategy usually depends on the starting point of the searcher. In general, if we allow the searcher to choose an arbitrary starting* *point, then the value of this game is $\mu/2$ if and only if there exists an Eulerian path (not necessarily closed) in $Q$. (If there exists such a path, $L$, then the searcher can keep the expected capture time $\leq \mu/2$ by an analogous strategy to $\hat{s}$ of Theorem 3.11, choosing the starting point randomly among the two end nodes of $L$. If there exists no Eulerian path in $Q$, then the hider can keep the expected capture time above $\mu/2$ by using $h_\mu$.)*

We now establish an upper bound for the value, which holds for all networks.

**Definition 3.14** *A closed trajectory that visits all the points of $Q$ and has minimal length will be called a minimal tour (or a Chinese postman tour) and is usually denoted by $L$. Its length will be denoted by $\bar{\mu}$.*

**Lemma 3.15** *Any minimal tour satisfies $\bar{\mu} \leq 2\mu$. Equality holds only for trees.*

**Proof.** Consider a network $Q_2$ obtained from $Q$ as follows. To any arc $b$ in $Q$, add another arc $\bar{b}$ that connects the same nodes and has the same length as $b$. Since every node of $Q_2$ has even degree, it follows that $Q_2$ has an Eulerian tour $L_2$ of length $\mu(Q_2) = 2\mu(Q) = 2\mu$. If we now map the network $Q_2$ into the original network $Q$ such that both arcs $b$ and $\bar{b}$ of $Q_2$ are mapped into the single arc $b$ of $Q$, then the tour $L_2$ is mapped into a tour $L$ of $Q$ with the same length $2\mu$. If $Q$ is not a tree, it contains a circuit $C$. If we remove all new arcs $\bar{b}$ in $Q_2$ corresponding to this circuit, then the resulting network is still Eulerian and contains $Q$ but has total length less than $2\mu$. ■

Finding a minimal tour for a given network is called the *Chinese postman problem*. This problem can be reformulated for any given network $Q$ as follows. Find a set of arcs, of minimum total length, such that when these arcs are duplicated (traversed twice in the tour), the degree of each node becomes even. This problem was solved by Edmonds (1965) and Edmonds and Johnson (1973) using a matching algorithm that uses $O(n^3)$ computational steps, where $n$ is the number of nodes in $Q$. This algorithm can be described as follows. First compute the shortest paths between all pairs of odd-degree nodes of $Q$. Then, since the number of odd degree nodes is even, partition them into pairs so that the sum of lengths of the shortest paths joining the pairs is minimal. This can be done by solving a weighted matching problem. The arcs of $Q$ in the paths identified with arcs of the matching are the arcs that should be duplicated (i.e., traversed twice). The algorithm is also described by Christofides (1975) and Lawler (1976). (An updated survey on the Chinese postman problem is presented by Eiselt et al., 1995.)

Once an Eulerian network is given, one can use the following simple algorithm for finding an Eulerian tour (see Berge, 1973). Begin at any node $A$ and take any arc not yet used as long as removing this arc from the set of unused arcs does not disconnect the network consisting of the unused arcs and incident nodes to them. Some algorithms, which are more efficient than this simple algorithm were presented by Edmonds and Johnson (1973). Actually, it is possible to slightly modify Edmond's algorithm in order to obtain a trajectory (not necessarily closed), which visits all the points of $Q$ and has minimal length. This trajectory is a minimax search trajectory, and its length is the minimax value of the game.

**Example 3.16** *Consider the graph in Figure 3.1 (having the same structure as Euler's Köninsberg bridge problem) in which all the four nodes have odd degrees.*

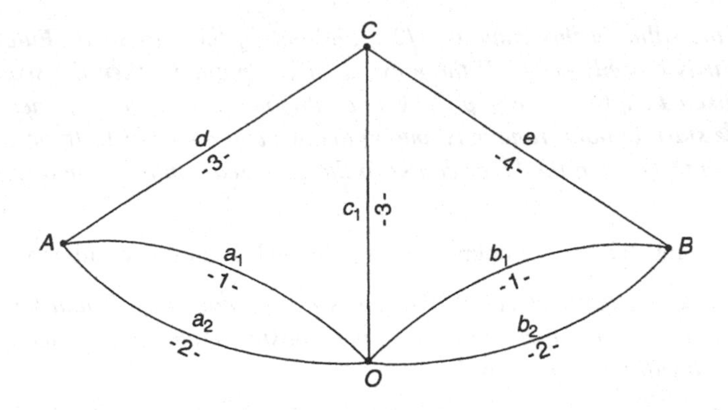


**Figure 3.1.**

*The duplicated arcs in the minimal tour can be either* $\{a_1, b_1, c_1\}$ *(based on the partition* $\{AB, OC\}$*) or* $\{b_1, d\}$ *(based on* $\{OB, AC\}$*) or* $\{a_1, e\}$ *(based on* $\{OA, BC\}$*). The corresponding sum of the lengths of the arcs is 5 or 4 or 5. Thus, the minimal tour duplicates the arcs* $b_1$ *and* $d$*. The minimal tour can be traversed by the following trajectory* $S_1$

$$S_1 = Ob_1Bb_1Ob_2BeCdAa_1Oa_2AdCc_1O \tag{3.2}$$

*with length* $\bar{\mu} = 20$*. A minimax search trajectory is based on duplicating arcs (having minimal total length) in order to make all the node degrees even except for the starting point O plus another node. It can be easily seen that the arcs that have to be duplicated are* $a_1$ *and* $b_1$ *leading to the following minimax trajectory*

$$Ob_1Bb_1Ob_2BeCdAa_1Oa_1Aa_2Oc_1C$$

*with length 18.*

Using the length $\bar{\mu}$ of the minimal tour, we now derive an upper bound for the value of the search game, with immobile hider, in a network in terms of the length of its minimal tour.

**Definition 3.17** *The search strategy* $\bar{s}$ *that encircles L equiprobably in each direction, will be called the random Chinese postman tour.*

**Lemma 3.18** *For any network Q, the random Chinese postman tour,* $\bar{s}$*, finds any point H in expected time not exceeding* $\bar{\mu}/2$*. Consequently* $v \le \bar{\mu}/2$*.*

**Proof.** For any hiding point $H$, if a path of $\bar{s}$ reaches $H$ at time $t$, then the opposite path reaches it not later than $\bar{\mu} - t$. Consequently, $\bar{s}$ finds $H$ in an expected time not exceeding

$$\frac{1}{2}(t) + \frac{1}{2}(\bar{\mu} - t) \le \frac{\bar{\mu}}{2}.$$

(Note that such a search strategy was shown to be optimal for Eulerian networks, in the beginning of Theorem 3.11. However, this strategy need not be optimal for other networks.) ■

A random Chinese postman tour of Figure 3.1 is to equiprobably follow $S_1$ (see (3.2)) or the same path in the opposite direction:

$$Oc_1CdAa_2Oa_1AdCeBb_2Ob_1Bb_1O.$$

Combining Theorems 3.3 and 3.11 and Lemmas 3.15 and 3.18 we obtain the following result. (The last statement of the theorem will be proven in the next section in Theorem 3.21.)

**Theorem 3.19** *For any network $Q$, the value $v$ of the search game with an immobile hider satisfies*

$$\mu/2 \leq v \leq \bar{\mu}/2 \leq \mu.$$

*The lower bound is attained if and only if $Q$ is Eulerian. The upper bound $\mu$ is attained if and only if $Q$ is a tree.*

## 3.3 Search on a Tree

We now consider the search game on a tree. Our main findings (Theorem 3.21) are that the value of such a game is simply the total length of the tree ($v = \mu$) and that a random Chinese postman tour is optimal for the searcher. The optimal strategy for the hider is to pick among the terminal nodes according to a certain recursively generated probability distribution, which will be explicitly described.

The fact that $v \leq \mu$ is an immediate consequence of Theorem 3.19. The reverse inequality is more difficult to establish. First observe that if $x$ is any point of the tree other than a terminal node, the subtree $Q_x$ (the connected component, or components, of $Q - \{x\}$, which doesn't contain the starting point $O$) contains a terminal node $y$. Since no trajectory can reach $y$ before $x$, hiding at $y$ strictly dominates hiding at $x$. So we may restrict our hiding strategies to those concentrated on terminal nodes.

To motivate the optimal hiding distribution over the terminal nodes, we first consider a very simple example. Suppose that $Q$ is the union of two trees $Q_1$ and $Q_2$ that meet only at the starting node $O$. Let $\mu_i$ denote the total length of $Q_i$. Let $p_i$ denote the probability that the hider is in the subtree $Q_i$. Assume that the searcher adopts the strategy of first using a random Chinese postman tour of $Q_1$ and then at time $2\mu_1$ starts again from $O$ to use a random Chinese postman tour of $Q_2$. The expected capture time $\hat{T}$ resulting from such a pair of strategies can be obtained as in the proof of Lemma 3.18, giving

$$p_1\mu_1 + p_2(2\mu_1 + \mu_2) = \mu_1 + p_2(\mu_1 + \mu_2).$$

Conducting the search in the opposite order gives an expected capture time of

$$\mu_2 + p_1(\mu_1 + \mu_2).$$

Consequently, if the $p_i$ are known, the searcher can ensure an expected capture time of

$$\min[\mu_1 + p_2(\mu_1 + \mu_2), \mu_2 + p_1(\mu_1 + \mu_2)].$$

Since the two expressions in the bracket sum to $2(\mu_1 + \mu_2)$, it follows that the hider can ensure an expected capture time of at least $\mu_1 + \mu_2$ only if these expressions are equal, or

$$p_1 = \frac{\mu_1}{\mu_1 + \mu_2}, \quad p_2 = \frac{\mu_2}{\mu_1 + \mu_2}.$$

This analysis shows that if $v = \mu$, then an optimal hider strategy must hide in each subtree with a probability proportional to its total length.

In general, an optimal hiding strategy will be constructed recursively by the following algorithm.

**Algorithm for hiding in a tree**

First recall our above argument that the hiding probabilities are positive only for the terminal nodes of $Q$. We start from the origin $O$ with $P(Q) = 1$ and go toward the leaves. In any branching we split the probability of the current subtree proportionally to the measures of subtrees corresponding to the branches. When only one arc remains in the current subtree we assign the remaining probability, $p(A)$, to the terminal node $A$ at the end of this arc. We illustrate this method for the tree depicted in Figure 3.2.

From $O$ we branch into $A_1, C, O_1$, and $O_2$ with proportions 1, 3, 6, and 3, respectively. Thus, the probabilities of the corresponding subtrees are $\frac{1}{13}, \frac{3}{13}, \frac{6}{13}$, and $\frac{3}{13}$, respectively. Since $A_1$ and $C$ are leaves we obtain $p(A_1) = \frac{1}{13}$ and $p(C) = \frac{3}{13}$. Continuing toward $O_1$ we split the probability of the corresponding subtree, $\frac{6}{13}$, with proportions $\frac{1}{5}, \frac{1}{5}$, and $\frac{3}{5}$ between $B_1, B_2$, and $C_1$ so that

$$p(B_1) = \frac{6}{65}, p(B_2) = \frac{6}{65} \quad \text{and} \quad p(C_1) = \frac{18}{65}.$$

Similarly,

$$p(A_2) = \frac{3}{13} \times \frac{1}{2} = \frac{3}{26} \quad \text{and} \quad p(A_3) = \frac{3}{13} \times \frac{1}{2} = \frac{3}{26}.$$

In order to show that $v = \mu$ we shall demonstrate that the above described hiding strategy $\bar{h}$ is optimal for trees, i.e., guarantees an expected capture time of at least $\mu$. This proof begins with the following result.

**Lemma 3.20** *Consider the two trees $Q$ and $Q'$ as depicted in Figure 3.3. The only difference between $Q$ and $Q'$ is that two adjacent terminal branches $BA_1$ of length $a_1$ and $BA_2$ of length $a_2$ (in $Q$) are replaced by a single terminal branch $BA'$ of length*

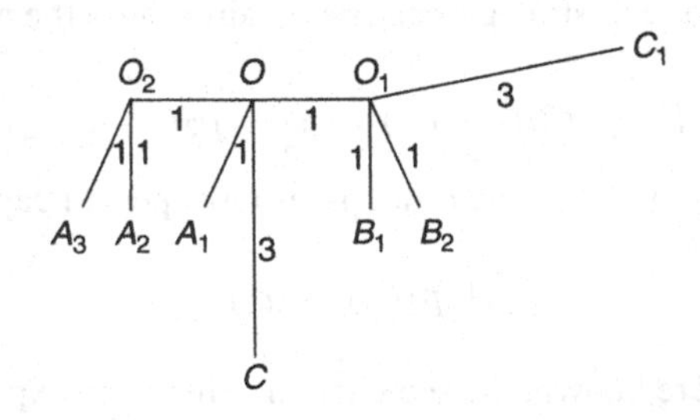


**Figure 3.2.**

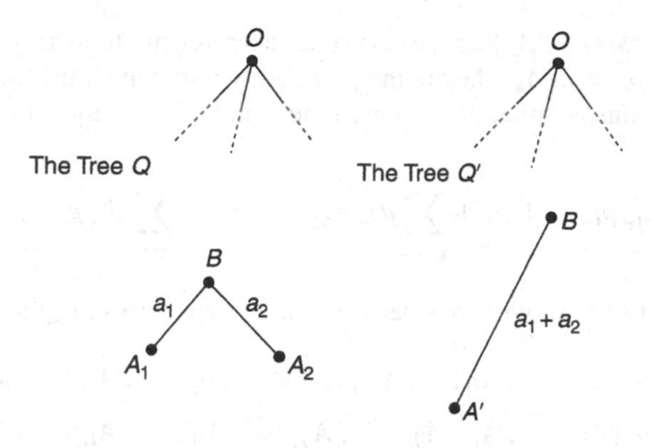


**Figure 3.3.**

*$a_1 + a_2$ (in $Q'$). Let $v$ be the value of the search game in $Q$ and let $v'$ be the value of the search game in $Q'$. Then $v \geq v'$.*

**Proof.** Let $h'$ be an optimal hiding strategy for the tree $Q'$, so that

$$c(S', h') \geq v', \text{ for any pure trajectory } S'. \tag{3.3}$$

We may assume, as explained above, that $h'$ is concentrated on terminal nodes. Given $h'$, we construct a hiding strategy in $h$ in the network $Q$ as follows. For any node other than $A_1$ or $A_2$, the hiding probability is the same for $h$ and $h'$. The probabilities $p_1 = h(A_1)$ and $p_2 = h(A_2)$ of choosing $A_1$ and $A_2$ when using $h$ are given by the formulae

$$p_1 = \frac{a_1}{a_1 + a_2} p' \quad \text{and} \quad p_2 = \frac{a_2}{a_1 + a_2} p', \tag{3.4}$$

where $p' = h'(A')$ is the probability of $A'$ being chosen by $h'$, and $a_1, a_2$ are the lengths defined in the statement of the lemma (see Figure 3.3). We shall show that $v \geq v'$ by proving that for any search trajectory $S$ in $Q$ we have

$$c(S, h) \geq v'. \tag{3.5}$$

In order to prove (3.5) we proceed as follows. Since the hider uses the strategy $h$ that chooses its hiding point at terminal nodes only, it is best for the searcher to use a search trajectory, which has the following characteristics. Starting from the root $O$, it moves via the shortest route to some terminal node, then moves via the shortest route to another terminal node, and so on until all the terminal nodes have been visited. More precisely, we can say that any search trajectory is dominated by one that visits the terminal nodes in the same order (of first visits) and has the above "shortest route" property. Consequently, there is a one-to-one correspondence (denoted by $\sim$) between the set of undominated search trajectories and the permutations of the terminal nodes. Bearing that in mind and assuming (without loss of generality) that the search strategy $S$ visits the terminal node $A_1$ before visiting $A_2$, $S$ can be represented by the following permutation of terminal nodes:

$$S \sim (A_{i_1}, \ldots, A_{i_I}, A_1, A_{j_1}, \ldots, A_{j_J}, A_2, A_{l_1}, \ldots, A_{l_L}).$$

Let $d_k = \min\{t : S(t) = A_k\}$ denote the time taken for the trajectory $S$ to reach the node $A_k$, and let $p_k = h(A_k)$ denote the probability with which the mixed strategy $h$ chooses $A_k$. With this notation, the required inequality (3.5) is equivalent to

$$\sum_{m=1}^{I} d_{i_m} p_{i_m} + d_1 p_1 + \sum_{m=1}^{J} d_{j_m} p_{j_m} + d_2 p_2 + \sum_{m=1}^{L} d_{l_m} p_{l_m} \geq v'. \tag{3.6}$$

In order to prove (3.6), we will consider two search trajectories in $Q'$:

$$S_1' \sim (A_{i_1}, \ldots, A_{i_I}, A', A_{j_1}, \ldots, A_{j_J}, A_{l_1}, \ldots, A_{l_L}) \quad \text{and}$$
$$S_2' \sim (A_{i_1}, \ldots, A_{i_I}, A_{j_1}, \ldots, A_{j_J}, A', A_{l_1}, \ldots, A_{l_L}).$$

It follows from (3.3) that

$$c(S_1', h') \geq v' \quad \text{and} \tag{3.7}$$
$$c(S_2', h') \geq v'. \tag{3.8}$$

Note that the nodes of $Q'$ are $A', A_3, A_4, \ldots$. For $i = 1, 2$ and $k \geq 3$, let $d_{ik}$ denote the time taken for the trajectory $S_i'$ to reach the node $A_k$, and let $d_i'$ denote the time for $S_i'$ to reach $A'$. It is easy to see that the following relations hold:

$$\begin{aligned} d_{1i_m} &= d_{2i_m} = d_{i_m} \\ d_{1j_m} &= d_{j_m} + 2a_2; \quad d_{2j_m} \leq d_{j_m} - 2a_1 \\ d_{1l_m} &\leq d_{l_m}; \quad d_{2l_m} \leq d_{l_m} \\ d_1' &= d_1 + a_2; \quad d_2' \leq d_2 - a_1. \end{aligned} \tag{3.9}$$

It follows from (3.7) and (3.8) that

$$\frac{a_1}{a_1 + a_2} c(S_1', h') + \frac{a_2}{a_1 + a_2} c(S_2', h') \geq v'$$

Consequently,

$$\frac{a_1}{a_1 + a_2}\left(\sum_{m=1}^{I} d_{1i_m} p_{i_m} + d_1' p' + \sum_{m=1}^{J} d_{1j_m} p_{j_m} + \sum_{m=1}^{L} d_{1l_m} p_{l_m}\right)$$
$$+ \frac{a_2}{a_1 + a_2}\left(\sum_{m=1}^{I} d_{2i_m} p_{i_m} + \sum_{m=1}^{J} d_{2j_m} p_{j_m} + d_2' p' + \sum_{m=1}^{L} d_{2l_m} p_{l_m}\right) \geq v'.$$

Using (3.9), we obtain

$$\sum_{m=1}^{I} d_{i_m} p_{i_m} + \sum_{m=1}^{J} d_{j_m} p_{j_m} + \sum_{m=1}^{L} d_{l_m} p_{l_m} + \frac{a_1}{a_1 + a_2} p' d_1 + \frac{a_2}{a_1 + a_2} p' d_2 \geq v', \tag{3.10}$$

and now the required inequality (3.6) immediately follows from (3.4) and (3.10), completing the proof. ■

Using Lemma 3.20, the next theorem uses induction on the number of terminal nodes in order to show that the hiding strategy $\bar{h}$ indeed guarantees an expected capture time $\geq \mu$ for any tree.

**Theorem 3.21** *Let $Q$ be a tree with total length $\mu$. Then*

*(i) The optimal search strategy is the random Chinese postman tour.*

*(ii) An optimal hiding strategy can be constructed recursively using the Algorithm for Hiding in a Tree.*

*(iii)* $v = \mu$.

*If $Q$ is a network which is not a tree, then $v < \mu$.*

**Proof.** First we show that $v = \mu$. We know from Theorem 3.19 that $v \leq \mu$, so if the theorem is false there is some tree $Q$ with a minimal number of nodes, for which $v < \mu$. Clearly, $Q$ cannot consist of a single arc $OA$, since in that case the hider ensures a capture time of at least $\mu$ by hiding at $A$. In all other cases, we can apply Lemma 3.20 to $Q$ to obtain a tree $Q'$ with fewer nodes and a search value $v'$ satisfying $v' < v$ and hence also $v' < \mu$. This contradicts the assumed minimality of the counter-example $Q$ and thus proves that for any tree $v = \mu$.

To prove that $v = \mu$ only for trees, note that by Theorem 3.19 $v \leq \bar{\mu}/2 \leq \mu$. Thus, $v = \mu$ implies $\bar{\mu} = 2\mu$, which implies, by Lemma 3.15, that $Q$ is a tree. ■

## 3.4 When is the Random Chinese Postman Tour Optimal?

In the case that the network $Q$ is neither Eulerian nor a tree, it follows from Theorem 3.19 that $\mu/2 < v < \mu$. Yet it may happen, for some networks, that the random Chinese postman tour is an optimal search strategy (as in the cases of Eulerian networks and trees). In this section we analyze such networks. In Section 3.4.1 we present, as a starter, a family of networks for which the random Chinese postman tour is optimal, and in Section 3.4.2 we present the widest family of networks with this property.

### 3.4.1 Searching weakly cyclic networks

**Definition 3.22** *A network is called weakly cyclic if between any two points there are at most two disjoint paths.*

An equivalent requirement, presented in [205], is that the network has no subset topologically homeomorphic with a network consisting of three arcs joining two points.

The difficulty in solving search games for the three-arcs network is illustrated in Section 3.5. Note that an Eulerian network may be weakly cyclic (e.g., if all the nodes have degree 2) but need not be weakly cyclic (e.g., 4 arcs connecting two points).

It follows from the definition that if a weakly cyclic network has a closed curve $\Gamma$ with arcs $b_1, \ldots, b_k$ incident to it, then removing $\Gamma$ disconnects $Q$ into $k$ disjoint

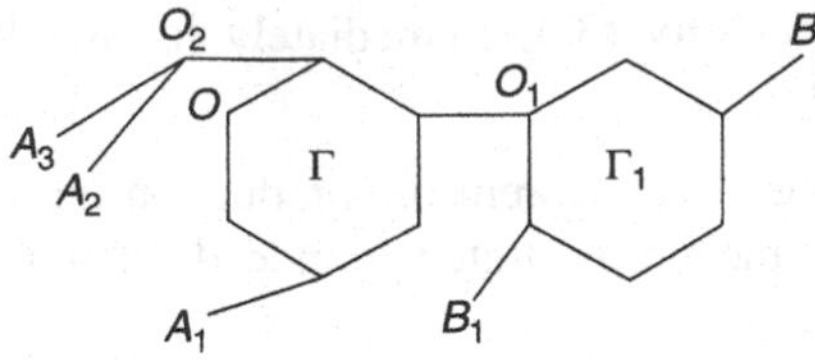

Figure 3.4.

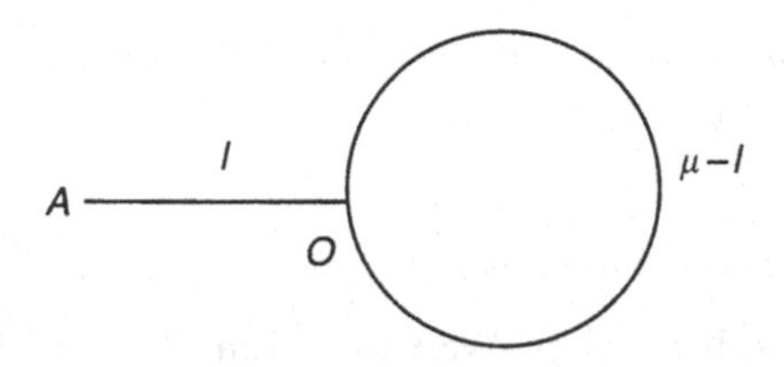


Figure 3.5.

networks $Q_1, \ldots, Q_k$ with $b_i$ belonging to $Q_i$. (If $Q_i$ and $Q_j$ were connected then the incidence points of $\Gamma$ with $b_i$ and with $b_j$ could be connected by three disjoint paths.) Thus, any weakly cyclic network can be constructed by starting with a tree (which is obviously weakly cyclic) and replacing some of its nodes by closed (simple) curves as, for example, $\Gamma$ and $\Gamma_1$ in Figure 3.4 (all edges have length 1).

We now formally define the above operation:

**Definition 3.23** *Let $\Gamma$ be a connected subnetwork of a network $Q$. If we replace the network $Q$ by a network $Q'$ in which $\Gamma$ is replaced by a point $B$ and all arcs in $Q - \Gamma$ that are incident to $\Gamma$ are incident to $B$ in $Q'$, we shall say that $\Gamma$ is shrunk and $B$ is the shrinking node.*

It is easy to see that if $Q$ contains a set of disjoint closed (simple) curves $\Gamma_1, \ldots, \Gamma_k$ such that shrinking them transforms $Q$ into a tree, then $Q$ is weakly cyclic.

In order to obtain a feeling about optimal solutions for such networks, we consider a simple example in which $Q$ is a union of an interval of length $l$ and a circle of circumference $\mu - l$ with only one point of intersection as depicted in Figure 3.5. Assume, for the moment, that the searcher's starting point, $O$, is at the intersection.

Note that the length of the Chinese postman tour is $\bar{\mu} = \mu + l$. (Remember that the trajectory has to return to $O$.) We now show that the value of the game satisfies $v = (\mu + l)/2 = \bar{\mu}/2$ and the optimal search strategy, $\bar{s}$, is the random Chinese postman tour.

The random Chinese postman tour guarantees capture time of at most $\bar{\mu}/2$ by Lemma 3.18. The following hiding strategy, $\bar{h}$, guarantees (at least) $\bar{\mu}/2$: hide with probability $\bar{p} = 2l/(\mu + l)$ at the end of the interval (at $A$) and with probability $1 - \bar{p}$ uniformly on the circle. It can be easily checked that if the searcher either goes to $A$, returns to $O$, and then goes around the circle or encircles and later goes to $A$, then the expected capture time is equal to $(\mu + l)/2$. Also, any other search trajectory yields a larger expected capture time.

Now assume that the starting point is different from $O$. In this case the value and the optimal search strategy remain the same, but the optimal hiding strategy remains $\bar{h}$ only if the starting point is anywhere on the circle. As the starting point moves from $O$ to $A$, the probability of hiding at $A$ decreases from $2l/(\mu + l)$ to 0.

The solution to search games on weakly cyclic networks was presented by Reijnierse and Potters (1993). They showed that $v = \bar{\mu}/2$ and presented an algorithm for constructing optimal hiding strategies. (The optimal search strategy is the random Chinese postman tour.)

We now present a simpler version of Reijnierse and Potters' algorithm. In our construction we transform the network $Q$ into an "equivalent" tree $\bar{Q}$ as follows: Shrink each closed curve $\Gamma_i$, with circumference $\gamma_i$, and replace it by an arc $c_i$ of length $\gamma_i/2$ that connects a new leaf (terminal node) $C_i$ to the shrinking node $B_i$. All other arcs and nodes remain the same. Let $\bar{h}$ be the optimal hiding strategy for the tree $\bar{Q}$. Then the optimal hiding strategy for $Q$ is obtained as follows:

- For a leaf of $Q$ (which is also a leaf of $\bar{Q}$) hide with the probability assigned to it by $\bar{h}$.
- For a curve $\Gamma_i$ (represented by leaf $C_i$ in $\bar{Q}$) hide uniformly along it with overall probability assigned by $\bar{h}$ to leaf $C_i$.
- No other arcs and nodes are ever chosen as hiding places.

We now use the above construction for the network $Q$ depicted in Figure 3.4, in the beginning of the subsection. The equivalent tree is depicted in Figure 3.2 (Section 3.3).

Note that the curves $\Gamma$ and $\Gamma_1$ are replaced by arcs $OC$ and $O_1C_1$. Thus, the optimal hiding probability is the same for the leaves of $Q$ and (replacing the leaves $C$ and $C_1$ by $\Gamma$ and $\Gamma_1$) hiding, uniformly, on $\Gamma$ with probability $\frac{3}{13}$ (i.e., probability density $\frac{3}{78}$) and on $\Gamma_1$ with probability $\frac{18}{65}$ (i.e., probability density $\frac{18}{390}$).

### 3.4.2 Searching weakly Eulerian networks

Reijnierse and Potters (1993) conjectured that their algorithm for constructing the optimal hiding strategy for the weakly cyclic network as well as the result $v = \bar{\mu}/2$ hold for the wider family of *weakly Eulerian networks*, i.e., networks obtained from a tree by replacing some nodes with Eulerian networks. This conjecture was shown to be correct by Reijnierse (1995). They also conjectured that $v = \bar{\mu}/2$ implies that the network is weakly Eulerian.

Gal (2000) provided a simple proof for the first conjecture and also showed that their second conjecture is correct. In order to present these results we first formally define the networks in question.

**Definition 3.24** *A network is called weakly Eulerian if it contains a set of disjoint Eulerian networks* $\Gamma_1, \ldots, \Gamma_k$ *such that shrinking them transforms* $Q$ *into a tree.*

An equivalent definition is that removing all the (open) arcs that disconnect the network (the "tree part") leaves a subnetwork(s) with all nodes having an even (possibly zero) degree. (Note that in particular removing an arc leading to a terminal node leaves the end node.) Obviously, any Eulerian network is also weakly-Eulerian. A weakly-Eulerian network has the structure illustrated in Figure 3.6.

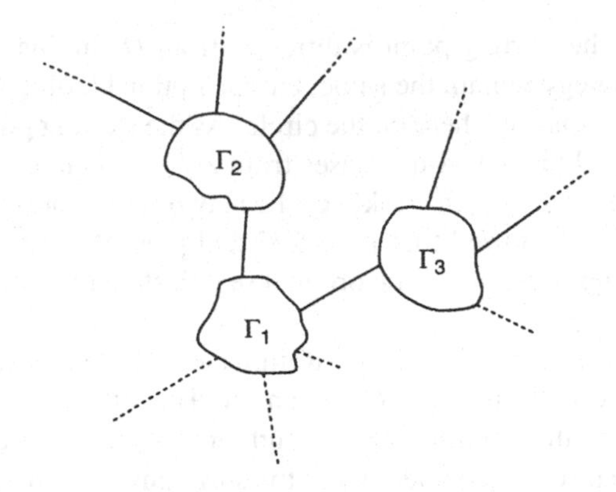


**Figure 3.6.**

**Theorem 3.25** *Let $Q$ be a weakly Eulerian network and let $\bar{\mu}$ denote the length of the Chinese postman tour of $Q$. Then*

1. $v = \bar{\mu}/2$.
2. *A random Chinese postman tour is an optimal search strategy.*
3. *An optimal hiding strategy for $Q$ is obtained as follows: First construct a tree $\bar{Q}$ by shrinking all the Eulerian subnetworks of $Q$, and adding at each shrinking node a new leaf with half the length of the associated Eulerian subnetwork. Hide in every terminal node of $Q$ according to the optimal probability for hiding in that node of the tree $\bar{Q}$. Hide uniformly on each Eulerian subnetwork of $Q$ with a total probability equal to the optimal probability of hiding at the end of the associated "new leaf" of $\bar{Q}$.*

**Proof.** We already know from Theorem 3.19 that the value $v$ of the search game on $Q$ satisfies $v \leq \bar{\mu}/2$. We will show that the stated hiding strategy ensures an expected capture time $\geq \bar{\mu}/2$, so that claims 1 and 3 will be established. Then claim 2 will then follow from Lemma 3.18.

We first show that the length of a Chinese postman path in $\bar{Q}$ is equal to $\bar{\mu}$, the same as that for the original network $Q$. To see this, partition $Q$ into the Eulerian parts $E$ (union of the Eulerian subnetworks) and the remaining treelike part $T$. Since $E$ is a union of networks, we have

$$\bar{\mu} = \bar{\mu}(Q) = \mu(E) + 2\mu(T). \tag{3.11}$$

If we denote the union of the "new leaves" of $\bar{Q}$ by $E'$ and recall that $2\mu(E') = \mu(E)$, we calculate the length $\bar{\mu}(\bar{Q})$ of the Chinese postman tour of $\bar{Q} = E' \cup T$ as

$$\begin{aligned}\bar{\mu}(\bar{Q}) &= 2\mu(E') + 2\mu(T)\\ &= \mu(E) + 2\mu(T)\\ &= \bar{\mu}(Q) = \bar{\mu}, \quad \text{by (3.11)}.\end{aligned}$$

Since $\bar{Q}$ is a tree, this implies that its total length is $\bar{\mu}/2$. Consequently by Theorem 3.21, there is an optimal hiding strategy $h^*$ on $\bar{Q}$, which guarantees an expected capture time of at least $\bar{\mu}/2$. We will show how to adapt this to $Q$ without changing this time. We now assume that lengths of all the arcs of the Eulerian subnetworks are rational. (We can do this because the value is a continuous function of the arc lengths.) Thus we can find an arbitrarily small $\varepsilon > 0$ such that these lengths are even integer multiples of $\varepsilon$. On such an arc of length $2k\varepsilon$, add $k$ additional nodes of degree two (called *grid points*), equally spaced at distances $\varepsilon, 3\varepsilon, \ldots, (2k-1)\varepsilon$ from an end node. Do this for all the arcs in the Eulerian subnetworks. The optimal hiding strategy $\bar{h}$ for $Q$ will be concentrated on the terminal nodes of $Q$ and on these new grid points.

Before defining $\bar{h}$, we introduce a new tree $\hat{Q}$ as follows: Shrink all the Eulerian subnetworks, and at each shrinking node $B$ add leaves of length $\varepsilon$ whose total length equals half the length of the corresponding Eulerian subnetwork. Identify the grid points of the Eulerian subnetwork with the terminal nodes of these leaves. The tree $\hat{Q}$ is the same as the tree $\bar{Q}$ except that instead of a single "new leaf" of length say $\gamma/2$ for each shrunk Eulerian network of length $\gamma$, there are $\gamma/2\varepsilon$ leaves of length $\varepsilon$. Observe that the optimal hiding strategy for a tree will put the same total probability (spread out equally) at the ends of the $\gamma/2\varepsilon$ small leaves of $\hat{Q}$, as on the single end of the corresponding larger leaf in $\bar{Q}$.

An optimal hiding strategy $\bar{h}$ for $Q$ can now be induced on $Q$ from the optimal hiding strategy $\hat{h}$ on $\hat{Q}$, with the probability of each grid point of $Q$ equal to that of the corresponding end of a small leaf. Suppose now that some search strategy $S$ on $Q$ visits the grid and terminal nodes of $Q$ in such a way that the expected time (relative to $\bar{h}$) to reach such a node is less than $\bar{\mu}/2$. Let $\hat{S}$ be the search strategy on $\hat{Q}$ that visits its corresponding terminal nodes in the same order, moving between consecutive nodes in least time. Note that our construction of $\hat{Q}$ ensures that the distance between any pair of grid or terminal nodes in $Q$ is at least as large as the distance between their corresponding terminal nodes in $\hat{Q}$. This ensures that the search path $\hat{S}$ will not get to any terminal node of $\hat{Q}$ later than $S$ gets to its corresponding node in $Q$. Hence the expected time (relative to $\hat{h}$) for $S$ to reach the hiding point is also less than $\bar{\mu}/2$, contradicting the optimality of $\hat{h}$ (and the value $v = \bar{\mu}/2$ for $\hat{Q}$). Consequently no search strategy $S$ on $Q$ finds an object hidden according to $\bar{h}$ is expected time less than $\bar{\mu}/2$, establishing our claim. ■

The strategy $\bar{h}$ given here is not uniquely optimal. For example, the simpler strategy stated in the theorem, of hiding uniformly in each Eulerian subnetwork, is also optimal. (Or that of hiding in the middle of each new arc.)

We now illustrate the result by an example: Let $Q$ be the union of an Eulerian network $\gamma$, of measure $\gamma$, and two arcs of lengths 1 and 2, respectively, leading to leafs $C_1$ and $C_2$ (see Figure 3.7).

If $O \in \Gamma$ then, $\bar{Q}$ would be a star with three rays of lengths 1, 2, and $0.5\gamma$, respectively. Thus $\bar{h}$ hides at $C_1$ with probability $1/(0.5\gamma+3)$ at $C_2$ with probability $2/(0.5\gamma+3)$ and uniformly on $\Gamma$ with overall probability $0.5\gamma/(0.5\gamma+3)$. If the

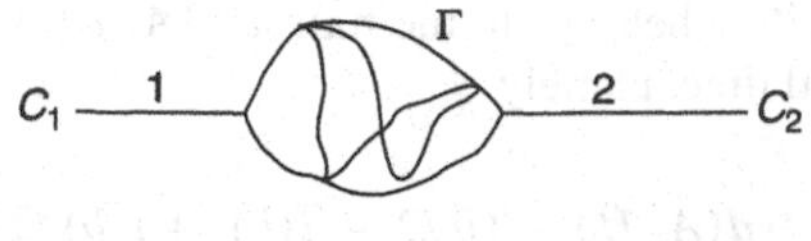

**Figure 3.7.**

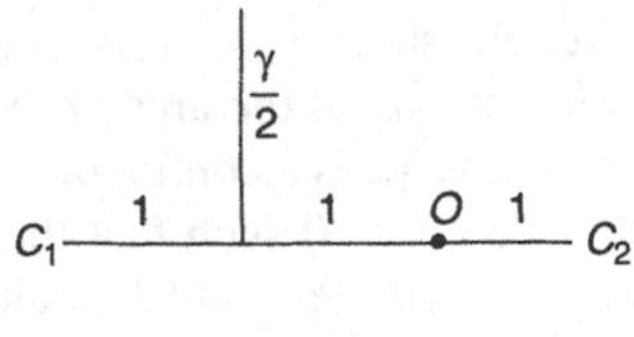


**Figure 3.8.**

starting point is on the arc leading to $C_2$ with distance 1 from $C_2$, then $\bar{Q}$ would be the tree depicted in Figure 3.8. Thus, the corresponding optimal hiding probabilities for $C_1$, $C_2$, and $\Gamma$ would be $(0.5\gamma+2)/(0.5\gamma+3)\times 1/(0.5\gamma+1)$, $1/(0.5\gamma+3)$, and $(0.5\gamma+2)/(0.5\gamma+3)\times 0.5\gamma/(0.5\gamma+1)$, respectively.

We now prove the second conjecture of Reijnierse and Potters (1993), i.e., that all networks that are not weakly Eulerian have value strictly smaller than $\bar{\mu}/2$.

**Theorem 3.26** *For any network $Q$, if $v=\bar{\mu}/2$, then $Q$ is weakly Eulerian.*

**Proof.** Let $L$ be a Chinese postman tour of $Q$ (with length $\bar{\mu}$). Let $Q_i$, $i=1,2$, be the subnetworks of $Q$ determined by the arcs traversed $i$ times by $L$. Note that $Q_1$ is the union of Eulerian networks: all its nodes have even degree. Let $Q'$ be the network consisting of $Q_1$ and a version of $Q_2$ in which all its arcs are doubled. The resulting network $Q'$ is Eulerian and has total length $\bar{\mu}$.

We shall show that under the hypothesis of the theorem, the removal from $Q$ of (the interior of) any arc $b$ of $Q_2$ results in a disconnected network $Q-b$. Consequently, the removal of all such arcs leaves the network $Q-Q_2=Q_1$, which has all even degree nodes. Thus the original network $Q$ will satisfy the equivalent definition given above for a weakly Eulerian network.

Let $b$ be any arc of the subnetwork $Q_2$ with length $l(b)>0$, and let $h_\varepsilon$ be a mixed hider strategy for $Q$ guaranteeing a capture time of at least $\bar{\mu}/2-\varepsilon$. We will establish that $Q-b$ is not connected by showing that the alternative assumption of connectedness lead to a contradiction of the $\varepsilon$-optimality of $h_\varepsilon$.

Since $Q-b$ is assumed to be connected, so is the network $Q''=Q'-b-b'$, where $b'$ is the added arc parallel to $b$. Since $Q''$ is Eulerian and has total length $\bar{\mu}-2l(b)$, its Eulerian tour produces a Chinese postman tour $L'$ of $Q-b$ with length $\bar{\mu}-2l(b)$. The random Chinese postman tour $s'$ based on $L'$ finds all points of $Q-b$ in expected time not exceeding $\bar{\mu}/2-l(b)$. Denote the endpoints of $b$ by $A$ and $C$ and its midpoint by $B$. Since $b$ cannot be a leaf of $Q$ (because its terminal node would become disconnected from the rest of $Q$), both of its endpoints $A$ and $C$ are visited by $L'$. Consequently, we may extend $L'$ to a Chinese postman tour of $Q$ in two ways $L_A$ and $L_B$. The tour $L_A$ traverses $b$ in both directions after its first arrival at $A$, the tour $L_C$ does the same when it reaches $C$. The search strategies $s_A$ and $s_C$ based on random Chinese postman tours of $L_A$ and $L_C$ each guarantee an expected capture time no more than $\bar{\mu}/2$ for all hiding points $H$ in $Q$. However, if $H$ belongs to the half arc $[A,B]$, then it will be found by $s_A$ with a smaller expected time, namely

$$c(s_A,H)=c(s',A)+d(A,H)\leq(\bar{\mu}/2-l(b))+l(b)/2=\bar{\mu}/2-l(b)/2.$$

It follows that the probability $p = h_\varepsilon([A, B])$ of hiding in $[A, B]$ under the $\varepsilon$-optimal strategy $h_\varepsilon$ satisfies

$$\bar{\mu}/2 - \varepsilon < c(s_A, h_\varepsilon) \leq p\left(\frac{\bar{\mu}}{2} - \frac{l(b)}{2}\right) + (1-p)\left(\frac{\bar{\mu}}{2}\right) = \frac{\bar{\mu} - p\,l(b)}{2},$$

and consequently

$$p < \frac{2\varepsilon}{l(b)}.$$

Using $s_C$ instead of $s_A$ we can show that the same estimate holds for the probability of hiding in $[B, C]$. Consequently, the probability of hiding in $b$ when using $h_\varepsilon$ is less than $4\varepsilon/l(b)$. Now consider the search strategy $s'_A$ on $Q$, which first follows $s'$, goes in the shortest route to $A$, and then traverses the arc $b$. For any $H \in Q - b$, we have $c(s'_A, H) = c(s', H) \leq \bar{\mu}/2 - l(b)$, and for any $H \in b$ we have $c(s'_A, H) \leq \bar{\mu} + d(O, A) + d(A, C) < 2\bar{\mu}$. Hence for sufficiently small $\varepsilon$, we have

$$c(s'_A, h_\varepsilon) \leq \left(\frac{\bar{\mu}}{2} - l(b)\right) \times \left(1 - \frac{4\varepsilon}{l(b)}\right) + (2\bar{\mu})\left(\frac{4\varepsilon}{l(b)}\right) < \frac{\bar{\mu}}{2} - \frac{l(b)}{2},$$

which contradicts the $\varepsilon$-optimality of $h_\varepsilon$ for $\varepsilon < l(b)/2$. ■

Combining Theorems 3.25 and 3.26 we have the following summarizing result:

**Summary 3.27** *If $Q$ is weakly Eulerian, then $v = \bar{\mu}/2$; otherwise, $v < \bar{\mu}/2$.*

An equivalent statement is that random Chinese postman tour is an optimal search strategy if and only if the network Q is weakly Eulerian.

**Remark 3.28** *Note that the value is $\bar{\mu}/2$, and the optimal search strategy is a random Chinese postman tour, independently of the specific starting point. If $Q$ is not weakly Eulerian, then the value may depend on the starting point $O$. (For example, in the next section we show that the value does depend on $O$ in the three-arcs network, see Remark 3.33.)*

*We conjecture that the independence of the value on the starting point holds only for weakly Eulerian networks.*

Searching a network which is not weakly Eulerian is expected to lead to rather complicated optimal search strategies. We shall show in the next section that even the "simple" network with two nodes connected by three unit length arcs, requires a mixture of infinitely many trajectories for the optimal search strategy.

## 3.5 Simple Networks Requiring Complicated Strategies

In the previous sections we solved the search game for weakly Eulerian networks. (Remember that this family includes Eulerian networks and the trees as "extreme" cases.) We have shown that the random Chinese postman tour is optimal, and $v = \bar{\mu}/2$, only for this family. We now present a simple network, which is not weakly Eulerian and hence has value strictly less than $\bar{\mu}/2$.

In this example, the network $Q$ consists of $k$ distinct equal length arcs, $b_1, \ldots, b_k$, connecting two points $O$ and $A$. An immediate consequence of the scaling lemma presented in Chapter 2 is that it is sufficient to consider the case in which all the arcs have unit length. This example will also be considered in the next chapter, where we deal with a mobile hider. If the number $k$ of arcs is even, then the network $Q$ is Eulerian, and the solution of the game is simple. On the other hand, it turns out that the solution is surprisingly complicated in the case that $k$ is an odd number greater than 1, even if $k$ is equal only to 3. For this network $\bar{\mu} = k + 1$, so we know from the last section that $v < (k+1)/2$. We will prove the stricter inequality:

**Lemma 3.29** *If $Q$ is a set of $k$ non-intersecting arcs of unit length, which join $O$ and $A$, and $k$ is an odd number greater than* 1, *then*

$$v < \frac{k}{2} + \frac{1}{2k}. \tag{3.12}$$

**Proof.** We will use a natural search strategy, $\hat{s}$, based on the following definition:

**Definition 3.30** *For a given set consisting of $k$ elements, choosing each element with probability* $1/k$ *will be called an equiprobable choice.*

We now proceed in establishing (3.12) by presenting the following search strategy $\hat{s}$. Starting from $O$ make an equiprobable choice among the $k$ arcs, and move along the chosen arc to $A$. Then make an equiprobable choice among the $k-1$ remaining arcs, independently of the previous choice, and move along this arc back to $O$. Then move back to $A$ and so on until all the arcs have been visited. Since $k$ is odd, any such path ends at $A$.

Let $H$ be any pure strategy (i.e., a point in $Q$) and assume that its distance from $A$ is $d$, so that its distance from $O$ is $1-d$. Let $m = (k-1)/2$, let $E_i$, $i = 1, \ldots, m$, be the event that the hider is discovered during the time period $(2(i-1), 2i]$, and let $E_f$ be the event that the hider is discovered during the time period $(k-1, k)$. Then

$$\begin{aligned} c(\hat{s}, H) &= \sum_{i=1}^{m} (2i-1) Pr(E_i) + (k-d) Pr(E_f) \\ &= \sum_{i=1}^{m} (2i-1) \frac{2}{k} + (k-d) \frac{1}{k} \\ &= \frac{k}{2} + \frac{1}{2k} - \frac{d}{k} < \frac{k}{2} + \frac{1}{2k}. \end{aligned}$$

■

The case where $Q$ consists of an odd number of arcs that connect two points has been used as an example for situations in which $v < \bar{\mu}/2$, but this case is interesting by itself. It is amazing that the solution of the game is simple for any even $k$ (and also, as will be demonstrated in the next chapter, for any odd or even $k$ if the hider is mobile), but it is quite complicated to solve this game even for the case that $k$ is equal only to 3. The reasonable symmetric search strategy $\hat{s}$, used in proving Lemma 3.29, which is optimal for an even $k$, can assure the searcher an expected search time less than $k/2 + 1/2k$.

We shall immediately show that $\hat{s}$ is not an optimal strategy for an odd number of arcs. (Incidentally, the fact that the search strategy $\hat{s}$ is not optimal or even $\varepsilon$-optimal, can be easily deduced from the following argument. If the searcher uses $\hat{s}$, then the hider can guarantee a payoff that is close to $k/2 + 1/2k$ only by hiding near $A$ with probability $1 - \varepsilon$. However, it can easily be verified that the payoff guaranteed by such a hiding strategy does not exceed $1 + \delta$, where $\delta$ is small. Thus, the value of the game has to be smaller than $k/2 + 1/2k$, which implies that the strategy $\hat{s}$ cannot be optimal or even $\varepsilon$-optimal.)

In order to demonstrate the complexity of this problem, we now consider the case of $k = 3$. In this case, the symmetric search strategy $\hat{s}$ satisfies $v(\hat{s}) = 3/2 + 1/6 = 5/3$, but we now present a strategy $\tilde{s}$, originally suggested by D. J. Newman, which satisfies

$$v(\tilde{s}) = (4 + \ln 2)/3 < 5/3.$$

The strategy $\tilde{s}$ is a specific choice among the following family $\{s_F\}$ of search strategies:

**Definition 3.31** *The family $\{s_F\}$ of search strategies is constructed as follows: Consider a set of trajectories $S_{ij\alpha}$, where $i$ and $j$ are two distinct integers in the set $\{1, 2, 3\}$ and $0 \le \alpha \le 1$. The trajectory $S_{ij\alpha}$ starts from $O$, moves along $b_i$ to $A$, moves along $b_j$ to the point $A_\alpha$ that has a distance of $\alpha$ from $A$ (see Figure 3.9), moves back to $A$, moves to $O$ along $b_m$, where $m \in \{1, 2, 3\} - \{i, j\}$, and then moves from $O$ to $A_\alpha$ along $b_j$.*

*Let $F(\alpha)$ be a cumulative probability distribution function of a random variable $\alpha$ $(0 \le \alpha \le 1)$. Then the strategy $s_F$ is a probabilistic choice of a trajectory $S_{ij\alpha}$, where $i$ is determined by an equiprobable choice in the set $(1, 2, 3\}$, $j$ is determined by an equiprobable choice in the set $\{1, 2, 3\} - \{i\}$, and $\alpha$ is chosen independently, using the probability distribution $F$.*

Note that the symmetric strategy $\hat{s}$ is a member of the family $\{s_F\}$ with the random variable $\alpha$ being identically zero.

We now show that there exists a search strategy $s = \tilde{s}$ in $\{s_F\}$ with value less than $v(\hat{s})$. The distribution function $\tilde{F}$ that corresponds to the strategy $\tilde{s}$ is the following:

$$\tilde{F}(\alpha) = \begin{cases} 0 & \alpha < 0, \\ \dfrac{1}{2} + \dfrac{e^\alpha}{4} & 0 \le \alpha \le \ln 2, \\ 1 & \alpha > \ln 2. \end{cases} \tag{3.13}$$

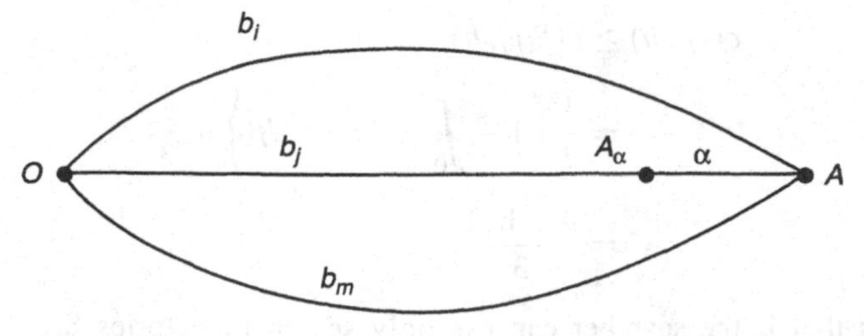


**Figure 3.9.**

(Note that $F$ has a probability mass of 3/4 at $\alpha = 0$.) Let $H_\beta$ be any hiding point with distance $\beta$ from $A$. Then it is easy to see that

$$c(\tilde{s}, H_\beta) = \frac{1}{3}(1-\beta) + \frac{1}{3}\left(1 + 2\int_0^{\ln 2} \alpha d\tilde{F}(\alpha) + \beta\right)$$
$$+ \frac{1}{3}\left(1 + \int_0^{\beta} (2\alpha + 2 - \beta) d\tilde{F}(\alpha) + \beta(1 - \tilde{F}(\beta))\right).$$

It is easy to check that the derivative of $c(\tilde{s}, H_\beta)$ with respect to $\beta$ is equal to zero for $0 < \beta < \ln 2$ and to $-1/3$ for $\ln 2 < \beta < 1$. Thus, it is sufficient to calculate $c(\tilde{s}, H)$ for $\beta = \varepsilon$, where $\varepsilon$ is small, and the calculation readily shows that

$$v(\tilde{s}) = \frac{4 + \ln 2}{3} < \frac{5}{3} = v(\hat{s}),$$

where $\hat{s}$ is the symmetric strategy.

We now show that $\tilde{s}$ is the best search strategy in the family $\{s_F\}$ by presenting a hiding strategy $\tilde{h}$ that satisfies for all $s_F \in \{s_F\}$,

$$c(s, \tilde{h}) \geq \frac{4 + \ln 2}{3} (\doteq 1.56).$$

**Definition 3.32** *The hiding strategy $\tilde{h}$ is presented as follows. Make an equiprobable choice of an arc $b_i$, $i \in \{1, 2, 3\}$, and hide at the point of $b_i$ that has a distance $\beta$ from $A$, where $\beta$ is a random variable that has the probability density $2e^{-\beta}$ for $0 \leq \beta \leq \ln 2$ and zero otherwise.*

It is easy to see that for all $S_{ij\alpha}$ used by any search strategy $s_F$:

$$\frac{d}{d\alpha} c(S_{ij\alpha}, \tilde{h}) = 2\left(\frac{1}{3} + \frac{1}{3}\int_\alpha^{\ln 2} 2e^{-\beta} d\beta\right) - \frac{2}{3} 2e^{-\alpha} = 0 \quad \text{for } 0 < \alpha < \ln 2,$$

and

$$= \frac{2}{3} \quad \text{for } \ln 2 < \alpha < 1.$$

Thus,

$$c(s_F, \tilde{h}) \geq c(S_{ij0}, \tilde{h})$$
$$= \frac{1}{3}\left(1 - \int_0^{\ln 2} 2\beta e^{-\beta} d\beta\right) + \frac{4}{3}$$
$$= \frac{4 + \ln 2}{3}.$$

It follows that if the searcher can use only search trajectories $S_{ij\alpha}$, then $\tilde{s}$ and $\tilde{h}$ presented previously are optimal strategies and the value of the restricted game is

$(4 + \ln 2)/3$. It is very plausible that $\tilde{s}$ and $\tilde{h}$ are optimal even if the searcher is not restricted to $S_{ij\alpha}$ trajectories. In order to show that, one would have to show that

$$c(S, \tilde{h}) \geq \frac{4 + \ln 2}{3} \quad \text{for all } S.$$

(We have just proved this inequality for $S = S_{ij\alpha}$).

Numerically, verifying that the value is indeed $(4 + \ln 2)/3$ (within any desired accuracy level) can be done by dynamic programming, as will be described in the next section. However, formally proving that fact has required a lot of effort: Bostock (1984) solved a discrete version of the three-arcs game, allowing the searcher to choose his trajectory more extensively than $S_{ij\alpha}$ (but still limited to a subfamily of trajectories). By letting the number of grid points tend to infinity he showed that $\tilde{s}$ and $\tilde{h}$ are optimal under a weaker assumption on the search strategies. Finally, L. Pavlovic (1993a, 1993b, and 1995b) has succeeded in proving that the value is indeed $(4 + \ln 2)/3$ and that $\tilde{s}$ is the optimal search trajectory, under no restrictions on the search trajectories. Pavlovic also presented the optimal solution of the search game for any odd number $k$ of arcs. The optimal search strategy is randomly (equiprobably) choose the traversed arc, among the untraversed arcs, until only three untraversed arcs remain; then use $\tilde{s}$ for these three arcs.

**Remark 3.33** *Unlike the weakly Eulerian networks, the value of the three-arcs network depends on the starting point of the searcher. For example,*[3] *assume that the searcher starts at the middle of arc $b_1$. Then, by using the uniform hiding strategy, $h_\mu$, the hider can guarantee an expected capture time $\geq 19/12 > (4 + \ln 2)/3$. (It can be verified that the searcher's best response is to go to one of the nodes, search $b_2$ and $b_3$, and finally retrace $b_1$ in order to search the unvisited half of $b_1$.)*

**Remark 3.34** *Solving the search game in a network with an arbitrary starting point for the searcher, is an interesting problem, which has not been investigated yet. Let $\tilde{v}$ be the value of this game. It is easy to see that $\tilde{v} \leq \tilde{\mu}/2$, where $\tilde{\mu}$ is the minimal length of a path (not necessarily closed) that visits all the points of $Q$. Indeed, sometimes $\tilde{v} = \tilde{\mu}/2$ as happens for networks with $\tilde{\mu} = \mu$ because the hider can guarantee expected capture time $\geq \tilde{\mu}/2$ by using the uniform strategy $h_\mu$. (Such an example is the three-arcs game, which is much more tractable with an arbitrary starting point than under the usual assumption of a fixed starting point known to the hider.) However, it is not clear for which family of networks the equality $\tilde{v} = \tilde{\mu}/2$ holds. For example, does it hold for trees?*

**Remark 3.35** *The problem of finding the optimal search strategy for a (general) network has been shown by von Stengel and Werchner (1997) to be NP-hard. However, they also showed that if the time of search is limited by a (fixed) bound, which is logarithmic in the number of nodes, then the optimal strategy can be found in polynomial time. The search game on a network can, in general, be formulated as an infinite-dimensional linear program. This formulation and an algorithm for obtaining its (approximate) solution is presented by Anderson and Armendia (1990).*

[3] This observation was made by Steve Alpern.

The difficult problem of finding an optimal (minimax) search strategy for a (non weakly Eulerian) network could be somewhat simplified if the following property holds.

**Conjecture 3.36** *Optimal strategies for searching for an immobile hider on any network never use trajectories that visit some arcs (or parts of arcs) more than twice.*

Note that this property does not hold for best responses to all hiding strategies. For example, in the three-arcs network, if the hiding probability is $\frac{1}{3} - \varepsilon$ for each point $B_i, i = 1, 2, 3$, having distance $\frac{1}{4}$ from $A$, and $\varepsilon$ for $C_i, i = 1, 2, 3$, having distance $\frac{1}{3}$ from A, then for a small $\varepsilon$ the optimal search strategy is to go to $A$, then to one of the unvisited $B_i$, then through $A$ to the unvisited $B_j$, continue to $C_j$ and finally from $C_j$ to $C_i$ through $B_i$. Thus, the segment $AB_i$ is traversed three times. Moreover, in the linear search problem, described in Section 8.1, there exist hiding distributions against which the best response search trajectories visit the unit interval an infinite number of times. Still, we conjecture that such a situation cannot occur against an optimal hiding strategy. This property holds for the weakly Eulerian networks and also for the three-arcs network.

## 3.6 Using Dynamic Programming for Finding Optimal Response Search Trajectories

Dynamic programming ($DP$) is a useful numerical optimization scheme developed by Bellman and others around 1960. It is based on the following recursive principle, which we apply to search problems. Assume that we wish to minimize the expected search time for an object that is hidden in one of $n$ possible points with *known* hiding probabilities. Any search trajectory is determined by a series of actions $a_1, a_2, \ldots, a_n$ such that each corresponds to going from the current location to a location that has not been searched before. Rather than trying to determine all the optimal actions together, $DP$ determines one action in each step minimizing the sum of the expected immediate cost plus the expected future cost. The optimality equation has the following form:

$$F(Z) = \min_{a \in A(Z)} \left[ \acute{C}(Z, a) + \left[ 1 - \frac{p(Z, a)}{P(Z)} \right] \times F(\phi(Z, a)) \right] \qquad (3.14)$$

where

- $Z$ is the *state* of the search (i.e., the current "unsearched" part of the space and the location of the searcher),
- $F(Z)$ is the current expected remaining search cost under an optimal trajectory,
- $A(Z) = \{a\}$ is the current set of actions $a$ available (i.e., what locations could be searched next),
- $P(Z)$ is the probability that, at state $Z$, the object has not yet been found,
- $\acute{C}(Z, a)$ is the cost of action $a$ at state $Z$ (e.g. the time needed to go to a specific next location),

- $p(Z, a)$ is the probability of that the object is in this location ($p(Z, a)/P(Z)$ is the probability that the object is found at the present stage under the condition that it has not been found before),
- $\phi(Z, a)$ is the "new" state after using action $a$, (i.e., the updated unsearched part and searcher's location), and
- $F(\phi(Z, a))$ is the new (minimal) expected remaining search time, after using action $a$ under the condition that the object has not been found at the present stage.

Note that at each step the number of unsearched locations of the state space is reduced by 1. Thus, the new search problem is reduced in some sense.

If the number of possible hiding locations is $n$ then we will have to use the recursion (3.14) for $n$ steps and compute $F(Z)$ for all the intermediate possible values of $Z$ finally obtaining the value of the optimal solution, $F(Z_0)$, where $Z_0$ is the initial state, i.e., the unsearched part being the whole space and the searcher at the origin $O$. The optimal action for each state is obtained from the *argmin* of the right side of (3.14).

Such a technique would obviously be inefficient if the number of possible states is an exponential function of $n$ (e.g., search in a complete network with $n$ nodes). However, there are many interesting examples where the structure of the problem keeps the number of possible states manageable (e.g., polynomial in $n$ with a small degree). For example, assume that we search the real line, starting from the origin, looking for an object with $n$ possible locations (on both sides of the origin). Then the number of possible states is bounded by $n^2$ because at each stage the current undiscovered part of the search space is determined by the two extreme right and left locations visited by the searcher. We will present the $DP$ algorithm for the search on the line in Section 8.7.

It is usually more convenient to use the recursive formula for the *contribution* $f(Z)$ of the current undiscovered part of the search space to the remaining expected search time, $f(Z) = P(Z) \times F(Z)$. In other words, $f$ is the sum of the original probabilities of the unsearched locations multiplied by the remaining time to reach them by the optimal search plan. Multiplying both sides of (3.14) by $P(Z)$, the corresponding recursive equation for $f$ is

$$f(Z) = \min_{a \in A(Z)} \{P(Z) \times \acute{C}(Z, a) + f(\phi(Z, a))\}, \tag{3.15}$$

where, as in (3.14), we compute $f(Z)$ for all the intermediate possible values of $Z$. In the final step we obtain the value of the optimal solution, $f(Z_0)$, where $Z_0$ is the initial state, i.e., the unsearched part is the whole space and the searcher is at the origin $O$. The optimal action for each state is obtained from the *argmin* of the right side of (3.15). Here, there is no need to compute new conditional probabilities at each stage; the same location probabilities can be used all the time.

We now illustrate the $DP$ approach for the three-arcs problem of the previous section. We would like to verify that the minimal expected search time against the hiding strategy $\tilde{h}$ (given in Section 3.5) $v(\tilde{h})$ is $(4 + \ln 2)/3$. The numerical value of $v(\tilde{h})$ can be approximated for any desired accuracy level using the following scheme. For each arc divide the segment starting from $A$ with length $\ln 2$ into $m$ intervals of equal length $\varepsilon = \ln 2/m$. For each interval, $j$, replace the (continuous) probability density of $\tilde{h}$ over this interval by one probability mass $p_j$ concentrated at the midpoint of the interval at distance $(j - \frac{1}{2})\varepsilon$ from $A$. (Choosing the center of gravity rather than the midpoint would lead to more accurate result, with the same complexity. For convenience, we present here the simpler version.) In the new (approximated) hiding strategy the object can be at any one of $n = 3m$ locations. For each stage, the current state space depends on the number of arcs with unsearched locations.

In the first time period, before $A$ has been visited, there are three such arcs so that the state of the search can be described by the 4-tuple $(i, k_1, k_2, k_3)$, where for each $j, 1 \le j \le 3; k_j, 1 \le k_j \le m$ is the number of unsearched locations (near $A$) on arc $j$; and $i$, $1 \le i \le 3$ is the arc number where the searcher is now located (at distance $\varepsilon \times \left(k_i + \frac{1}{2}\right)$ from $A$). The next location to be searched is either the next point on arc $i$ or the left point of one of the other arcs. Thus, for $i = 1$, the recursive formula is

$$f(1, k_1, k_2, k_3) = \min \begin{cases} P(k_1, k_2, k_3) \times \varepsilon + f(1, k_1 - 1, k_2, k_3), \\ P(k_1, k_2, k_3) \times [2 - (k_1 + k_2)\varepsilon] + f(2, k_1, k_2 - 1, k_3), \\ P(k_1, k_2, k_3) \times [2 - (k_1 + k_3)\varepsilon] + f(3, k_1, k_2, k_3 - 1) \end{cases}$$

where $P(k_1, k_2, k_3)$ is the sum of the probabilities over the unsearched locations.

A similar recursion holds for $i = 2$ and $i = 3$ (see Figure 3.10).

After one of the $k_j$ reduces to 0 (say $k_3$) then unsearched locations of the two remaining arcs would be in two segments $1 \le l_1 \le k_1$ and $1 \le l_2 \le k_2$ with the searcher either on arc 1 at distance $\left(l_1 - \frac{3}{2}\right)\varepsilon$ from $A$ (in this case $1 < l_1$) or on arc 2 at distance $\left(l_2 - \frac{3}{2}\right)\varepsilon$ from $A$ (with $1 < l_2$). The corresponding two arcs recursive formula used for $i = 1$ (similar recursion holds for $i = 2$) is

$$\varphi(1, k_1, l_1; k_2, l_2) = \min \begin{cases} P(k_1, l_1; k_2, l_2) \times \varepsilon + \varphi(1, k_1, l_1 + 1; k_2, l_2), \\ P(k_1, l_1; k_2, l_2) \times (l_1 + l_2 - 2)\varepsilon + \varphi(2, k_1, l_1; k_2, l_2 + 1) \end{cases}$$

where $\varphi$ takes the place of $f$ in (3.15) and $P(k_1, l_1; k_2, l_2)$ is the sum of probabilities over the unsearched locations (on the two remaining arcs).

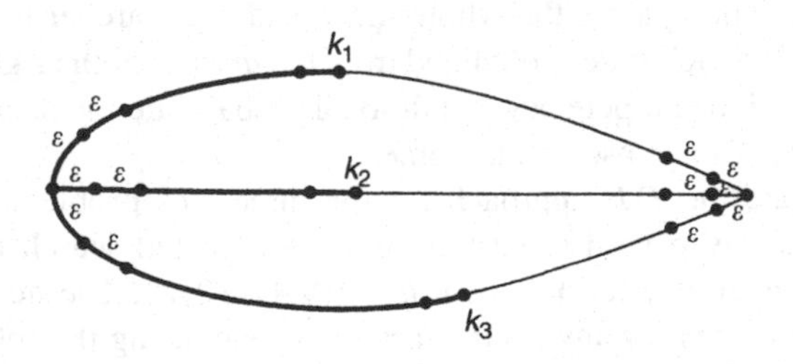


**Figure 3.10.**

Note that the recursive formulas for the transition between three and two arcs has the following form (e.g., when $k_1$ and $k_2 > 1$ and $k_3 = 1$):

$$f(1, k_1, k_2, 1) = \min \begin{cases} P(k_1, k_2, 1) \times \varepsilon + f(1, k_1 - 1, k_2, 1); \\ P(k_1, k_2, 1) \times [2 - (k_1 + k_2)\varepsilon] + f(2, k_1, k_2 - 1, 1); \\ P(k_1, k_2, 1) \times [2 - (k_1 + 1)\varepsilon] \\ \quad + \min \begin{cases} P(k_1, 1; k_2, 1) \times \varepsilon + \varphi(1, k_1, 2; k_2, 1), \\ P(k_1, 1; k_2, 1) \times \varepsilon + \varphi(2, k_1, 1; k_2, 2). \end{cases} \end{cases}$$

When $l_i$ becomes $k_i$ (i.e., arc $i$, say $i = 2$, has been completely searched) and only one arc remains, the following boundary condition is used:

$$\varphi(i, k_1, l_1; k_2, k_2) = \min \begin{cases} (k_2 - 1) \times \varepsilon \sum_{j=l_1}^{k_1} p_j + \sum_{j=l_1}^{k_1} j p_j \times \varepsilon; \\ (2 - (k_2 - 1) \times \varepsilon) \sum_{j=l_1}^{k_1} p_j - \sum_{j=l_1}^{k_1} j p_j \times \varepsilon. \end{cases}$$

Note that in each of the above formulae the number of unsearched locations corresponding to the right side states is smaller than the number of unsearched locations in the right side. This enables us to recursively calculate the $\varphi$ values ($O(n^4)$) and then the $f$ values ($O(n^3)$). The minimal search time is obtained by

$$f(Z_0) = 1 - \left(m - \tfrac{1}{2}\right) \times \varepsilon + f(1, m - 1, m, m).$$

(Due to symmetry we can assume that the search starts along arc 1.) The optimal search trajectory is obtained step by step by using the alternative that produced the minimal value at any given state in order to move into the next state.

For the location probability based on $\tilde{h}$, given Definition 3.32 we should obtain an expected capture time approximately $(4 + \ln 2)/3$. This would demonstrate, numerically, that the hider can indeed achieve $(4 + \ln 2)/3$. Since we have shown that the searcher can guarantee capture time not exceeding $(4 + \ln 2)/3$, it would follow that it is indeed the value of the three-arcs game and that $\tilde{s}$ and $\tilde{h}$ are optimal strategies.[4]

The $DP$ algorithm will be used later in Sections 8.7 and 16.8 (search on the line and rendezvous search on the line).

## 3.7 Search in a Multidimensional Region

When considering search games with a mobile or an immobile hider in multidimensional compact regions, we would like to avoid unnecessary complications, and thus we shall make a rather weak assumption about the region $Q$. We shall use the following definition.

[4] Such a computation has been carried out by Victoria Ptashnikov, who also removed several bugs from the formulae.

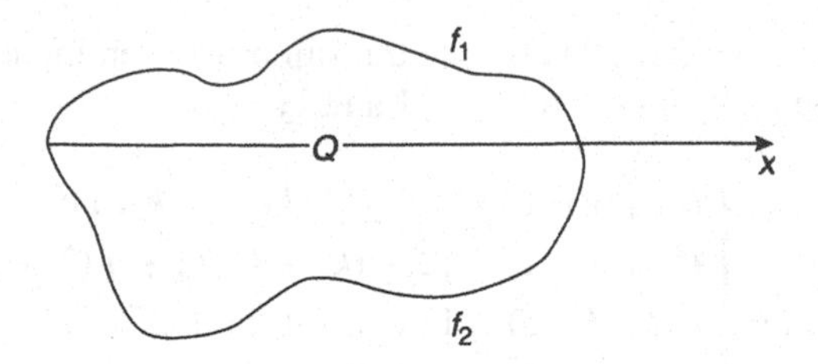


**Figure 3.11.**

**Definition 3.37** *A compact region $Q$ is called simple if the boundary of $Q$ can be represented as the union of two continuous single-valued functions in some coordinate system (see Figure 3.11).*

We shall usually make the following assumption.

**Assumption 1**. The compact region $Q$ is the union of a finite number of simple regions with disjoint interiors.

Note that we allow multiply connected regions with a finite number of "holes" in them.

When considering search games in multidimensional regions, we assume that the hider is captured at the first instant in which the distance between him and the searcher is less than or equal to $r$, where the discovery radius $r$ is a small positive constant. (The case in which $r$ is not a constant is considered in the next section.)

In this section, we are concerned with an immobile hider. Since Theorem 3.3 holds for any search space $Q$, it follows that by using the uniform hiding strategy, $h_\mu$, the hider can guarantee that the expected discovery time is greater than or equal to $\mu/2\rho$, (neglecting $O(r^2)$ terms), where $\mu$ is the Lebesgue measure of $Q$ and $\rho$ is the maximal discovery rate. Thus, the value $v$ of the game satisfies $v \geq \mu/2\rho = \mu/4r$.

An upper bound for $v$ can be derived by extending the notion of a *tour* (used for networks) to multidimensional region:

**Definition 3.38** *A closed curve $L$ that passes inside $Q$ is called a tour if for any $Z \in Q$ there exists $Z' \in L$ such that $d(Z, Z') \leq r$.*

It has been shown by Isaacs (1965, sec. 12.3) that if $L$ is a tour with length $\tau$, then the searcher can guarantee an expected capture time not exceeding $\tau/2$ by choosing each one of the directions, of encircling $L$ with probability 1/2. (We used such a search strategy for Eulerian networks.) Thus, if we could find a tour whose length $\tau$ satisfies

$$\tau \leq (1+\varepsilon)\mu/\rho \tag{3.16}$$

then we would have $v \sim \mu/2\rho = \mu/4r$, and in this case the strategies of the searcher and the hider already described would be $\varepsilon$-optimal. We now show that for any two-dimensional region which satisfies a rather weak condition, it is possible to find a tour whose length satisfies inequality (3.16).

**Lemma 3.39** *Let $Q$ be a two-dimensional compact region that satisfies Assumption 1 with the boundary of each one of the simple regions composing $Q$ having a finite length.*

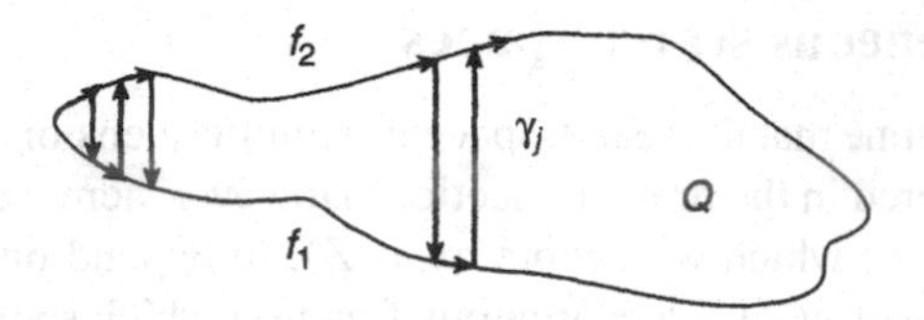


**Figure 3.12.**

*Then for any $\varepsilon > 0$ there is an $r_\varepsilon$ such that for any $r < r_\varepsilon$ there exists a tour with length less than $(1+\varepsilon)\mu/2r$.*

**Proof.** Let $Q = \bigcup_{i=1}^{m} Q_i$, where $Q_i$, $1 \le i \le m$, is a simple region with area $\mu_i$. If for each $1 \le i \le m$ we could find a line that covers $Q_i$ with length less than $(1+\varepsilon/2)\mu_i/2r$, then we could "link" these curves together and form a tour of $Q$ by adding $m$ arcs. The length of such a tour for $Q$ would not exceed

$$\left(1+\frac{\varepsilon}{2}\right)\frac{\mu}{2r} + mD = \left(1+\frac{\varepsilon}{2}+\frac{2rmD}{\mu}\right)\frac{\mu}{2r},$$

where $D$ in the diameter of $Q$. Obviously, if $r$ is small enough, then $2rmD/\mu < \varepsilon/2$, so that we would have the required tour for $Q$. Thus, we may assume that $Q$ is already a simple region. Let $f_1$ and $f_2$ be the graphs of the continuous functions that bound $Q$. We can now cover $Q$ by parallel strips of width $2r$ each, as depicted in Figure 3.12.

Let $y_1, y_2, \ldots$ be the lengths of the vertical line segments formed by these strips. Then by moving alternately along these line segments and along the boundary of $Q$, we form a covering *line* (not necessarily closed) for $Q$ with length less than $\sum y_j + \gamma$, where $\gamma$ is the length of the boundary of $Q$. Due to the fact that $f_1$ and $f_2$ are Riemann integrable, it follows that for any $\varepsilon > 0$, if $r$ is small enough, then $2r\sum y_i < (1+\varepsilon/2)\mu$. Thus, the length of the covering line is less than

$$\left(1+\frac{\varepsilon}{2}+\frac{2r\gamma}{\mu}\right)\frac{\mu}{2r} < (1+\varepsilon)\frac{\mu}{2r}$$

for a small enough $r$. ■

In proving Lemma 3.39 we used the fact that two-dimensional regions can be covered by narrow strips with little overlap. The analog construction for three dimensions would require covering the region with narrow cylinders, but in this case the overlap would not be negligible. Thus, by using the uniform strategy, the hider can keep the expected capture time above $\beta\mu/\rho$, where $\beta > 1/2$. It should be noted, though, that in the next chapter we will show that the value of a search game with a mobile hider in a multidimensional region satisfies

$$v < (1+\varepsilon)\mu/\rho, \quad \text{where } \varepsilon \to 0 \text{ as } r \to 0,$$

and this bound is obviously applicable for the search game with an immobile hider.

**Remark 3.40** *The result of this section extends to the case of $J$ searchers. Thus, if $\rho$ is the total rate of discovery of the searchers (see Corollary 3.6), then for two-dimensional regions $v \sim \mu/2\rho$.*

### 3.7.1 Inhomogeneous search spaces

In this section, we assume that the search space is a multidimensional compact region of the same type considered in the previous section. However, here we allow the maximal velocity of the searcher, which we denote by $v(Z)$, to depend on the location of the searcher. We assume that $v(Z)$ is a continuous function which satisfies

$$0 < v_1 < v(Z) < v_2, \tag{3.17}$$

where $v_1$ and $v_2$ are constants. We also allow the detection radius to depend on the location of the hider. In this case, the detection radius is a function $r(Z)$ whereby if the hider is located at a point $Z$, then he can be seen from any point $Z_1$ that satisfies $d(Z_1, Z) \leq r(Z)$. We assume that

$$r(Z) = \bar{r}\eta(Z),$$

where $\bar{r}$ is a small number and $\eta(Z)$ is a continuous function that satisfies $\min_{Z \in Q} \eta(Z) > 0$.

The discovery rate of the searcher, $\rho(Z)$, is defined as $2v(Z)r(Z)$ for two-dimensional regions, $\pi v(Z) r^2(Z)$ for three-dimensional regions, and so on. The capture time $c(S, H)$ is given by

$$c(S, H) = \min\{t: d(S(t), H(t)) \leq r(H(t))\}.$$

Such a search space will be referred to as an *inhomogeneous search space*. We will show that results established in the first part of the section can be extended in a natural way. We first extend Theorem 3.3 to the inhomogeneous case. In this case, the natural randomization of the hider is given by choosing his location using a probability density which is proportional to $1/\rho(Z)$.

**Lemma 3.41** *Let $\tilde{h}$ be the hiding strategy that uses the probability density $1/\tau\rho(Z)$, where*

$$\tau = \int_Q \frac{d\mu(Z)}{\rho(Z)} \tag{3.18}$$

*and $\mu$ is the Lebesgue measure. Then $\tilde{h}$ satisfies*

$$v(\tilde{h}) \geq \frac{1}{1+\varepsilon} \frac{\tau}{2}. \tag{3.19}$$

**Proof.** A simple continuity argument (for details see Gal, 1980) implies that for any search trajectory $S$, the Lebesgue measure of the strip that is swept during a small time interval $\Delta t$ is less than $(1 + \varepsilon/2)\rho(S(t))\Delta t$. If the hider uses $\tilde{h}$, then the probability mass of such a strip is less than $(1 + \varepsilon)\Delta t/\tau$, where $\tau$ is given by (3.18). Thus, $Pr(T \leq t) < (1 + \varepsilon)t/\tau$, so that

$$c(S, \tilde{h}) = \int_0^\infty Pr(T > t)dt \geq \int_0^\infty \max\left(0, 1 - \frac{1+\varepsilon}{\tau} t\right) dt = \frac{1}{1+\varepsilon} \frac{\tau}{2}.$$

■

We now extend the result established in the first part of the section and show that the value of the search game with an immobile hider in a two-dimensional compact region satisfies $v \sim \tau/2$, where $\tau$ is given by (3.18) with

$$\rho(Z) = 2v(Z)r(Z).$$

The inequality $v > (1/1+\varepsilon)\tau/2$ is an immediate consequence of Lemma 3.41.

An upper bound for the value can be obtained by constructing a tour (see Definition 3.38), such that the time required to encircle it is less than

$$(1+\varepsilon)\int_Q \frac{d\mu(Z)}{2v(Z)r(Z)}.$$

This can be done by dividing $Q$ into several regions $Q_1, \ldots, Q_n$ such that the variation of $\eta(Z)$ in $Q_i i = 1, \ldots, n$, is small and then using the technique adopted in Lemma 3.39.

**Remark 3.42** *All the results presented for the networks in the previous sections can easily be extended to the case in which the maximal velocity of the searcher,* $v(z)$, *depends on his location in the network* $Q$. *(The radius of detection is assumed to be zero, as before.) In this case, the Lebesgue measure of* $Q$, $\mu$, *should be replaced by*

$$\tau = \int_Q \frac{d\mu(Z)}{v(Z)}.$$

# Chapter 4

# Search for a Mobile Hider

## 4.1 Introduction

The interest in search games with a mobile hider was motivated by the presentation of the *Princess and Monster game* described by Isaacs (1965, sec. 12.4). In this game, the Monster searches for the Princess in a totally dark room $Q$ (both of them being cognizant of its boundary). Capture occurs when the distance between the Monster and the Princess is less than or equal to $r$, where $r$ is small in comparison with the dimension of $Q$. As a stepping stone for the general problem, Isaacs suggested a simpler problem in which $Q$ is the boundary of a circle. The Princess and Monster game on the boundary of a circle was solved several years later by Alpern (1974), Foreman (1974), and Zelikin (1972), under the additional assumption that the maximal velocity $w$ of the hider, satisfies $w \geq 1$ (recall that the maximal velocity of the searcher is usually taken as 1). Their formulation is a little different from the one that we usually adopt, in the sense that the searcher does not start from a fixed point $O$. Instead, they assume that at $t = 0$ the initial relative position of the searcher with respect to the hider has a known probability distribution. (Alpern, 1974; Zelikin, 1972, assume a uniform distribution, while Foreman, 1974, considers a general probability distribution.) A discretized version of this problem was solved by Wilson (1972). An attempt to develop a technique for obtaining approximate solutions of some discrete games with a mobile hider by restricting the memory of the players was presented by Worsham (1974). Some versions of this game with a fixed termination time were considered by Foreman (1977). A general solution of the Princess and Monster game in a convex multidimensional region, was presented by Gal (1979, 1980). The solution will be presented in detail in Section 4.5. An important intermediate step is solving the search game with a mobile hider on a network consisting of $k$ arcs connecting two points, which will be presented in Section 4.2. Some extensions were made by Garnaev (1991) (the velocity vectogram of the searcher is a rhombus-type set) and (1992) (the probability of detection depending on the distance between the players). A new type of search strategy presented by Lalley and Robbins (1988), will be described in Section 4.6.1.

Anderson and Armendia (1992) formulated the search game in a network as an infinite-dimensional linear program. They obtained an optimality condition for the strategies and an improvement technique (if the currently used strategies are not optimal).

We now present the general framework of search games with a mobile hider, to be considered in this chapter. We assume that the search space $Q$ is either a network or a compact multidimensional region. A pure strategy of the searcher is a continuous trajectory $S(t), t \geq 0$ inside $Q$ which satisfies

$$S(0) = O \quad \text{and for any } 0 \leq t_1 < t_2\colon d(S(t_1), \quad S(t_2)) \leq t_2 - t_1.$$

A pure strategy of the hider is a trajectory $H(t), t \geq 0$, inside $Q$, with $H(0)$ an arbitrary point chosen by the hider, which satisfies

$$\text{for any } 0 \leq t_1 < t_2\colon d(H(t_1), \quad H(t_2)) \leq w(t_2 - t_1),$$

where $w$ is the maximal velocity of the hider.

The capture time $T$ is the first instant $t$ with $d(S(t), H(t)) \leq r$, or infinity if no such $t$ exists. As usual, we take $r = 0$ for the search in a network and assume that $r$ is a small positive number in the case that $Q$ is a multidimensional region.

When considering the role of mixed search strategies, one immediately observes that their advantage over pure search strategies is much greater in the case of a mobile hider than in the case of an immobile hider. In order to see this fact, consider the case of an *immobile* hider in an Eulerian network of length $\mu$. In this case, by using a trajectory (pure strategy) that traces an Eulerian curve, the searcher can guarantee an expected capture time not exceeding $\mu$, whereas the use of mixed strategies (tracing the curve equiprobably in each direction) enables the searcher to guarantee an expected capture time not greater than $\mu/2$. Thus, the use of mixed strategies against an immobile hider yields an improvement by a factor of 2.

On the other hand, if a *mobile* hider can move on a circle with maximal velocity $w \geq 1$, then no pure search strategy can guarantee capture. This is in fact true for any network other than the line segment. However, Alpern and Asic (1985) showed that if mixed strategies are allowed, then the search value is finite (see Section 4.4).

In the case of the circle (see Section 4.3), the use of a mixed search strategy enables the searcher to achieve expected capture time not exceeding $\mu$. Thus, the advantage of using mixed strategies is much greater in the case of a mobile hider.

**Remark 4.1** *The problem of finding the minimal speed advantage of the searcher, which guarantees capture using pure strategies only was investigated for a given network by Fomin (1999) and for rectangular domains by Avetisyan and Melikyan (1999a, 1999b). Note that having a search trajectory that guarantees capture actually guarantees it even if the (mobile) hider can see the searcher (but, as usual, the searcher cannot see the hider unless their distance is within the capture radius).*

**Remark 4.2** *If only pure search strategies are used then, except for trivial cases, several searchers are needed to guarantee capture. The interesting problem of determining the minimal number of such searchers in a given network, called the search number, was considered by Parsons (1978a, 1978b) and Megiddo and Hakimi (1978). This problem* *has attracted much research. An important result for this game was obtained by Megiddo et al. (1988), who showed that the problem of computing the search number is NP-hard for general graphs but can be solved in linear time for trees.*

As in the discussion in Section 3.1, it is worthwhile to consider the unrestricted game (which is actually a discrete version of the search game) in which the physical constraint of the continuity of the trajectories of the players is disregarded. In this game, the set $Q$ is divided into $n$ "cells" $Q_1, \ldots, Q_n$, each with measure $\mu/n$. The searcher and the hider can move from one cell to another at the time instants $t_i = i\Delta t$, $i = 0, 1, \ldots, n$, where $\Delta t = \mu/n\rho$. The time interval $\Delta t$ may be regarded as the time required by the searcher to sweep one cell. We assume that capture occurs at the end of the first time interval in which both the searcher and the hider occupy the same cell. It is easy to see that the value of the unrestricted game is $\mu/\rho$, because both the searcher and the hider can guarantee this value by using completely random strategies, choosing each cell with equal probability, $1/n$, independently of previous choices, at each time instant $t_i$, $i = 0, 1, \ldots, n$.

It is worthwhile to estimate the probability of capture after time $t$ in the unrestricted game, both in the case of a mobile and an immobile hider (assuming optimal play on both sides). If the hider is immobile then, under optimal strategies, the probability of capture after time $t$ satisfies

$$Pr(T > t) \sim 1 - \rho t/\mu, \quad 0 \le t < \mu/\rho.$$

On the other hand, if the hider is mobile, then

$$Pr(T > t) \sim e^{-\rho t/\mu}, \quad 0 \le t < \infty.$$

In contrast to the unrestricted game, we shall be mainly concerned with games in which both the searcher and the hider are restricted to move along continuous trajectories, but it is interesting to note that in spite of that restriction, the value is still $(1 + \varepsilon)\mu/\rho$ for most games to be considered in this chapter. The first search space we discuss is a network consisting of $k$ arcs connecting two points (Section 4.2). Solving the search game for the $k$ arcs can give us a useful insight, which helps us to understand how to solve the Princess and Monster game in multidimensional regions (Sections 4.5 and 4.6).

A remarkable property of most of the search games solved in this chapter is the following. There exists a function $P(t)$, which decreases exponentially in $t$, such that for all $t$ both the searcher and the hider can keep the probability of capture after time $t$ around $P(t)$. In practice, this property may sometimes be more useful than the expected capture time, since the cost function need not be $T$ itself but may place a different (e.g., heavier) penalty on larger values of $T$. Indeed, this property can be used to show that the optimal ($\varepsilon$-optimal) strategies obtained for the games in which the capture time serves as a cost function are still optimal ($\varepsilon$-optimal) even if the capture time is replaced by a more general cost function (Section 4.6.8). In other words, the optimal ($\varepsilon$-optimal) strategies of these games are uniformly optimal for all reasonable cost functions. (Strategies which guarantee $P(t)$ are called *uniformly optimal* in Book II.) An exception to the existence of uniformly optimal strategies was investigated by Alpern and Asic (1986) (see Remark 4.16).

**Remark 4.3** *The Princess and Monster game was also used in some biological models. See Meyhofer et al. (1997) and Djemai et al. (2000).*

## 4.2 Search on $k$ Arcs

Solving the $k$ arcs search is important because it gives an insight into the nature of optimal search strategies for a mobile hider in more general domains and in particular for multidimensional regions. Here we consider the following search game. The search space $Q$ is a set of $k$ non-intersecting unit length arcs $b_1, \ldots, b_k$, joining two points $O$ and $A$, as depicted in Figure 4.1.

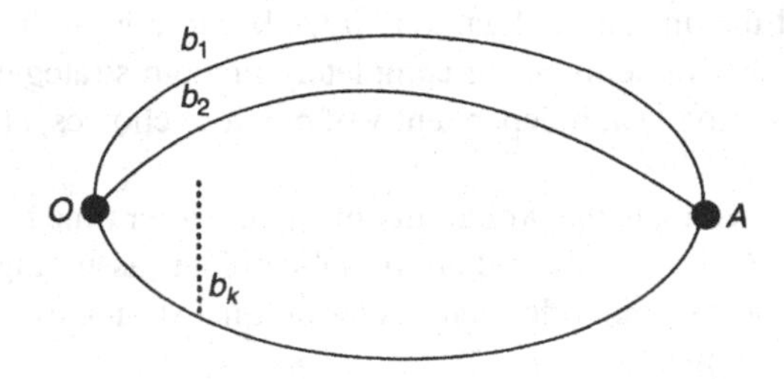


**Figure 4.1.**

(The solution of this problem in the case that the arcs have a common length different from 1, is easily obtained by using the scaling lemma of Chapter 2.) The searcher has to start moving from $O$ with maximal velocity equal to 1. The hider can choose an arbitrary starting point, and from this point he can move along any continuous trajectory in $Q$ with maximal velocity $w$. In this chapter, we shall usually assume that $w \geq 1$. The capture time $T$, which is the loss of the searcher (or the gain of the hider), is the time elapsed until the searcher reaches a point that is occupied at the same time by the hider. In order to avoid unnecessary complications, we make the following assumption.

**Non-Loitering Assumption**. The searcher can pass but not stay at the points $O$ or $A$.

This assumption is not needed for $k = 2$ (search on a circle) as we shall see in Section 4.3.

We shall show that the optimal strategies of the searcher and of the hider are both of the type $U(Z, \tau)$ to be defined.

**Definition 4.4** *Let $Z$ be either the point $O$ or the point $A$ and let $0 \leq \tau < \infty$ be any nonnegative number. Then the (random) trajectory $U(Z, \tau)$ is defined by the following rules. At time $\tau$, starting from the point $Z$, choose an integer $i$ equiprobably from the integers $(1, 2, \ldots, k)$ and move along the arc $b_i$ with unit velocity to the end point of this arc ($O$ or $A$). Then (at time $\tau + 1$) make another equiprobable choice of an integer $j \in \{1, 2, \ldots, k\}$, independently of $i$, and move along $b_j$ until the other end point ($A$ or $O$) is reached, and so on.*

Using Definition 4.4, we state the following theorem.

**Theorem 4.5** *For the search game on k arcs, an optimal strategy for the searcher is $U(O, 0)$. An $\varepsilon$-optimal strategy for the hider is to start at A, stay there until time $1 - \varepsilon$, and then use the strategy $U(A, 1 - \varepsilon)$. The value of the game is $k(= \mu/\rho)$.*

The theorem will be proved using three lemmas. The following *sweeping lemma* is the fundamental one and will also be used in the next sections.

**Lemma 4.6 (Sweeping Lemma)** *Let Q denote the network consisting of two nodes O and A, connected by k unit length arcs. Define k "sweepers," which move from O to A at unit velocity during the time interval $0 \leq t \leq 1$, each along a different arc. Then*

*(a) Any (continuous) hider trajectory $H(t)$ will meet at least one sweeper by time $t = 1$.*

*(b) If the hider's trajectory $H(t)$ has velocity not exceeding one, then it will not meet more than one sweeper by time $t = 1$.*

**Proof.**

(a) Let $f(t) = d(H(t), O)$ denote the distance (in the network $Q$) of the hider from the searcher starting node $O$ at time $t$. Since $H$ is continuous, so is the function $H$. The function $t - f(t)$ is continuous on the interval $[0, 1]$ and is negative for $t = 0$ and non-negative for $t = 1$. So the Intermediate Value Theorem ensures that for some $t_0$ with $t_0 \leq 1$ we have $t_0 - f(t_0) = 0$. At time $t_0$ the hider is therefore at distance $t_0$ from $O$ along one of the $k$ arcs – but so is one of the sweepers.

(b) Let $t_0$ be the first time when the hider meets one of the sweepers – say, $i$. In order to meet another sweeper before time 1, he must move to a different arc. If he does this via the node $O$, he cannot catch the sweeper on that arc before time 1, as he has the same speed. So he would have to move to another arc via $A$. However, he cannot reach $A$ before sweeper $i$, hence not before time 1.

■

**Lemma 4.7** *If the searcher uses the strategy $U(O, 0)$ (see Definition 4.4), then for any hiding strategy H*

$$c(U(O, 0), H) \leq k.$$

**Proof.** Strategy $U(O, 0)$ for the searcher means: At time $t = 0$, the searcher makes an equiprobable choice from the sweepers of the sweeping lemma and moves with the chosen sweeper to $A$; then at time $t = 1$ he makes an independent equiprobable choice from $k$ such sweepers that run from $A$ to $O$, etc. We may look upon the hider as the fugitive, and thus the sweeping lemma implies that if the searcher uses $U(O, 0)$ and if the hider has not been captured by time $j - 1$, then the probability of capture in the time period $j - 1 < t \leq j$ is at least $1/k$ (independently of the previous part of the trajectory). Thus,

$$Pr(T > j) \leq \left(\frac{k-1}{k}\right)^j. \tag{4.1}$$

It follows that for any hiding trajectory $H$ we have

$$c(U(O,0),H) \le \sum_{j=1}^{\infty} Pr(j-1<T\le j)\times j$$

$$= \sum_{j=1}^{\infty} Pr(j-1<T) \le \sum_{j=1}^{\infty}\left(\frac{k-1}{k}\right)^{j-1} = k.$$

■

**Lemma 4.8** *Let $R_A$ be the strategy of the hider described as follows. Stay at point $A$ until time $t = 1-\varepsilon$ and then use $U(A, 1-\varepsilon)$ (see Definition 4.4). Then for any search trajectory $S$,*

$$c(S, R_A) \ge k - \varepsilon.$$

**Proof.** Using strategy $U(A, 1-\varepsilon)$ for the hider means that he chooses each of the sweepers (note that it is the hider who now moves with one of the sweepers) with probability $1/k$ (and independently of previous choices) at the time instants $1-\varepsilon, 2-\varepsilon, \ldots$.

Now the searcher takes the role of the fugitive of the sweeping lemma. We shall use the fact that the velocity of the searcher does not exceed the velocity of the hider and the non-loitering assumption, which states that the searcher cannot wait for the hider at either of the points $O$ or $A$. Thus it can be assumed that at any one of the time instants $1-\varepsilon, 2-\varepsilon, \ldots$, the probability that the searcher is either at $O$ or at $A$ is zero (this can be achieved by the hider by using $\varepsilon$ as a random variable uniformly distributed in any small interval). Thus the condition of part ($b$) of the sweeping lemma is satisfied so that, for any trajectory of the searcher, if the hider has not been captured by time $t = j-\varepsilon$, then the probability of capture in the time period $j-\varepsilon < t \le j+1-\varepsilon$ is equal to $1/k$. Thus, the probability that capture will occur in the time period $j-\varepsilon < t \le j+1-\varepsilon$ is equal to $(k-1/k)^{j-1} \times 1/k$. Hence

$$c(S, R_A) \ge \sum_{j=1}^{\infty}\left(\frac{k-1}{k}\right)^{j-1} \times \frac{1}{k}(j-\varepsilon) = k-\varepsilon.$$

■

Theorem 4.5 is an immediate consequence of Lemmas 4.7 and 4.8. Actually, if $k > 1$, then $c(U(O,0),H) < k$ for any hiding trajectory $H$. The above statement holds because if $H(1) = A$, then $c(U(O,0),H) = 1$, while if $H(1) \ne A$, then the capture time is equal to $\alpha$, $\alpha < 1$, with probability $1/k$. Thus

$$c(U(O,0),H) \le \frac{\alpha}{k} + \frac{k-1}{k}(1+k) < k.$$

Therefore, any mixed Hiding strategy $h$ satisfies $c(U(O,0),h) < k$. Since $v = k$, it follows that the hider does not have an optimal strategy (but obviously has $\varepsilon$-optimal strategies) while the searcher does have an optimal strategy. This result is in accordance with the results presented in Appendix A.

**Remark 4.9** *It follows from the proof of Lemmas 4.7 and 4.8 that the searcher can keep the probability of capture after t below* $(k-1/k)^{\lfloor t\rfloor}$ *($\lfloor t\rfloor$ is the integer part of t), while the hider can keep the probability of capture after t above* $(k-1/k)^{\lfloor t+\varepsilon\rfloor}$. *Thus, if instead of choosing the capture time T as the cost function, we consider a more general cost function W(T), where W is any monotonic nondecreasing function, then the searcher can keep the value below*

$$f(W)=\sum_{i=1}^{\infty}W(i)\times\frac{1}{k}\left(\frac{k-1}{k}\right)^{i-1} \tag{4.2}$$

*while (for any* $\varepsilon>0$*) the hider can keep the value above*

$$f_{\varepsilon}(W)=\sum_{i=1}^{\infty}W(i-\varepsilon)\times\frac{1}{k}\left(\frac{k-1}{k}\right)^{i-1}.$$

*If W is continuous on the left at the points* $T=1,2,3,\ldots$, *then* $f_{\varepsilon}(W)\to f(W)$ *as* $\varepsilon\to 0$. *In other words, if the cost is a continuous nondecreasing function of the capture time, then by using* $U(O,0)$, *the searcher can guarantee the value* $f(W)$, *while the hider can guarantee* $(1-\varepsilon)f(W)$. *It follows that in this case, the value of the game is* $f(W)$ *and the optimal (ε-optimal) strategies of the searcher and the hider are still the same as those described in Lemmas 4.7 and 4.8.*

It is interesting to note that the value of the game is equal to $k$ irrespective of the maximal speed of the hider $w$, as long as it is not less than unity. If $w$ is less than unity, then the optimal strategies of the searcher and the hider may be quite complicated. However, using an argument similar to the one previously presented in this section, it seems to us that if $k$ is large and $w$ is not too small, then the value of the game should be approximately $k$, even for the case $0<w<1$, because the hider can achieve a value of $(1-\varepsilon)k$ by randomly choosing one of the arcs and moving along it with speed $w$ from $A$ to $O$, then using an independent equiprobabilistic choice of another arc and moving along the chosen arc with maximal speed from $O$ to $A$, and so on. If $m=\lceil 1/w\rceil$ and the searcher moves with maximal speed (unity), then, when reaching each of the points $O$ or $A$, the maximal amount of information, which may be available to the searcher is that at that moment the hider is not located on any of the $m$ last arcs visited by the searcher. Thus, even if the searcher could rule out $m$ arcs out of $k$ each time he reaches $O$ or $A$, then his gain would be to increase the probability of capture for each period $j<t\leq j+1$ from $1/k$ to $1/(k-m)$. Thus the expected capture time would decrease at most to $k-\lceil 1/w\rceil$, so that if $k$ is large in comparison with $1/w$, the value obtained in Theorem 4.5 would remain about the same even for $w<1$.

Note that for an immobile hider (see Sections 3.2 and 3.5) the value $v$ of the search game on $k$ parallel arcs is equal to $k/2$ for an even $k$ and is approximately equal to $k/2$ for an odd $k$. Thus, for an immobile hider, $v\sim\mu/2\rho$. For a mobile hider, $v$ is doubled and becomes $\mu/\rho$, and this is due to the fact that, contrary to the case of an immobile hider, the searcher cannot rule out the arcs previously visited by him. We shall have the same phenomenon for the case of a two-dimensional search space.

## 4.3 Search on a Circle

In this section, we consider searching for a mobile hider on a circle. As in the previous section, we shall make the assumption that the maximal velocity $w$ of the hider is greater than or equal to 1. By denoting the (known) starting point of the searcher by $O$ and its antipode by $A$, the problem reduces to a special case of the search game solved in the previous section with $k = 2$ so that the value of the game is the length of the circle, $\mu$. However, we must be aware of the fact that, in the previous section, we restricted the strategies of the searcher by the non-loitering assumption which requires that the searcher cannot wait at $O$ or at $A$. We now show that the result for $k = 2$ remains valid even without this assumption. The searcher's strategy $U(O, 0)$ (see Definition 4.4), which guarantees an expected capture time $\mu (= 2)$, remains the same, but the hider's strategy needs a little modification. At time $t = 0$, the hider should choose a random point, $A'$, uniformly distributed around $A$ with distance $\leq \varepsilon$ (a small number). It is easy to see that by staying at $A'$ until $t = 1 - \varepsilon$ and then using $U(A', 1 - \varepsilon)$, i.e., moving from $A'$ to $O'$ randomly, the hider can make sure that the capture time will exceed $2 - \varepsilon (= \mu - \varepsilon)$ because the probability of capture at $O'$ or at $A'$ is zero, so that the argument used in Lemma 4.8 is valid here as well.

It should be noted that the result $v = \mu$ depends on the assumption that the initial starting point of the searcher is known ($S(0) = O$). This result is no longer valid if one changes this assumption. For example, if one assumes that $S(0)$ has a uniform distribution on the circle, as was done by Alpern (1974) and Zelikin (1972), then the value is $\frac{3}{4}\mu$. This result is demonstrated by the following theorem, which is taken from Alpern (1974). (The optimal strategy below was given the name *coin half tour* by Foreman, 1974.)

**Theorem 4.10** *Consider the search game with a mobile hider on the circle of circumference $\mu$, assuming that $S(0)$ and $H(0)$ are uniformly and independently distributed on the circle. Then*

$$v = \tfrac{3}{4}\mu \tag{4.3}$$

*An optimal strategy for each player is to oscillate at speed 1 between his initial point and its antipode, each time making an equiprobable choice between the clockwise and counterclockwise directions, independently of previous choices.*

**Proof.** Assume for convenience that $\mu = 2$. For any $t > 0$, let $I = \lfloor t \rfloor$ and $\alpha = t - \lfloor t \rfloor$. Let

$$L(t) = \frac{1}{2^I} - \frac{\alpha}{2^{I+1}}. \tag{4.4}$$

First we show that under the search strategy $\bar{s}$, described by Theorem 4.10, the probability of capture after time $t$ satisfies

$$Pr(T > t) \leq L(t) \tag{4.5}$$

for any hiding trajectory $H$.

We prove (4.5) as follows. Assume without loss of generality that $I$ is even. Denote the following events

$$E: d(O, H(t)) \leq \alpha \tag{4.6}$$

$$\tilde{E}: \text{the complement of } E. \tag{4.7}$$

Then the same considerations used in Lemma 4.7 lead to the inequalities

$$Pr(T > t|E) \leq 1/2^{I+1} \tag{4.8}$$

and

$$Pr(T > t \mid \tilde{E}) \leq 1/2^{I}. \tag{4.9}$$

Since $S(0)$ is uniformly distributed on the circle, it can be shown that regardless of the hider's motion $H$

$$Pr(E) = \alpha. \tag{4.10}$$

Now (4.5) readily follows from (4.8) to (4.10). Thus,

$$c(\bar{s}, H) = \int_0^\infty Pr(T > t)\,dt$$

$$= \sum_{I=0}^{\infty} \int_0^1 Pr(T > I + \alpha)\,d\alpha \leq 3/2 \text{ (by(4.4) and (4.5))}.$$

The optimality of the hider's strategy $\bar{h}$, described by Theorem 4.10, is established rather similarly, using Lemma 4.8 instead of Lemma 4.7, by proving that for any trajectory $S$, if the hider uses $\bar{h}$, then $Pr(T > t) = L(t)$, and thus $c(S, \bar{h}) = 3/2$. ■

Note that the optimal search strategy is the same as in the case that $S(0)$ is a fixed known point, while the hiding strategy has to be modified due to the fact that the hider does not know $S(0)$.

The value would still be $\frac{3}{4}\mu$ in the case that both $S(0)$ and $H(0)$ can be chosen by the players because each one of them can guarantee an expected capture time not worse than $\frac{3}{4}\mu$ by choosing a uniformly distributed starting point and then using the Coin Half Tour described above.

In contrast to the foregoing discussion, we shall see in Section 4.5 that in the case of multidimensional search, the value of the game is not sensitive to any assumptions about the location of $S(0)$.

## 4.4 Quickly Searched Networks

In general, solving search games in a Network is difficult. Still, Alpern and Asic (1985) (part a) found the following upper bound, demonstrating that the search value is finite:

**Theorem 4.11** *Let $Q$ be a Network with $m$ edges, $n$ nodes, diameter $D$, and minimal tour length $\bar{\mu}$. Then for any searcher starting point the value of the search game with* *mobile hider satisfies*

(a) $v \leq 6mD$. *(If no arc has length exceeding D, then* $v \leq 4mD$*) and*

(b) $v \leq (n+2)(D+\bar{\mu}/2)$. *(This bound requires loitering strategies.)*

**Proof.** Consider the following mixed searcher strategy. Randomly pick one of the arcs $e$. Move so as to arrive equiprobably at one of the endpoints of $e$ at time $D$. Observe that no arc can have length exceeding $2D$, since otherwise the distance from its midpoint to an end would exceed $D$. So it will always be possible to traverse the arc so as to arrive at the opposite end at time $3D$. (If $e$ is a loop, then it should be traversed equiprobably in either direction.) If the hider has not been found by time $3D$, then repeat the process in each time interval $[3iD, 3(i+1)D]$, making choices independently of previously choices. Observe that in each time period, the probability that the searcher will encounter the hider is at least $p = 1/2m$. This is because the hider will start the time period on some arc $e$ and hence will meet the agent of the searcher starting at one of the ends of $e$. The expected capture time is therefore not more than $3D/p = 6mD$. In the common case that no arc has length exceeding the diameter $D$, the period $3D$ can be reduced to $2D$ and the value to $4mD$.

If $Q$ has many arcs but few nodes (for example, many multiple edges or loops), and loitering strategies are allowed, then another strategy may be better. From time $0$ to $D$ move with probability $n/(n+2)$ to one of the nodes and wait there until time $D+\bar{\mu}/2$ choosing among them equiprobably. With probability $2/(n+2)$ wait until time $D$ and then traverse a minimal tour of $Q$ halfway around, equiprobably in either direction. Repeat this process, with independent randomization, in each period of length $D+\bar{\mu}/2$. Whether or not the hider moves, he will meet the searcher in a given period with probability $1/(n+2)$, either when the searcher is waiting at a node or when he is searching. Consequently the expected capture time does not exceed $(n+2)(D+\bar{\mu}/2)$. ■

For example, if $Q$ is the graph with an even number $m$ of unit length arcs connecting $n = 2$ nodes, then $D = 1$ and $\mu = \bar{\mu} = m$. So the two estimates are $4m$ (part a) and $4(1+m/2)$ (part b). In this case estimate (b) is better for $m > 2$. For the complete graph on $n$ nodes there are $m = n(n-1)/2$ unit length arcs and we have $D = 1$ and $\bar{\mu} \geq m$. In this case the two estimates are $3n(n-1)$ (part a) and at least $(n+2)(1+n(n-1)/4)$ (part b). For $n > 9$ the estimate in part (a) is better.

What can we say about a lower bound for the value? In Chapter 3 we proved Theorem 3.19, which stated that the value of the search game with an immobile hider lies between $\mu/2$ and $\mu$, where $\mu$ is the sum of the lengths of the arcs of the network $Q$. The lower bound is achieved for Eulerian networks, while the upper bound is achieved for trees. It would be interesting to ask similar questions concerning the search for a mobile hider. For example,

*Assuming that* $O = S(0)$ *is known to the hider, are there any networks with* $v < \mu$? *(Recall that* $v = \mu$ *holds for the search on the circle or on k arcs under the non-loitering Assumption.)*

It is not too easy to find such a network. (In fact, it was conjectured in Gal, 1980, that $v \geq \mu$ always hold.) However, Alpern and Asic (1985) showed that $v < \mu$ for the *figure eight* network presented as follows.

### 4.4.1 The figure eight network

The "figure eight" network (see Figure 4.2) consists of two circles, each of unit circumference and only one point in common, $O$. Thus $\mu = 2$.

Denote the antipodes of $O$ by $A$ for the left circle, and $B$ for the right circle and denote the arcs from $A$ to $O$ by $a_1$ and $a_2$, and from $B$ to $O$ by $b_1$ and $b_2$. Each of these arcs has length 1/2.

We shall now show that if the searcher starts at the center $O$ of the figure eight network, then

$$v \leq \frac{15}{8} = \frac{15}{16}\mu.$$

Suppose the searcher begins by traversing the four possible arcs out of $O$ equiprobably, until arriving at one of the endpoints $A$ or $B$ at time 1/2. By this time he will have captured the hider with probability at least 1/4. If capture has not occurred and the hider has met only one of the possible paths of the searcher (the worst case from the searcher's point of view), then the game that remains at time 1/2 is the following: The searcher starts at one of the ends, say $A$, with probability 1/3, and at the other end with probability 2/3. The probability at each end is known to the hider. Call the value of this game (starting at time 0) $v'$. For example, if the hider starts to the left of $O$ and is not caught by time 1/2, he knows that at this time the hider is at $A$ with probability 1/3. If he started to the right of $O$ then $B$ will be the low probability starting point in the subgame. Assuming the searcher plays optimally in the subgame, this strategy ensures an expected capture time (and hence an estimate of $v$) given by

$$v \leq \left(\frac{1}{4}\right)\frac{1}{2} + \left(\frac{3}{4}\right)\left(\frac{1}{2} + v'\right) = \frac{1}{2} + \frac{3}{4}v'. \tag{4.11}$$

The estimation of $v'$ is carried out in the following.

**Lemma 4.12** $v' \leq \frac{11}{16}$.

**Proof.** Assume that $A$ is the low (1/3) probability end. Suppose the searcher starts by going at speed 1 from his starting point ($A$ or $B$) equiprobably along either the top pair of arcs ($a_1b_1$) or the bottom pair of arcs ($a_2b_2$), arriving at time 1 at the other end ($B$ or $A$). Regardless of subsequent play, the optimal response of the hider is to meet a

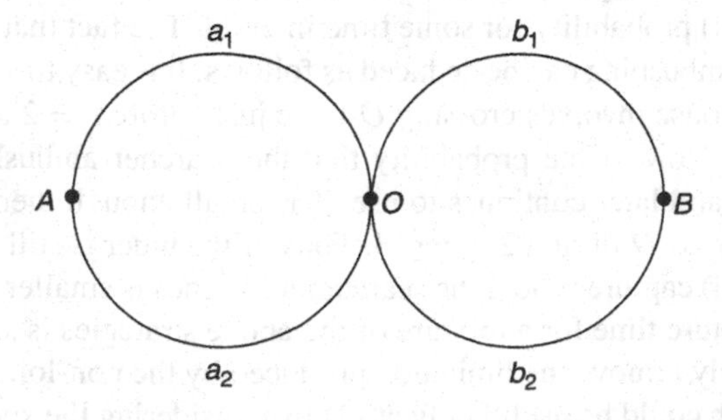


Figure 4.2.

possible searcher path from the low probability ($A$) side of $O$ at time just before $\frac{1}{2}$. If he has not been captured by time $\frac{1}{2}$, then it is even more likely than at the beginning that the searcher started at $B$, so he should go close to $A$, meeting a searcher path starting from B just before time 1. If he has not been captured by time 1, the remaining game is the same as the original one, with the starting probabilities on the ends reversed, so with optimal play the expected *remaining* capture time is again $v'$. Consequently, the hider will be captured just before time $\frac{1}{2}$ with probability $(\frac{1}{3})(\frac{1}{2}) = \frac{1}{6}$ and just before time 1 with probability $(\frac{2}{3})(\frac{1}{2}) = \frac{2}{6}$. With the remaining probability $\frac{1}{2}$ the same game will be played again (with $A$ and $B$ reversed) starting at time 1. Consequently, we have the estimate

$$v' \leq \left(\frac{1}{6}\right)\frac{1}{2} + \left(\frac{2}{6}\right)1 + \left(\frac{1}{2}\right)(1 + v').$$

Solving the above inequality for $v'$ gives

$$v' \leq \frac{11}{6}.$$

■

Substituting this value of $v'$ into the inequality (4.11) gives $v' \leq \frac{15}{8}$. Writing this in terms of the length $\mu$ of the network, we obtain the desired result.

**Theorem 4.13** *The value of the search game with mobile hider on the figure eight network with the searcher starting at the center satisfies*

$$v' \leq \frac{15}{16}\mu.$$

*This can be achieved by a strategy that is consistent with the non-loitering assumption.*

**Remark 4.14** *Note that if the starting point of the searcher is at a known antipode, say A, then the value of the search game is* $2 = \mu$. *(The hider can use the* $\varepsilon$*-optimal strategy for the graph that consists of the two arcs* $a_1 \cup b_1$ *and* $a_2 \cup b_2$ *with a random small* $\varepsilon$*. Since the non-loitering assumption prevents the searcher from ambushing at* $O$*, this strategy is as effective as in the two arcs game as described in Section 4.2.) The advantage of starting at* $O$ *is that at time* $\frac{1}{2}$ *the hider does not know whether the searcher sweeps from A to B or from B to A.*

Can the expected capture time be reduced below $\frac{15}{16}\mu$? Yes, if we do not limit the searcher by the non-loitering assumption and allow the searcher to *ambush* at the node $O$, with some (small) probability for some time interval. The fact that the searcher can actually gain from ambushing can be deduced as follows. It is easy to see that, against $\bar{s}$, the hider's best response involves crossing $O$ once just before $t = 2$ and an additional time before $t = 3$. Now, if the probability that the searcher ambushes at $O$ for the time interval [2,3] (and later continues to use $\bar{s}$) is small enough then the hider's best response is still to cross $O$ during $2 \leq t \leq 3$. Thus, if the hider is still at large at $t = 2$, then the (conditional) capture time if the searcher ambushes is smaller than if he uses $\bar{s}$. So the expected capture time for a mixture of the above strategies is smaller than $\frac{15}{16}\mu$.

We can effectively remove the limitation produced by the non-loitering assumption (note that an ambush could be useful only at $O$) by considering the *spectacles* network given by Figure 4.3.

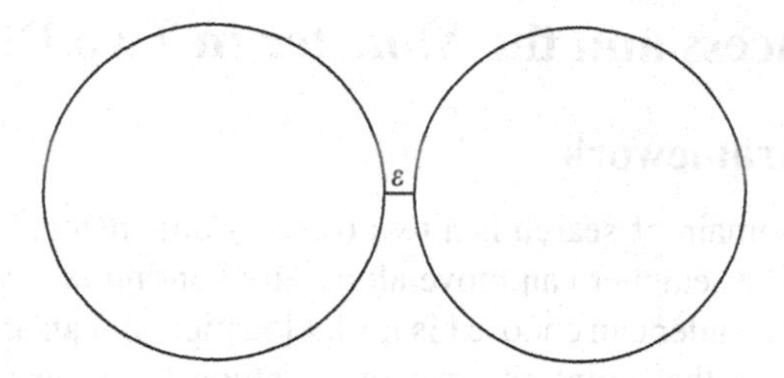


**Figure 4.3.**

Instead of the central node, $O$, of the figure eight graph, we have here a very short central interval. Thus, for the spectacles network, even with the non-loitering assumption, the searcher can "ambush" by slowly moving in the central interval, say during $2 \leq t \leq 3$.

Alpern and Asic (1985) conjectured that the spectacles graph (with an infinitesimal central interval) achieves the minimum possible $v/\mu$, under the non-loitering assumption.

They also conjectured that when loitering is allowed, the quickest graph to find the hider in is the $n$-leafed clover ($n \to \infty$) consisting of a central vertex, where the searcher starts, to which $n$ loops of equal length are attached. Both of the above conjectures are still open problems.

**Remark 4.15** *Analyzing the effect of ambush strategies in a star consisting of a central vertex $O$ to which $n$ line segments of equal length are attached, seems a "neat" problem. However, even for this simple graph, finding the optimal search strategy, allowing ambushing at the central node, seems quite complicated. (A reasonable search strategy is, for each stage, either to visit a segment among those that were not inspected lately, or, with a small probability, ambush at $O$).*

**Remark 4.16** *Alpern and Asic (1986) investigated the effect of removing the non-loitering assumption in the k-arcs game (see Section 4.2). They showed that allowing the searcher to ambush at vertices ($O$ or $A$) in the k-arcs game reduces the value below $k$ for $k > 3$. If $k = 3$, then the value remains 3, but the optimal strategies are not "uniformly optimal" (see Remark 4.9). The analysis is quite complicated (probably because of the lack of uniformly optimal strategies).*

It should also be noted that if we allow the searcher an arbitrary starting point, then the lower bound for the value decreases to $\mu/2$ (= lower bound for an *immobile* hider). This lower bound, $v = \mu/2$, is obtained in the case that $Q$ is any line segment $[a, b]$ and the searcher can choose his starting point. (The searcher can guarantee an expected capture time not exceeding $(b-a)/2$ by choosing $S(0) = a$ with probability $\frac{1}{2}$ and $S(0) = b$ with probability 1/2, and moving with maximal speed to the other end of the segment, while the hider can guarantee an expected capture time greater, or equal to $(b-a)/2$ by choosing a point $H$ that is uniformly distributed in $[a, b]$, and staying at $H$.)

## 4.5 The Princess and the Monster in Two Dimensions

### 4.5.1 General framework

In this chapter, the domain of search is a two (or >2) dimensional convex[1] region in a Euclidean space. The searcher can move along any continuous trajectory that starts from the origin $O$. The hider can choose his initial location and an arbitrary continuous trajectory starting from that point and can move along his trajectory with maximal speed $w$. In contrast to the search on $k$ arcs, we will not require that $w \geq 1$ but assume that $w$ is not too small. (The exact formulation of this condition will be presented in Section 4.5.3.)

The notations that will frequently be used in this chapter are $\mu$, the Lebesgue measure of relevant search space; $D$, its diameter; and $\rho$, the maximal rate of discovery of the searcher (which is equal to $2r$ for two-dimensional sets, $\pi r^2$ for three-dimensional regions, etc.). The radius of detection $r$ will be assumed to be small in relation to the magnitude of $Q$. We shall proceed under the assumption that the detection radius is a constant, and the maximal velocity of the searcher is 1, but in Section 4.6 we show how to extend the results to the case in which both the radius of detection and the maximal velocity of the searcher depend on the location inside $Q$.

For the case of an immobile hider in a two-dimensional region, it has been shown in Section 3.7 that the value, $v$, of the search game, satisfies $v \sim \mu/2\rho$. For the case of a mobile hider, we show in Section 4.5.2 that the searcher can guarantee an expected capture time not exceeding $(1+\varepsilon)\mu/\rho$. The dual result is presented in Section 4.5.3. We show there that the hider can make sure that the expected capture time will exceed $(1-\varepsilon)\mu/\rho$. Thus, we demonstrate that the value of the Princess and Monster game in a multidimensional region satisfies $v \sim \mu/\rho$, independently of the shape of the search space, and we present $\varepsilon$-optimal strategies for both players. Specifically, we prove the following theorem.

**Theorem 4.17** *For a convex two-dimensional region, $Q$, the value, $v$, of the search game with a mobile hider satisfies*

$$v = (1+\varepsilon)\frac{\mu}{2r}$$

*where $\varepsilon \to 0$ as $r \to 0$.*

This result can be extended to non-convex regions as well using a mild modification of the optimal strategies, as discussed in Section 4.6.2. We conduct the detailed proofs for two-dimensional regions, but in Section 4.6.3 we also show how to extend the results to higher-dimensional regions and also discuss some other extensions, including the result that our strategies remain $\varepsilon$-optimal even if a more general cost function is used.

[1] In Section 4.6.2 we show that our results also hold for nonconvex regions. We make this assumption in order to simplify the presentation.

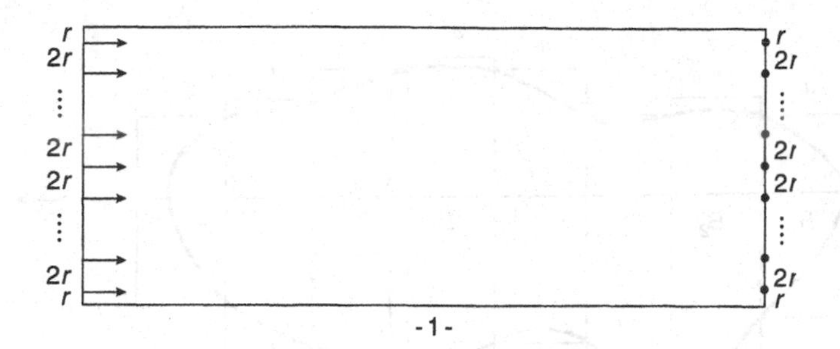


**Figure 4.4.**

## 4.5.2 Strategy of the searcher

In this section, we shall prove that for any bounded two-dimensional convex[2] region $Q$, if the detection radius $r$ is small, then the searcher has a strategy $\bar{s}$ that makes sure that the expected capture time does not exceed $(1+\varepsilon)\mu/2r$, where $\varepsilon$ is small.

As an easy "starter" consider searching (for a mobile hider) in a narrow rectangle with width 1 and height $\varepsilon = 2rk$ ($k$ being any positive integer and $r$ very small). Cover $Q$ by $k$ parallel strips, each of height $2r$. Consider $k$ sweepers (each of them assigned to a different strip) that move with speed 1 in the middle of the strip from one (narrow) side to the other in the time segment $\tau \le t \le \tau + 1$ (see Figure 4.4).

Then, just as in the $k$ arcs network, at least one of these sweepers must meet the hider. The problem now is very similar to the $k$ arcs network, except for a small perturbation. Assume that the searcher starts from one of the narrow edges. He can choose any sweeper (with equal probability) to reach him by time $\tau_0 = 2kr$ and join him during the time segment $\tau \le t \le \tau + 1$. Then, at time $\tau_1 = \tau_0 + 1 + 2kr$, he can (randomly) join another sweeper, etc. Just as in Lemma 4.7, the expected capture time would satisfy

$$\tilde{T} \le 2kr + k(1 + 2rk) = \varepsilon + k(1+\varepsilon) \sim k = \frac{\mu}{2r}.$$

We now construct the search strategy for any convex region $Q$. The search strategy $\bar{s}$ to be considered has the following general structure. The region $Q$ is covered by a set of parallel congruent rectangles with height much smaller than the width; a rectangle, chosen randomly, is entered and examined by moving $N$ times from one narrow side to the other along randomly chosen heights; then another rectangle is randomly chosen, etc. The number $N$ should be large enough to "absorb" the effect of the time spent in going from one rectangle to another, but on the other hand, $N$ must not be so large that too much time is spent in one rectangle. Having this idea in mind we proceed as follows.

Let $Q_a \supset Q$ be the set with minimal area that is a union of the rectangles $B_l, l = 1, \ldots, L, \ldots$, where each $B_l$, with height $\gamma_l$, and width $2a$, is parallel to the $x$ axis as shown in Figure 4.5.

[2] We have already noted that convexity is not really needed (see Section 4.6.2).

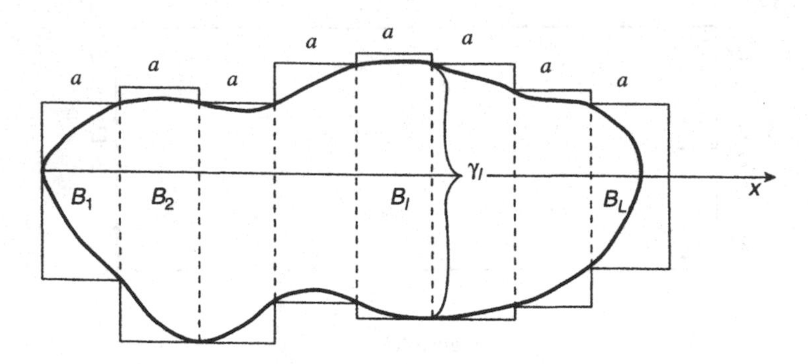


Figure 4.5.

Let $\mu$ be the area of $Q$ and $\mu_a$ be the area of $Q_a$. Then

$$\mu_a = (1 + \phi_a)\mu, \tag{4.12}$$

where, by the convexity of $Q$,

$$\phi_a \downarrow 0 \text{ as } a \downarrow 0. \tag{4.13}$$

For any $Z_1, Z_2 \in Q_a$, let $d(Z_1, Z_2)$ be the minimal length of a path that connects $Z_1$ and $Z_2$ and passes inside $Q_a$. As usual, $D$ will denote the diameter of the relevant search space ($Q_a$).

We shall give a constructive proof of the following theorem.

**Theorem 4.18** *Let $r$ satisfy*

$$r = \varepsilon \frac{a^2}{2D} \tag{4.14}$$

*and assume that*

$$\delta = \varepsilon^{1/4} \ll 1. \tag{4.15}$$

*Then there exists a search strategy $\bar{s}$ in $Q_a$ such that for any evading trajectory $H$ used by the hider, the expected capture time $c(\bar{s}, H)$ satisfies*

$$c(\bar{s}, H) \le (1 + 4\delta)\frac{\mu_a}{2r} = (1 + 4\delta)(1 + \phi_a)\frac{\mu}{2r}. \tag{4.16}$$

We use the following construction. Let

$$\alpha = a\delta^2. \tag{4.17}$$

We first divide each rectangle $B_l, l = 1, \ldots, L$, of size $\gamma_l \times 2a$ (see Figure 4.5) by horizontal lines, into narrow rectangles, $Q_m$, so that all of these rectangles except possibly one, have a width $\alpha$, while the upper one has a width $\alpha' \le \alpha$; then split each rectangle $Q_m$ into two halves, each of length $a$, by a vertical line in the middle and denoted by $Q_{m0}$ – the left rectangle and $Q_{m1}$ – the right rectangle, as depicted in Figure 4.6.

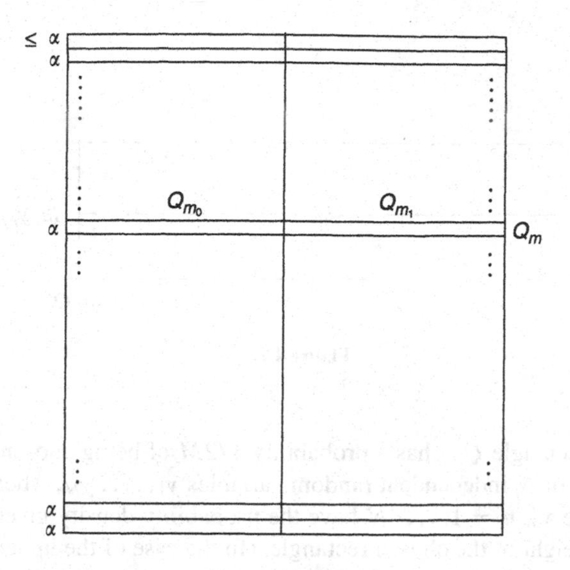


**Figure 4.6.**

Thus, each rectangle $Q_{mi}, i = 0, 1; m = 1, \ldots, M$, has a height $\leq \alpha$. Since $\mu_a = \sum_l 2a\gamma_l$ (see Figure 4.5), then the number $2M$ of such rectangles satisfies

$$2M = 2\sum_{l=1}^{L} \lceil \gamma_l/\alpha \rceil \leq 2\sum_{l=1}^{L} \frac{\gamma_l}{\alpha} + 2L \leq \frac{\mu_a}{\alpha a} + 2\frac{D+a}{a}$$

$$= \frac{\mu_a}{\alpha a}\left(1 + 2\frac{\alpha(D+a)}{\mu_a}\right) < \frac{\mu_a}{\alpha a}\left(1 + \frac{\delta}{2}\right) \tag{4.18}$$

by (4.17) and (4.15).

Let $N$ be a positive integer defined by

$$N = \left\lceil \frac{D}{\delta a} \right\rceil. \tag{4.19}$$

Let $y$ be a random variable $y$ with the density

$$\begin{aligned} f(y) &= \frac{2}{\alpha + 2r} && \text{for } 0 \leq y \leq r \\ &= \frac{1}{\alpha + 2r} && \text{for } r < y < \alpha - r \\ &= \frac{2}{\alpha + 2r} && \text{for } \alpha - r \leq y \leq \alpha \\ &= 0 \quad \text{elsewhere.} \end{aligned} \tag{4.20}$$

The search strategy $\bar{s}$ is composed of independent repetitions of the following step. At time $t = 0$ make a random choice out of the narrow rectangles $Q_{10}, Q_{11}, \ldots Q_{M0}, Q_{M1}$

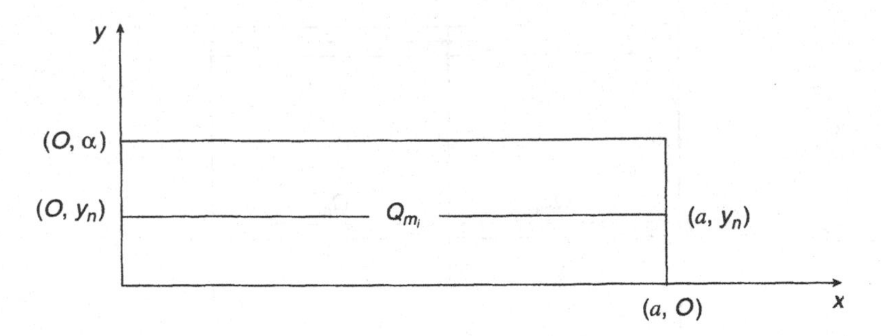


**Figure 4.7.**

such that each rectangle $Q_{mi}$ has a probability $1/2M$ of being chosen, and also make a random choice of $N$ independent random variables $y_1, \ldots, y_N$, where $N$ is given by (4.19) and all the $y_n, n = 1, \ldots, N$ have the probability density given by (4.20), and where $\alpha$ is the height of the chosen rectangle. (In the case of the upper rectangle, with height $\alpha' < \alpha$, then $\alpha'$ replaces $\alpha$ in (4.20); if $\alpha' < 2r$ we define $y$ to be identically $\alpha'/2$.) In order to describe the motion of the searcher within $Q_{mi}$, we shall use a coordinate system with origin at the lower left corner of $Q_{mi}$ as depicted in Figure 4.7. At time $t = 0$, the searcher starts moving as fast as possible to the point $(ia, y_1)$ (i.e., on the left vertical edge for $i = 0$ and on the right edge for $i = 1$). He rests at that point until $t = D$ and then moves with maximal velocity *horizontally* to the other vertical edge (to $(a, y_1)$ if $i = 0$ and to $(0, y_1)$ if $i = 1$) and reaches it at time $t = D + a$, then he moves *vertically* to $y = y_2$, i.e., to the point $(a, y_2)$ if $i = 0$ and to $(0, y_2)$ if $i = 1$. He rests there until $t = D + a + \alpha$, and at that moment he starts moving horizontally to the other vertical edge, etc.

The important feature of the movement of the searcher, using $\bar{s}$, is that at the time segments

$$\tau_n = [D + (n-1)(a+\alpha), D + (n-1)(a+\alpha) + a], \quad n = 1, \ldots, N,$$

he moves, *horizontally*, along the intervals that join $(0, y_n)$ to $(a, y_n)$. We shall show that for this kind of movement, the following proposition holds.

**Proposition 4.19** *If the searcher uses $\bar{s}$, then, for any hider strategy, the probability $p$ of capture during the time segment $0 \le t \le D + N(a+\alpha)$ satisfies*

$$p \ge \frac{1}{2M}\left(1 - \left(\frac{\alpha}{\alpha + 2r}\right)^N\right). \tag{4.21}$$

**Proof.** Consider a specific time segment $\tau_n$ given by

$$\tau_n = \{t: D + (n-1)(a+\alpha) \le t \le D + (n-1)(a+\alpha) + \alpha\} \tag{4.22}$$

where $1 \le n \le N$.

We shall distinguish between two cases. If $n$ is odd and $i = 0$ or if $n$ is even and $i = 1$, then for any $t \in \tau_n$, we define $G_{mi}(t), m = 1, \ldots, M$, as the vertical line segment of width $\alpha$, which has a distance $d(t)$ from the *left* vertical edge of $Q_{mi}$, where

$$d(t) = t - (D + (n-1)(a + \alpha)) \tag{4.23}$$

so that $G_{mi}(t)$ is given by

$$G_{mi}(t) = \{(d(t), y), 0 \leq y \leq \alpha'\}. \tag{4.24}$$

If $n$ is even and $i = 0$ or if $n$ is odd and $i = 1$, then for any $t \in \tau_n$, we define $G_{mi}(t), m = 1, \ldots, M$, as the vertical line segment of width $\alpha'$, which has a distance $d(t)$ from the *right* vertical edge of $Q_{mi}$, i.e., in this case

$$G_{mi}(t) = \{(a - d(t), y), 0 \leq y \leq \alpha'\}. \tag{4.25}$$

In both cases we define

$$G(t) = \bigcup_{m=1}^{M} \bigcup_{i=0}^{1} G_m(t). \tag{4.26}$$

By an argument similar to those used in proving the sweeping lemma (4.6) of Section 4.2, one can show that the following lemma holds. ■

**Lemma 4.20** *If $H$ is any trajectory used by the hider, then for any $n$ there exists at least one time instant $t_n \in \tau_n$ (see (4.22)) such that the point $H(t_n)$ visited by the hider at time $t_n$ satisfies $H(t_n) \in G(t_n)$ (see (4.26)).*

It follows from the lemma that for any $n$ there exist $m$ and $i$ such that

$$H(t_n) \in G_{mi}(t_n). \tag{4.27}$$

Let $I_{mi}(n) = 1$ if (4.27) holds and zero otherwise, and define

$$I_{mi} = \sum_{n=1}^{N} I_{mi}(n). \tag{4.28}$$

Then it follows from the foregoing discussion that

$$\sum_{m=1}^{M} \sum_{i=0}^{1} I_{mi} \geq N \tag{4.29}$$

Assume that the searcher chooses the rectangle $Q_{mi}$. If $I_{mi}(n) = 1$, then it follows from the definition of the random variable $y_n$ that the probability of capture during the time segment $\tau_n$ is greater than or equal to the probability that at the time $t_n$ (see (4.27)) the random interval $Y_n$ given by

$$Y_n = [y_n - r, y_n + r] \cap [0, \alpha']$$

will contain the $y$-th coordinate of $H(t_n)$. Now, it follows from (4.20) that for any point $b$ in the interval $[0, \alpha']$ the probability that $b \in Y_n$ is greater than or equal to

$$\frac{2r}{\alpha' + 2r} \geq \frac{2r}{\alpha + 2r}.$$

Since the random variables $y_1, \ldots, y_N$ are independent, it follows that the probability of capture during the time segment $0 \leq t \leq D + N(a + \alpha)$ is greater than or equal to

$$1 - \left(1 - \frac{2r}{\alpha + 2r}\right)^{I_{mi}} \qquad \text{(see (4.28))}.$$

Since each rectangle $Q_{mi}$, $m = 1, \ldots, M$, $i = 0, 1$, is chosen with probability $1/2M$, it follows that the probability $p$ of capture during the time segment $0 \leq t \leq D + N(a + \alpha)$ satisfies

$$\begin{aligned} p &\geq \sum_{m=1}^{M} \sum_{i=0}^{1} \frac{1}{2M} \left(1 - \left(\frac{\alpha}{\alpha + 2r}\right)^{I_{mi}}\right) \\ &= 1 - \frac{1}{2M} \sum_{m=1}^{M} \sum_{i=0}^{1} \left(\frac{\alpha}{\alpha + 2r}\right)^{I_{mi}}. \end{aligned} \tag{4.30}$$

Since for any nonnegative integers $J$, $K$

$$\left(1 - \left(\frac{\alpha}{\alpha + 2r}\right)^{J}\right)\left(1 - \left(\frac{\alpha}{\alpha + 2r}\right)^{K}\right) \geq 0,$$

it follows that

$$\left(\frac{\alpha}{\alpha + 2r}\right)^{J} + \left(\frac{\alpha}{\alpha + 2r}\right)^{K} \leq 1 + \left(\frac{\alpha}{\alpha + 2r}\right)^{J+K}$$

so that (by (4.29))

$$\begin{aligned} \sum_{m=1}^{M} \sum_{i=0}^{1} \left(\frac{\alpha}{\alpha + 2r}\right)^{I_{mi}} &\leq 2M - 1 + \left(\frac{\alpha}{\alpha + 2r}\right)^{\sum_{m=1}^{M} \sum_{i=0}^{1} I_{mi}} \\ &\leq 2M - 1 + \left(\frac{\alpha}{\alpha + 2r}\right)^{N}. \end{aligned} \tag{4.31}$$

Thus, it follows from (4.30) and (4.31) that

$$p \geq \frac{1}{2M}\left(1 - \left(\frac{\alpha}{\alpha + 2r}\right)^{N}\right).$$

■

We can now proceed with the proof of Theorem 4.18. First we note that (4.14), (4.15), (4.17), and (4.19) imply that

$$\frac{2r}{\alpha} = \frac{2\delta^4 a^2}{2D\delta^2 a} = \delta^2 \frac{a}{D} \geq \frac{\delta}{N}.$$

Thus by (4.21), the probability $p$ of capture in the time segment $0 \leq t \leq D+N(a+\alpha)$ satisfies

$$p \geq \frac{1}{2M}\left(1 - \frac{1}{(1+(2r/\alpha))^N}\right) \geq \frac{1}{2M}\left(1 - \frac{1}{(1+(\delta/N))^N}\right) \tag{4.32}$$

$$\geq \frac{1}{2M}\left(1 - \frac{1}{1+\delta}\right) = \frac{\delta}{2M(1+\delta)}. \tag{4.33}$$

Now, since the search strategy $\bar{s}$ is composed of independent repetitions of the step described for the time segment $0 \leq t \leq D + N(a+\alpha)$, then for any hiding trajectory $H$, the probability $\bar{p}_K$ of capture after the time instant $t = K(D + N(a+\alpha))$ satisfies

$$\bar{p}_K \leq (1-p)^k. \tag{4.34}$$

Thus the expected capture time $c(\bar{s}, H)$ satisfies (see (2.5))

$$c(\bar{s}, H) \leq (D + N(a+\alpha)) \sum_{K=0}^{\infty} \bar{p}_K \leq \frac{(D + N(a+\alpha))}{p} \text{ (by (4.34))}$$

$$\leq \frac{2M(1+\delta)}{\delta} aN \left(1 + \frac{D}{aN} + \frac{\alpha}{a}\right) \text{(by (4.32))}$$

$$\leq \frac{\mu_a}{a\alpha}\left(1+\frac{\delta}{2}\right)\frac{1+\delta}{\delta}\left(\frac{D}{a\delta}+1\right)a(1+\delta+\delta^2) = \text{ (by (4.17)–(4.19))}$$

$$= \frac{1}{\delta^4}\frac{\mu_a D}{a^2}\left(1+\frac{\delta}{2}\right)(1+\delta)\left(1+\delta\frac{a}{D}\right)(1+\delta+\delta^2)$$

$$\leq \frac{\mu_a}{(a^2/D)\varepsilon}(1+4\delta) = \frac{\mu_a}{2r}(1+4\delta) \tag{4.35}$$

(by (4.14) and (4.15)). This completes the proof of Theorem 4.18.

The following corollary is important for establishing the $\varepsilon$-optimality of $\bar{s}$ for a more general cost function (see Section 4.6.8).

**Corollary 4.21** *If the searcher uses the strategy $\bar{s}$ described in the proof of Theorem 4.18, then for any hiding trajectory $H$ the probability that the capture time $T$ exceeds $t$ satisfies*

$$Pr(T > t) \leq (1+\varepsilon)\exp\left(\frac{-2rt}{(1+\varepsilon)\mu}\right)$$

*with $\varepsilon \to 0$ as $r \to 0$.*

**Proof.** Let

$$I_t = \left\lfloor \frac{t}{D+N(a+\alpha)} \right\rfloor.$$

Then it follows from (4.32) and (4.34) that if the searcher uses $\bar{s}$, then for any $H$

$$
\begin{aligned}
Pr(T > t) &\leq Pr(T > (D + N(a+\alpha))\, I_t) \\
&\leq (1-p)^{I_t} \leq \left(1 - \frac{\delta}{2M(1+\delta)}\right)^{I_t} \leq \exp\left(-\frac{\delta I_t}{2M(1+\delta)}\right) \\
&\leq \exp\left(-\frac{t\delta}{2M(1+\delta)(D+N(a+\alpha))} + \frac{\delta}{2M(1+\delta)}\right)
\end{aligned}
$$

by (4.12), (4.14), (4.15), and (4.17)–(4.19):

$$
\leq \exp\left(-\frac{2rt}{\mu(1+\phi_a)(1+4\delta)} + \frac{\delta}{2M(1+\delta)}\right).
$$

Since both $\delta \to 0$ and $\phi_a \to 0$ as $r \to 0$, we obtained the desired result. ■

It should be noted that if the searcher uses the strategy $\bar{s}$ presented in the proof, then a part of his trajectory, which is near the boundary of $Q_a$ might be slightly outside of the original search space $Q$. However, if $Q$ is convex, then we can make a slight modification in $\bar{s}$ and introduce a search strategy $\hat{s}$ that uses trajectories entirely inside $Q$ and still guarantees that the result (4.16) of Theorem 4.18, holds. The strategy $\hat{s}$ is defined as follows. If the chosen rectangle $Q_{mi}$ is inside $Q$, then the movement is identical to the one in $\bar{s}$. However, if a part of $Q_{mi}$ is outside $Q$, as depicted in Figure 4.8, we make the following modification.

Assume that in the strategy $\bar{s}$ the searcher moves from the point $(0, y_n)$ to $(a, y_n)$ in the time segment $\tau_n$ (see (4.22)) and then moves from the point $(a, y_{n+1})$ to $(0, y_{n+1})$ in the time segment $\tau_{n+1}$. The movement in $\hat{s}$ is as follows. The searcher moves from the point $(x_n, y_n)$ to $(x_n', y_n)$ (see Figure 4.8) in the time segment

$$
D + (n-1)(a+\alpha) + x_n \leq t \leq D + (n-1)(a+\alpha) + x_n',
$$

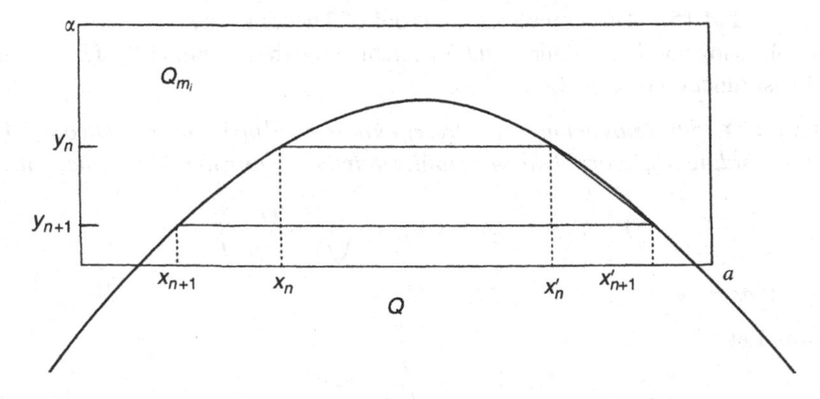


**Figure 4.8.**

then moves with maximal velocity in a straight line from $(x'_n, y_n)$ to $(x'_{n+1}, y_{n+1})$, and stays there until the time instant

$$t' = D + n(a + \alpha) + a - x'_{n+1}.$$

The searcher can arrive at $(x'_{n+1}, y_{n+1})$ before $t'$ because the length of the segment that joins $(x'_n, y_n)$ to $(x'_{n+1}, y_{n+1})$ does not exceed

$$\left(a - x'_n + \alpha + a - x'_{n+1}\right).$$

He then moves from $(x'_{n+1}, y_{n+1})$ to $(x_{n+1}, y_{n+1})$ in the time segment

$$t' \leq t \leq D + n(a + \alpha) + a - x_{n+1}, \text{ etc.}$$

By using exactly the same method of proving (4.16), it can be shown that for any hiding trajectory $H$, the expected capture time satisfies

$$c(\hat{s}, H) \leq (1 + 4\delta)\frac{\mu_a}{2r}$$

so that for a convex $Q$, the strategy $\hat{s}$ guarantees the desired result by moving only inside $Q$.

### 4.5.3 Strategy of the hider

The strategy $h_u$ of the hider to be considered in this section is defined as follows.

**Definition 4.22** *Choose a point $Z_1$ using a uniform probability distribution in $Q$, and stay there during the time period $0 \leq t \leq u$. At the time $t = u$, choose a point $Z_2$, that is uniformly distributed in $Q$ independent of $Z_1$, move toward $Z_2$ with velocity $w_1 = min(w, 1)$ in a straight line, and stay in $Z_2$ for a time period of length $u$. Then choose a point $Z_3$ uniformly distributed in $Q$ and independent of $Z_1$ and $Z_2$, move toward it, again with velocity $w_1$ in a straight line, and stay there for a time period of length $u$, and so on. The "resting time" $u$ should satisfy two conditions:*

*(1) It should not be too long, so that the area covered by the searcher in a time interval of length $u$ would be small relatively to $\mu$.*

*(2) On the other hand, in order to keep the probability of capture during motion relatively small, the hider should not move too frequently and thus $u$ should not be too short.*

Note that the hider need not randomly choose the time of changing his location; it is enough that he moves relatively rarely.

Assume that the hider uses a strategy $h_u$ as described by Definition 4.22. Let

$$E_i - \text{the event capture occur at point } Z_i, i = 1, 2, \ldots.$$

Now if it were possible to neglect the probability of capture during the motion from $Z_i$ to $Z_{i+1}$, then for any search trajectory $S$, the expected capture time would approximately satisfy

$$c(S, h_u) \gtrsim \sum_{i=1}^{\infty}(i-1)u \times Pr(E_i)$$

$$= u\sum_{n=1}^{\infty}\sum_{i=n+1}^{\infty} Pr(E_i) \geq u\sum_{n=1}^{\infty} Pr\left(\bigcup_{i=1}^{\bar{n}} E_i\right). \tag{4.36}$$

Since each $Z_i$ is uniformly distributed in $Q$ and the maximal rate of discovery is $2r$, it then follows that for each $Z_i$ the probability of being discovered at $Z_i$ is at most $2ru/\mu$. Thus, it follows from the independence of $Z_i$ that

$$Pr\left(\bigcup_{i=1}^{\bar{n}} E_i\right) \geq \left(1 - \frac{2ru}{\mu}\right)^n. \tag{4.37}$$

Thus, it would follow from (4.36) and (4.37) that for all $S$

$$c(S, h_u) \gtrsim u\sum_{n=1}^{\infty} Pr\left(\bigcup_{i=1}^{\bar{n}} E_i\right) \geq u\frac{1 - 2ru/\mu}{2ru/\mu} = \frac{\mu}{2r} - u, \tag{4.38}$$

so that if the parameter $u$ of the hider's strategy $h_u$ could be chosen to be small in comparison with $\mu/2r$, we would get the desired result.

We have presented the previous discussion in order to help the reader understand the motivation behind the definition of the strategy $h_u$ and the idea of the proof of Theorem 4.23, which follows.

We must also require that the maximal velocity of the hider, $w$, should not be too small because when $w$ reduces to zero, we approach the situation of an immobile hider considered in the previous chapter, and the value of the search game should then be approximately $\mu/2\rho(= \mu/4r)$. Such a condition is the following:

$$\Delta = \frac{D}{\mu w_1} r \ll 1 \tag{4.39}$$

where

$$w_1 = \min(w, 1).$$

In the next theorem we show for any two-dimensional convex set, that if (4.39) holds, then the hider can make sure that the expected capture time will exceed $(1-\varepsilon)\mu/2r$. The theorem is formulated and proved for two dimensions, but it can be easily extended to three or more dimensions.

**Theorem 4.23** *Let the search space $Q$ be any two-dimensional convex set. Denote*

$$\delta = \left[16\pi\left(\Delta + \frac{r^2}{\mu}\right)\right]^{1/4}, \tag{4.40}$$

*where $\Delta$ satisfies condition (4.39). If the hider uses the strategy $h_u$ presented in Definition 4.22, where*

$$u = \delta \frac{\mu}{2r}, \tag{4.41}$$

*then for any search trajectory $S$ the expected capture time satisfies*

$$c(S, h_u) \geq \frac{\mu}{2r}(1 - 3\delta). \tag{4.42}$$

The hiding points used in the proof $Z_i, i = 1, 2, \ldots,$ are described by Definition 4.22. Let $F_i$ and $F_i'$ be the following events:

$$F_i\text{: capture occurs while the hider is moving from } Z_i \text{ to } Z_{i+1} \tag{4.43}$$

$$F_i'\text{: capture does not occur before the hider leaves } Z_i \tag{4.44}$$

then (see (4.36) and (4.44))

$$\begin{aligned} Pr(F_n') &= 1 - Pr\left(\bigcup_{i=1}^{n} E_i \bigcup_{i=1}^{n-1} F_i\right) \\ &\geq 1 - Pr\left(\bigcup_{i=1}^{n} E_i\right) - Pr\left(\bigcup_{i=1}^{n-1} F_i\right) \\ &\geq Pr\left(\overline{\bigcup_{i=1}^{n} E_i}\right) - \sum_{i=1}^{n-1} Pr\left(F_i\right). \end{aligned} \tag{4.45}$$

The first stage of the proof is to establish an upper bound for $Pr(F_i)$ so as to show that it is negligible in comparison to the other relevant terms. The following general lemma concerning mixtures of uniform variables will be used.

**Lemma 4.24** *Let $Z_i$ and $Z_{i+1}$ be two independent random variables uniformly distributed in a convex (two dimensional) region, $Q$ (i.e., with probability density $1/\mu$ each). Let $Y_\theta$ be the random point*

$$Y_\theta = \theta Z_i + (1 - \theta) Z_{i+1},$$

*where $0 \leq \theta \leq 1$ is a fixed constant. Then the probability density of $Y_\theta$ does not exceed $4/\mu$.*

**Proof.** Assume that $\theta \geq 1/2$ (otherwise the role of $Z_i$ and $Z_{i+1}$ would be interchanged). Let $x$ be any point inside $Q$. Choose $\varepsilon > 0$ small enough so that the disc $\{y\colon d(y, x) \leq \varepsilon\}$ is inside $Q$. Now the event

$$d(Y_\theta, x) \leq \varepsilon \tag{4.46}$$

occurs only if

$$d\left(Z_i, \frac{1}{\theta}x - \frac{1-\theta}{\theta} Z_{i+1}\right) \leq \frac{1}{\theta}\varepsilon \leq 2\varepsilon. \tag{4.47}$$

Since $Z_i$ is a uniform random variable, it follows that for any $Z_{i+1}$ the conditional probability for the event (4.47) (and hence of (4.46)) does not exceed $4\pi\varepsilon^2$. Thus, the overall probability of (4.46) is at most $4\pi\varepsilon^2$ so that the probability density of $Y_\theta$, at any point $x$ inside $Q$, is at most $4/\mu$. ■

Then we establish the following proposition:

**Proposition 4.25** *$Pr(F_i) \le \delta^4$ (see (4.40) and (4.43)).*

**Proof.** The hider moves from point $Z_i$ to point $Z_{i+1}$ in a straight line with velocity $w_1 = \min(w, 1)$. Divide the time of movement into $J$ equal sections where

$$J = \left\lceil \frac{D}{rw_1} \right\rceil. \tag{4.48}$$

Then the time duration of each section $\le (d(Z_i, Z_{i+1}))/Jw_1 \le r$, and the distance traveled by the hider $\le rw_1$. Also, the location, $H(t_{ij})$, of the hider at the middle of the $j$-th section, $t_{ij}$, $j = 1, \ldots, J$, satisfies

$$H(t_{ij}) = \frac{J - j + (1/2)}{J} Z_i + \frac{j - (1/2)}{J} Z_{i+1}. \tag{4.49}$$

Let $S_i$ be the part of the searcher's trajectory that corresponds to the time the hider moves from $Z_i$ to $Z_{i+1}$. We also divide $S_i$ into $J$ arcs for $S_{i1}, \ldots, S_{iJ}$, where $S_{ij}$ is the part of the trajectory followed by the searcher during section $j$. Note that each section $S_{ij}$ is traversed during a time interval of length at most $r$.

Let $F_{ij}$ be the event:

$F_{ij}$: At some point of $S_{ij}$, the distance between the searcher and the hider does not exceed $r$. (4.50)

Obviously (see (4.43)),

$$Pr(F_i) = Pr\left(\bigcup_{j=1}^{J} F_{ij}\right) \le \sum_{j=1}^{J} Pr(F_{ij}). \tag{4.51}$$

Let

$$\alpha = \left(1 + \frac{1}{2} + \frac{w_1}{2}\right) r \le 2r. \tag{4.52}$$

Note that for each of the $J$ time sections the distance traveled during half of the section is at most $r/2$ for the searcher and $rw_1/2$ for the hider. Thus, a necessary condition for the validity of the event $F_{ij}$ (see (4.50)) is the event $M_{ij}$:

$$M_{ij}: d(S(t_{ij}), H(t_{ij})) \le \alpha, \tag{4.53}$$

where $t_{ij}$ is the middle point of $j$-th time segment.

We can now use Lemma 4.24 (and (4.52)) and get

$$Pr(M_{ij}) \le \frac{4\pi\alpha^2 r^2}{\mu} \le \frac{16\pi r^2}{\mu}$$

so that by (4.51), (4.48), (4.39) and (4.40)

$$Pr(F_i) \leq \sum_{j=1}^{J} Pr(M_{ij}) \leq \left(\frac{D}{rw_1} + 1\right) \frac{16\pi r^2}{\mu} = \delta^4.$$

■

We can now complete the proof of Theorem 4.23 as follows: For any search trajectory $S$ we now calculate the expected capture time, $c(S, h_u)$, with the "resting" parameter $u$ of the hider's strategy $h_u$ chosen as

$$u = \frac{\delta\mu}{2r}.$$

If $F_n'$ is defined, as in (4.43), to be the event that capture does not occur before the hider leaves the point $Z_n$, then inequality (2.6) leads to the inequality

$$c(S, h_u) \geq u \sum_{n=0}^{\infty} Pr(F_n') \geq u \sum_{n=0}^{N} Pr(F_n') \tag{4.54}$$

(where the integer $N$ will be determined later)

$$\geq u \sum_{n=1}^{N} \left( Pr\left(\overline{\bigcup_{i=1}^{n} E_i}\right) - \sum_{i=1}^{n-1} Pr(F_i) \right)$$

(see (4.45))

$$\geq u \left( \sum_{n=1}^{N} \left(1 - \frac{2ru}{\mu}\right)^n - \sum_{i=1}^{N} (n-1)\,\delta^4 \right)$$

(by (4.37) and Proposition 4.25)

$$\geq u \left( \frac{(1 - (2ru/\mu)) - (1 - (2ru/\mu))^{N+1}}{2ru/\mu} - N^2\delta^4 \right).$$

Now if we choose $N = \left\lfloor \frac{1}{\delta^2} \right\rfloor$ and use $u = \frac{\delta\mu}{2r}$, we obtain

$$c(S, h_u) \geq \frac{\delta\mu}{2r} \left( \frac{1 - \delta - (1-\delta)^{1/\delta^2}}{\delta} - 1 \right)$$

$$= \frac{\mu}{2r} \left(1 - 2\delta - (1-\delta)^{1/\delta^2}\right). \tag{4.55}$$

It can be easily seen that for $0 < \delta < 1$ as assumed,

$$(1-\delta)^{1/\delta^2} = \exp\left(\frac{1}{\delta^2} \ln(1-\delta)\right) < \exp\left(-\frac{1}{\delta}\right) < \delta$$

(because $\delta \ln \delta \geq -e^{-1} > -1$). Thus, by (4.55)

$$c(S, h_u) > \frac{\mu}{2r}(1 - 3\delta).$$

This completes the proof of Theorem 4.23.

The following corollary is important for establishing the $\varepsilon$-optimality of $h_u$ for a more general cost function (see Section 4.6.8).

**Corollary 4.26** *If the hider uses the strategy $h_u$ (with an appropriately chosen $u$), then for any bounded time interval $0 \leq t \leq \tau$ the probability that the capture time $T$ exceeds $t$ satisfies for any search trajectory*

$$Pr(T > t) \geq (1 - \varepsilon) \exp(-2rt/\mu)$$

*with $\varepsilon \to 0$ as $r \to 0$.*

**Proof.** For a (small) $\delta > 0$ assume that the hider uses $h_u$ with

$$u = \frac{\delta \mu}{2r}.$$

Let

$$I_t = \left\lceil \frac{t}{u} \right\rceil.$$

Then (4.45), (4.37), and Proposition 4.25 imply that, for any search trajectory, the capture time $T$ satisfies

$$\begin{aligned} Pr\,(T > t) &\geq Pr\left(\bigcup_{i=1}^{\bar{I}_t} E_i\right) - \sum_{i=1}^{I_t - 1} Pr\,(F_i) \\ &\geq \left(1 - \frac{2ru}{\mu}\right)^{(t/u)+1} - \frac{t}{u}\delta^4 \\ &= (1 - \delta)^{(2rt/\mu\delta)+1} - \frac{2rt\delta^3}{\mu}. \end{aligned} \tag{4.56}$$

Let $\alpha$ be a positive number that satisfies

$$e^{-(1+\alpha)} = (1 - \delta)^{1/\delta} \tag{4.57}$$

(note that $\alpha \to 0$ as $\delta \to 0$), and let

$$\varepsilon = \max\{2\left(1 - (1 - \delta)\, e^{-\alpha q}\right), 2qe^q\delta^3\}, \tag{4.58}$$

where

$$q = \frac{2r\tau}{\mu}.$$

Then (4.56)–(4.58) imply that for $0 \le t \le \tau$:

$$\begin{aligned} Pr(T > t) &\ge (1 - \delta)\, e^{-(1+\alpha)2rt/\mu} - \frac{2rt\delta^3}{\mu} \\ &\ge (1 - \delta)\, e^{-\alpha(2r\tau/\mu)} e^{-(2rt/\mu)} - \frac{2r\tau\delta^3}{\mu} \\ &= (1 - \delta)\, e^{-\alpha q} e^{-(2rt/\mu)} - q\delta^3 \\ &\ge \left(1 - \frac{\varepsilon}{2}\right) e^{-(2rt/\mu)} - \frac{\varepsilon}{2} e^{-q} \ge (1 - \varepsilon)\, e^{-(2rt/\mu)}. \end{aligned}$$

■

It follows from Corollaries 4.21 and 4.26 that both the searcher and the hider can keep the probability of capture before time $t$ close to $1 - \exp(-2rt/\mu)$. This is the expression obtained, in a different context, by Koopman (1956) for the probability of detection in a "random search" (see also Washburn, 1981).

## 4.6 Modifications and Extensions

### 4.6.1 A "non-localized" search strategy

Our $\varepsilon$-optimal search strategy, presented in Section 4.5.2, is characterized by extensively searching a small part of $Q$ for a relatively long period of time $\tau$, then moving to another part of $Q$ and so on. This type of strategy would not be efficient if the rules of the game were changed in such a way that the hider is informed about the position of the searcher from time to time. A search strategy that is robust to such a change of rules was presented by Lalley and Robbins (1988b) as follows.

Assume that $Q$ is a compact, convex region in $R^2$ with smooth boundary $\partial Q$ such that any line tangent to $\partial Q$ meets $Q$ in only one point. Let $\Theta_1, \Theta_2, \ldots$ be *i.i.d.* random variables with density $\sin\theta/2$, $0 < \theta < \pi$. Define a sequence of random points $Z_1, Z_2, \ldots$ on $\partial Q$ as follows. Let $Z_1$ be uniformly chosen on $\partial Q$. Having defined $Z_i$, draw the chord in $Q$ that makes an angle $\Theta_i$ with the tangent to $\partial Q$ at $Z_i$ and define $Z_{i+1}$ to be the second point of intersection of this chord with $\partial Q$. The trajectory of the searcher is obtained by following the chords $Z_1Z_2, Z_2Z_3, \ldots$ at unit speed.

Lalley and Robbins (1988b) proved that, as $r \to 0$, the above strategy guarantees expected capture time $\sim\mu/2r$ and capture probability $\sim 1 - \exp(-2rt/\mu)$ as in Section 4.5.2. They have also shown that this strategy is robust to partial information: even if the hider is given the position and direction of the searcher from time to time (not too often though) then the hider will not be able to predict its course for very long.

It should be noted, though, that Lalley and Robbins' strategy may be ineffective for non-convex regions while our search strategy, described in Section 4.5.2, is also effective for non-convex regions, as we show in the next subsection.

### 4.6.2 Search in non-convex regions

The main result of Section 4.5, $v \sim \mu/\rho$, has been shown to hold under the assumption that the search space $Q$ is convex. However, both the searcher's strategy $\bar{s}$ presented in Section 4.5.2 and the hider's strategy $\bar{h}$ presented in Section 4.5.3 can be easily modified for non-convex regions in such a way that the basic results still hold for that case as well.

Considering the searcher's strategy, it is easy to perturb $\bar{s}$ near the boundary of $Q$, so that all the trajectories will pass inside $Q$. With regard to the hider's strategy, we can choose the points $Z_1, Z_2, \ldots$ by the same method used by $h_u$, and the only problem is to make the movement from $Z_i$ to $Z_{i+1}$ in such a way that the probability of capture during this movement would still be negligible with respect to the probability of capture at $Z_i$. This requirement should be achievable if the detection radius is small enough. (Note that even the existence of "Narrow passages" inside $Q$ does not change the value, as we show in Section 4.6.6.)

### 4.6.3 Multidimensional extensions

The results stated and proved in Sections 4.5.2 and 4.5.3 can be extended to any number of dimensions greater than 2, by the same techniques used in Theorems 4.18 and 4.23. For example, if $Q$ is a three-dimensional compact region, then the construction of a search strategy that keeps the expected capture time below $(1+\varepsilon)\mu/\pi r^2$ can be made as follows. Cover $Q$ by a large number $2M$ of boxes of dimension $\alpha \times \alpha \times a$, where $r \ll \alpha$, randomly choose one of the boxes, and move along $N$ random horizontal segments that join the two faces of size $\alpha \times \alpha$, etc. A hiding strategy that can keep the expected capture time above $(1-\varepsilon)\mu/\pi r^2$ is the strategy $h_u$ presented by Definition 4.22 with a parameter $u$ that satisfies $1 \ll u \ll \mu/r^2$.

### 4.6.4 The case of several searchers

We now consider a game in which $J$ cooperative searchers seek a single mobile hider in a multidimensional region. We do not assume that all the searchers have the same characteristics so that each searcher $i, i = 1, \ldots, J$, may have a different maximal speed $v_i$ and a different discovery radius $r_i$. Let $\rho_i$ be the maximal discovery rate of the $i$-th searcher ($\rho_i = 2v_i r_i$ for two dimensions, $\pi v_i r_i^2$ for three dimensions, etc.). Then the total discovery rate $\rho$ is defined as

$$\rho = \sum_{i=1}^{J} \rho_i .$$

Under the assumption that $\max_{1 \le i \le J} r_i$ is small with respect to the dimension of $Q$ and that the maximal velocity of the hider is not too small, it is possible to establish a result that is analogous to the one obtained for a single searcher; i.e., that $v \sim \mu/\rho$. This follows from the fact that by adopting the strategy $h_u$, used in Section 4.5.3, the hider can keep the probability of capture after time $t$ close to the function $\exp(-\rho t/\mu)$. On the other hand, if each of the searchers adopts, independently, the strategy $\bar{s}$ used in Section 3.5.2, then the probability of capture after time $t$ will also be close to $\exp(-\rho t/\mu)$.

### 4.6.5 The case of an arbitrary starting point

In Section 4.3, which dealt with a search for a mobile hider on a circle, we noted that the result $v = \mu/\rho$ depends on the assumption that the searcher has to start from a fixed point known to the hider and that if the starting point of the searcher is arbitrary or if it is random, then the value would be substantially less than $\mu/\rho$. Obviously, that phenomenon does not occur in the multidimensional search because Theorem 4.17 is valid under any assumption about the location of the starting point, so that the result $v \sim \mu/\rho$ is not sensitive to the assumption $S(0) = O$.

### 4.6.6 Loitering strategies

Assume that the search region $Q$ is composed of two regions $Q_1$ and $Q_2$ that have the same area, $\mu/2$, and intersect at only one point, $A$. It may seem reasonable for the searcher to loiter at $A$ when the hider uses the strategy $h_u$, but the hider can easily modify his strategy as follows. He chooses each $Q_i$, $i = 1, 2$ with probability 1/2 and stays in $Q_i$ all the time using $h_u$ for this chosen region. Then, for each $t$, if the searcher spends time $t_1$ in $Q_1$ and $t - t_1$ in $Q_2$, then the probability that the capture time exceeds $t$ is

$$\frac{1}{2}\exp\left(-\frac{2rt_1}{\mu/2}\right)+\frac{1}{2}\exp\left(-\frac{2r(t-t_1)}{\mu/2}\right) \geq \exp\left(-\frac{2rt}{\mu}\right).$$

Thus, even though ambushing at $A$ seems natural it is actually ineffective.

Note that this is very different from the search in a network, e.g., the $k(\ k > 3)$ arcs network considered in Section 4.2 for which loitering gives an advantage to the searcher (see Remark 4.16).

### 4.6.7 Non-homogeneous search spaces

In Section 3.7.1 we have shown that the results established for the search game with an immobile hider can be extended to the case in which the maximal velocity of the searcher $\nu(Z)$ depends on the location of the searcher, and the discovery radius $r(Z)$ depends on the location of the hider. Using the framework and definitions of Section 3.7.1 (i.e., that $\nu(Z)$ is continuous and satisfies $0 < \nu_1 \leq \nu(Z) \leq \nu_2$ and $r(Z) = \bar{r}\eta(Z)$, where $\eta(Z)$ is continuous and positive and $\bar{r}$ is small). We now extend the results of Sections 4.5.2 and 4.5.3 to this case.

In order to simplify the presentation, we describe this extension for two-dimensional regions. In this case, if $\bar{r}$ is small, then the value of the game satisfies

$$v \sim q = \int_Q \frac{d\mu(Z)}{\rho(Z)} = \int_Q \frac{d\mu(Z)}{2\nu(Z)r(Z)}$$

where $\mu$ is the Lebesgue measure. The intuitive reason that the value should approximate $q$ is that both the searcher and the hider can guarantee that, except for negligible time periods, the probability of detection in a small time interval $\Delta t$ is $(1 \pm \varepsilon)\Delta t/q$. Since both the searcher and the hider can keep the probability of detection close to an exponential function, it follows that the expected capture time is approximately $q$.

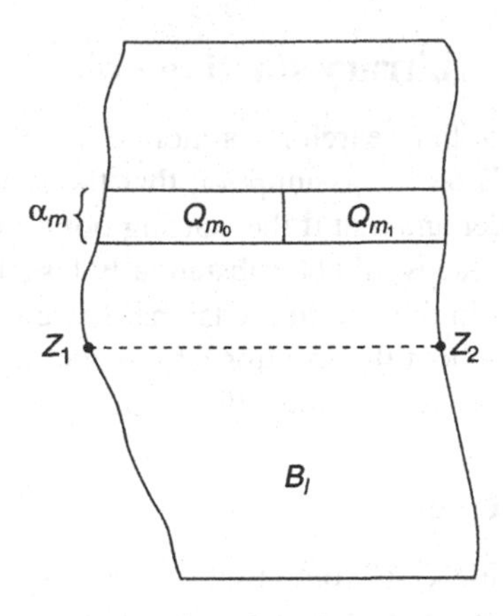


**Figure 4.9.**

In order to achieve the value $q$, the hider's strategy $h_u$ should be modified as follows. The hiding points $Z_1, Z_2, \ldots$ should be chosen, independently, using the probability density $1/(2\nu(Z)r(Z)q)$ instead of the uniform density that was originally used.

With regard to the searcher, we first consider the simpler case in which $\nu(Z)$ is a constant, and only $r(Z)$ depends on $Z$. In this case, the only modification needed in $\bar{s}$ is to construct the rectangles $Q_{mi}$ in such a way that the width $\alpha_{mi}$ of $Q_{mi}$ would be proportional to $\bar{r}_{mi}$, where $\bar{r}_{mi}$ is the average of $r(Z)$ in $Q_{mi}$. It is also required that $1 \gg a \gg \alpha_{mi} \gg \bar{r}$, where $a$ is the length of $Q_{mi}$. If $\nu(Z)$ is not a constant, then a substantial modification is needed in $\bar{s}$. In this case, the region $B_l$ would not be a rectangle, as in Figure 4.6. Instead, it would be a strip, as shown in Figure 4.9.

The distance between any point $Z_1$ on the left boundary line and the corresponding point $Z_2$ on the right boundary line (at the same level) should be equal to the maximal distance that can be traveled in a time interval of length $2a$ by a searcher moving from $Z_1$ in a horizontal direction. The region $B_l$ should then be divided into strips by parallel horizontal lines. The width $\alpha_m$ of the strip $Q_{m0} \cup Q_{m1}$ should be proportional to $r_m$, where $r_m$ is the average of $r(Z)$ in this strip. It should also be required that $1 \gg a \gg \alpha \gg \bar{r}$.

The fact that the above-described modifications of $h_u$ and $\bar{s}$ guarantee the value $(1 \pm \varepsilon)q$ can be established by a technique similar to that used in Section 4.5.

### 4.6.8 A general cost function

It follows from Corollaries 4.21 and 4.26 of Section 4.5 that each one of the players of the search game in a two-dimensional convex region can keep the probability of capture after time $t$ close to the function $\exp(-\lambda t)$, where

$$\lambda = \frac{2r}{\mu}. \tag{4.59}$$

This result can be used to obtain the solution of the game in the case that the cost function $W$ is a monotonic non-decreasing function of the capture time $T$, rather than

$T$ itself. In this case, one would expect that the value $v$ of the game would satisfy

$$v \sim \lambda \int_0^\infty W(T)e^{-\lambda T}\,dT. \tag{4.60}$$

Actually (4.60) holds, under the following assumption about $W$.

**Condition 4.27** *There exist constants $\alpha > 0$, $\gamma > 0$, and $\beta > 1$ such that for all $T \geq \gamma$*

$$W(\beta T) \leq \alpha W(T).$$

Condition 4.27 is satisfied for a wide variety of functions including all the polynomials, the bounded functions, and actually any reasonable function that does not increase faster than all polynomials.

**Theorem 4.28** *Let $Q$ be a (two-dimensional) search space. If the cost function $W(T)$ is a positive non-decreasing function of the capture time $T$ that satisfies Condition 4.27, then for any $\delta > 0$ there exists a detection radius $r_0$ so that for any search game (inside $Q$) with $r < r_0$ the strategies $\bar{s}$ and $h_u$ presented in Sections 4.5.2 and 4.5.3 satisfy*

$$v(\bar{s}) \leq (1+\delta)v$$

*and*

$$v(\bar{h}) \geq (1-\delta)v$$

*with*

$$v = \lambda \int_0^\infty W(T)e^{-\lambda T}\,dT$$

*where*

$$\lambda = \frac{2r}{\mu}.$$

The details of the proof are presented in Gal (1980).

$\mathcal{F}$ itself. In [illegible] of the [illegible] that the value of the game would satisfy

$$[illegible] \quad (4.69)$$

[illegible] By (4.69) [illegible] the following [illegible] about $V$.

**Corollary 4.27** [illegible] $d = 1$ [illegible]

$$V(\mathcal{Q}, T) [illegible]$$

[illegible] bounded functions, [illegible] functions [illegible] than all polynomials.

**Theorem 4.28** [illegible] $(T)$ [illegible] Section 4.2 and 4.5.3 [illegible]

$$[illegible]$$

and

$$[illegible]$$

[illegible]

$$[illegible] = \lambda \int_0 [illegible]$$

where

$$[illegible]$$

[5] The details of the proof are presented in [illegible] (1980) [illegible]

# Chapter 5

# Miscellaneous Search Games

This chapter presents several search games in compact spaces that are not included in the framework described in Chapter 2. These games are interesting and have attracted a considerable amount of research, some of it rather recently.

## 5.1 Search in a Maze

In Section 3.2 we examined the problem of searching for an immobile hider in a finite connected network $Q$ of total length $\mu$. We now make the searcher's problem more difficult by depriving him of a view of the whole network $Q$ and instead let him see (and remember) only that part of $Q$ that he has already traversed. We also let him remember the number of untried arcs leading off each of the nodes he has visited. Under these informational conditions, the network $Q$ is known as a *maze*. The starting point $O$ is known as the *entrance*, and the position $A$ of the hider is known as the *exit*. In this section we present the randomized algorithm of Gal and Anderson (1990) for minimizing the expected time for the searcher to find the exit in the worst case, relative to the choice of the network and the positioning of its entrance and exit. This strategy (called the *randomized Tarry algorithm*) may be interpreted as an optimal search strategy in a game in which the maze (network with entrance and exit) is chosen by a player (hider) who wishes to maximize the time required to reach the exit.

Finding a way for exiting a maze has been a challenging problem since ancient times. Deterministic methods that ensure finding the exit were already known in the 19-th century. Lucas (1882) described such an algorithm developed by Trémaux, and Tarry (1895) presented an algorithm which is now very useful in computer science for what is known as *depth-first search*. An attractive description of depth-first search is presented in Chapter 3 of Even (1979). The algorithms of Trémaux and Tarry mentioned above each guarantee that the searcher will reach the exit by time $2\mu$. Fraenkel (1970, 1971) presented a variant of Tarry's algorithm, that has an improved performance for some cases but has the same worst-case performance of $2\mu$.

The worst-case performance of any fixed search strategy cannot be less than $2\mu$. This can easily be seen by considering the family of "star" mazes consisting of $n$ rays of equal length $\mu/n$, all radiating from a common origin $O$. The least time required to visit all the nodes of this maze is $(n-1)(2\mu/n)+(\mu/n) = 2\mu - \mu/n$, which converges to $2\mu$. Can this worst-case performance be improved by using mixed strategies and requiring only that the *expected* time to find the exit is small? This question is related to the following game: For a given parameter $\mu$, Player II (the hider) chooses a maze with measure $\mu$ and specifies the entrance $O$ and the exit $A$. Player I (the searcher) starts from $O$ and moves at unit speed until the first time $T$ (the payoff to the maximizing player II) that he reaches $A$. We will show that this game has a value $v$ that is equal to the length $\mu$ of the maze. Obviously, $v \geq \mu$, because the hider can always choose the maze to be a single arc of length $\mu$ going from $O$ to $A$. On the other hand, we shall show that the searcher has a mixed strategy that guarantees that the expected value of $T$ does not exceed $\mu$. Consequently, $v = \mu$. Therefore this (optimal) search strategy achieves the best worst-case performance for reaching the exit of a maze. This result was obtained by Gal and Anderson (1990), on which the following discussion is based.

The optimal strategy generates random trajectories that go from node to node by traversing the intervening arc at unit speed without reversing direction. Consequently, it is completely specified by giving (random) rules for leaving any node that it reaches. This strategy is based in part on a coloring algorithm on the *passages* of the maze. A passage is determined by a node and an incident arc. Thus each arc has two passages, one at each end. We assume that initially the passages are uncolored, but when we go through a passage, we may sometimes color it either in yellow or in red. The strategy has two components *coloring rules* and *path rules*, as follows. Since this strategy is in fact a randomized version of Tarry's algorithm, we will refer to it as the *randomized Tarry algorithm*.

**Coloring rules**

1. When arriving at any node *for the first time*, color the arriving passage in *yellow*. (This will imply that there is at most one yellow passage at any node.)
2. When leaving any node, color the leaving passage in *red*. (This may require changing a yellow passage to red.)

**Path rules (how to leave a node)**

1. If there are uncolored passages, choose among them equiprobably.
2. If there are no uncolored passages but there is a yellow passage, choose it for leaving the node.
3. If there are only red passages, stop.

Obviously, when adopting the above algorithm to exit a maze, the searcher would use the obvious additional rule of stopping when the exit $A$ is reached. However, for analytical reasons we prefer to consider the full paths $\tau(t)$ produced when the searcher does not stop at $A$ but continues until the single stopping rule 3 is applied. We call these full paths *tours* and denote the set of all tours by $\mathcal{T}$. The set $\mathcal{T}$ is finite because there are only finitely many randomizations involved in the generation of the tours.

To illustrate how the randomized Tarry algorithm generates tours of the maze, consider the particular maze drawn in Figure 5.1 (all the arcs have unit length).

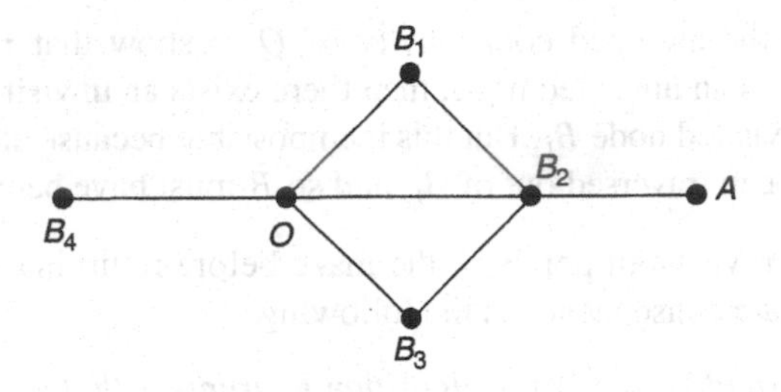


Figure 5.1.

All arcs have length 1 so that $\mu = 7$. For this maze a particular tour $\bar{\tau} \in \mathcal{T}$ is given by the node sequence (which determines the continuous path)

$$\bar{\tau} = O, B_1, B_2, O, B_2, B_3, O, B_4, O, B_3, B_2, A, B_2, B_1, O. \tag{5.1}$$

Note that the tour $\bar{\tau}$ stops at $O$ when all its passages become red. It traverses each arc exactly once in each direction and consequently takes 14 ($= 2\mu$) time units. The following lemma shows that this behavior is not specific to the maze or the particular tour chosen but *always* occurs.

**Lemma 5.1** *Let $\tau$ be any tour generated by the randomized Tarry algorithm for a particular maze $Q$ of total length $\mu$ and entrance $O$. Then $\tau$ finishes at $O$ at time $2\mu$, after traversing every arc of $Q$ exactly once in each direction.*

**Proof.** The path rules ensure that a node will never be left via a red passage, so coloring rule 2 guarantees that an arc will never be traversed more than once in either direction. Consequently, the path $\tau$ must end and can have length at most $2\mu$.

We now show that $\tau$ cannot end at any node $B$ other than $O$. Let $k$ denote the degree of $B$. Note that whenever $\tau$ is at $B$, it has arrived at $B$ one more time than it has left. If $\tau$ ends at $B$, all the passages at $B$ must be red, and hence it must have already left $B$ by all $k$ arcs. Hence it must have arrived at $B$ at least $k + 1$ times, which is impossible, since the $k$ incoming arcs at $B$ can each be traversed at most once in the incoming direction.

Next we show that at the time $\tau$ stops at $O$, all the arcs incident with any *visited* node have already been traversed exactly once in each direction. Let $B_1$ denote the node (other than $O$) that has been discovered first, $B_2$ the node discovered second, $B_3$ the node discovered third, and so on. When $\tau$ stops at $O$, the arc at $O$ that led to the discovery of $B_1$ has already been traversed into $O$ so that the yellow passage of node $B_1$ has been used. By path rule 2 this implies that all the passages out of $B_1$ have been used. Since the number of times the trajectory entered $B_1$ is equal to the number of times it left $B_1$, it follows that all the arcs incident with $B_1$ have been traversed in both directions, and coloring rule 2 ensures that an arc cannot be traversed more than once in any direction. Similarly, $B_2$ can be discovered by an arc from either $O$ or $B_1$. We have just shown that all the arcs incident with $O$ or with $B_1$ have been traversed in both directions so that the yellow passage of $B_2$ has been used so that all the arcs incident with $B_2$ have been traversed exactly once in each direction. The same argument applies to all nodes that have been reached by $\tau$.

Finally, we use the assumed connectivity of $Q$ to show that $\tau$ visits all nodes. Assuming that there is an unvisited node, then there exists an unvisited node (say, $\bar{B}$, ) that is adjacent to a visited node $B_i$. But this is impossible because all the arcs incident with node $B_i$ have been traversed out of $B_i$, and so $\bar{B}$ must have been discovered. ■

Since each tour $\tau$ visits all points of the maze before returning to the entrance at time $2\mu$, an immediate consequence is the following.

**Corollary 5.2** *The randomized Tarry algorithm guarantees that the time to reach the exit is smaller than* $2\mu$.

We have already observed that the set of tours is finite. The following lemma computes its cardinality, which is also the reciprocal of the probability of any particular tour.

**Lemma 5.3** *Let $Q$ be a maze with $I+1$ nodes. Denote by $d_0$ the degree of $O$ and by $d_i$ the degree of the remaining nodes $i$, $i = 1, \ldots, I$. Then the cardinality $K$ of the set $\mathcal{T}$ consisting of all tours $\tau$ is given by*

$$K = \#\mathcal{T} = d_0! \prod_{i=1}^{I} (d_i - 1)!$$

*Each tour $\tau$ is obtained with an equal probability of* $1/K$.

**Proof.** If the randomization that produces the tours $\tau \in \mathcal{T}$ is carried out in a more concentrated but equivalent manner, the calculation of $K$ is simplified, and the claimed formula easily obtained. Suppose that whenever a tour $\tau$ arrives at a new node $B$ that has degree $k$, the searcher labels the incoming passage $k$ and assigns the numbers $1, \ldots, k-1$ to the remaining passages at $B$ according to a random permutation. Then whenever the searcher has to leave $B$ he always chooses the lowest numbered passage that has not already been used for that purpose. There are $(k-1)!$ ways of making these choices at $B$. At the start of his search, he does the same thing with respect to *all* the $d_0$ passages at $O$. Consequently, the number of possible tours is the number $K$ given by the stated formula. Since every tour $\tau$ determines a particular permutation of this type at each node, it occurs with probability $1/K$. ■

We now consider the main problem of evaluating the expected time taken for the randomized Tarry algorithm to reach a point $A$ of the maze (which might be taken as the exit). For $\tau \in \mathcal{T}$ and any node $A$ of $Q$, define $T_A(\tau)$ to be the first time that the tour $\tau$ reaches the node $A$, that is, $T_A(\tau) = \min\{t : \tau(t) = A\}$. Consequently, the *expected* time $\tilde{T}_A$ for the randomized Tarry algorithm to reach $A$ is given by

$$\tilde{T}_A = \frac{1}{K} \sum_{\tau \in \mathcal{T}} T_A(\tau). \tag{5.2}$$

In order to estimate $\tilde{T}_A$ it will be useful to compare the performance of a tour $\tau$ with the performance of its *reverse tour* $\phi(\tau) = \tau'$ defined by

$$[\phi(\tau)](t) \equiv \tau'(t) = \tau(2\mu - t). \tag{5.3}$$

For example, the reverse tour of the tour $\bar{\tau}$ given in (5.1) is expressed in node sequence form as

$$\bar{\tau}' = O, B_1, B_2, A, B_2, B_3, O, B_4, O, B_3, B_2, O, B_2, B_1, O.$$

Observe that while the time $T_A(\bar{\tau}) = 11$ is greater than $\mu = 7$, when averaged with the smaller time of $T_A(\bar{\tau}') = 3$ for its reverse path, we get $(11 + 3)/2 = 7$. This motivates the following analysis. (The reversal mapping $\phi$ is a bijection of $\mathcal{T}$.) Consequently we can just as well calculate $\tilde{T}_A$ by the formula

$$\tilde{T}_A = \frac{1}{K} \sum_{\tau \in \mathcal{T}} T_A(\tau'), \text{ and hence also by}$$

$$\tilde{T}_A = \frac{1}{K} \sum_{\tau \in \mathcal{T}} \frac{T_A(\tau) + T_A(\tau')}{2}. \tag{5.4}$$

Observe that since $\tau(T_A(\tau)) = A$ by definition, and similarly we have by (5.3) that $\tau'[2\mu - T_A(\tau)] = A$.

If $\tau$ reaches $A$ exactly once (which is the same as $A$ having degree 1) then so does $\tau'$ and consequently $T_A(\tau') = 2\mu - T_A(\tau)$. If $\tau$ also reaches $A$ at some later time, then $\tau'$ also reaches $A$ at some earlier time $(2\mu - \max\{t\colon \tau(t) = A\})$, and hence $T_A(\tau') < 2\mu - T_A(\tau)$. In either case, we have that

$$T_A(\tau') \le 2\mu - T_A(\tau),$$

i.e.,

$$T_A(\tau') + T_A(\tau) \le 2\mu, \tag{5.5}$$

with equality if and only if $A$ is a node of degree 1. Combining (5.4) and (5.5), we have

$$\begin{aligned} \tilde{T}_A &= \frac{1}{K} \sum_{\tau \in \mathcal{T}} \frac{T_A(\tau) + T_A(\tau')}{2} \\ &\le \frac{1}{K} \sum_{\tau \in \mathcal{T}} \frac{2\mu}{2} = \mu. \end{aligned}$$

Again, we have equality if and only if $A$ has degree 1. Since in this argument $A$ was arbitrary, we have established the following.

**Theorem 5.4** *The randomized Tarry algorithm reaches every possible exit point in $Q$ in expected time not exceeding the total length $\mu$ of $Q$. Furthermore, except for nodes of degree 1, it reaches all points in expected time strictly less than $\mu$.*

Since we have already observed that the hider can choose a maze in which the exit lies at the end of a single arc of length $\mu$ (which would take time $\mu$ to reach), we have the following.

**Corollary 5.5** *The value of the game in which the hider chooses a maze of total length $\mu$, with entrance and exit, and the searcher moves to minimize the (Payoff) time required to find the exit, is equal to $\mu$.*

We conclude this section by giving some comparisons of the randomized Tarry algorithm with other possible methods of searching a maze. Notice that in the Tarry algorithm a tour (such as $\bar{\tau}$ given in (5.1) on the first return to $O$) may choose an arc that has already been traversed (in the opposite direction), even when an untraversed arc is available at that node. This may appear to be a foolish choice because the untraversed node may lead to a new part of the maze (maybe even straight to the exit). Yet a strategy that is based on giving priority to unvisited arcs leads to significantly inferior performance in the worst case: $2\mu$ instead of $\mu$, as shown in Gal and Anderson (1990).

The randomized Tarry algorithm requires only local information in each node. This type of locally determined strategy is similar to some policies that could be employed in a distributed computing environment to deal with incoming messages or queries directed to an unknown node of the computer network. Such a model was used, for example, by Golumbic (1987). Our strategy provides an alternative path finding mechanism that uses only local information. Other local information schemes of searching an unknown graph by several (decentralized) agents (ant-robots that leave chemical odor traces) are described and analyzed by Wagner, Lindenbaum and Bruckstein (1996, 1998, 2000).

## 5.2 High–Low Search

In this section we consider games that are usually played on an interval or a discrete subset of it. The hider chooses a point $H$ in the interval, and the searcher tries to approach it by a sequence of guesses $g_1, g_2, \ldots,$ of $H$, being told for each $g_i$ whether it is too high or too low.

### 5.2.1 Continuous high–low search

The continuous version of the high–low search game was introduced by Baston and Bostock (1985). In this version the hider picks a point $H$ in the unit interval $Q = [0, 1]$, which is also the hider's pure strategy space $\mathcal{H}$. In each time period $i = 1, 2, \ldots,$ the searcher announces a "guess" $g_i \in Q$ and is told the direction of the error. The searcher wishes to minimize the "sum of errors" cost function

$$c = \sum_{i=1}^{\infty} |g_i - H|.$$

Baston and Bostock found a pure search strategy that guaranteed that this sum was always less than 0.628. (Note that the natural bisection strategy can't guarantee cost less than 1 against hiding at an end of the interval.)

Subsequently, Alpern (1985) continued the analysis in the following way. The information feedback can be coded in a binary fashion as $a_i = 0$ if $g_i > H$ (too high) and $a_i = 1$ if $g_i < H$ (too low). We may assume that if the searcher is told if $g_i = H$, then all subsequent guesses will be the correct value $H$. The information available to the searcher before making guess $g_i$ is the feedback sequence $a_1, a_2, \ldots, a_{i-1}$, which may be uniquely coded as a dyadic rational number $d_i$ of rank $i$ (one with a power of 2 as its denominator, the power called the rank) by the formula

$$d_i = a_1 a_2, \ldots a_{i-1} 1 \text{ (base 2)}.$$

Thus the initial information is coded as $d_1 = 0.1 = 1/2$ and the information after the first guess is either $d_2 = 0.01_2 = 1/4$ (if the first guess was too high) or $d_2 = 0.11_2 = 3/4$ (if too low). Hence the pure strategy space $\mathcal{S}$ consists of all maps (pure strategies) $S : D \to [0, 1] = Q$, where $D$ is the set of all reduced proper dyadic rationals. For example, the "halving" strategy, which always guesses at the middle of the uncertainty interval (in which the hider is known to be), is given by the linear function $S(d) = d$. In general, $S$ and $H$ determine $d_i$ by the beginning of period $i$, and then the next guess is determined as $g_i = S(d_i) = S(d_i(S, H))$.

The cost function

$$c(S, H) = \sum_{i=1}^{\infty} |S(d_i(S, H)) - H|$$

is lower semicontinuous on the product space $\mathcal{S} \times \mathcal{H}$ if the usual distance on [0, 1] is used for $\mathcal{H}$ and the distance between two pure search strategies $S$ and $S'$ in $\mathcal{S}$ is given by

$$\rho(S, S') = \sum_{k=1}^{\infty} 2^{-k} |S(d(k)) - S'(d(k))|,$$

for some enumeration $d(1), d(2), \ldots$ of all the proper dyadic rational numbers. Since under this metric $\mathcal{S}$ is compact (it is homeomorphic to the Hilbert cube), the existence of a value for this game follows from the minimax theorem of the authors (Alpern and Gal, 1988, or Appendix A).

To eliminate some pure strategies from consideration, we say that $S$ is *effective* if it ensures that $g_i \to H$ and *undominated* if there is no other strategy that is never worse and sometimes better, in terms of the cost function $c(S, H)$. Observe that the dyadic rationals $D$ form a dense subset of [0, 1]. It is shown in Alpern (1985) that

**Theorem 5.6** *A pure strategy $S : D \to [0, 1]$ is effective and undominated if and only if it has an extension to an increasing homeomorphism*[1] *of* [0, 1] *onto itself. In particular, such strategies always make the next guess inside the current interval of uncertainty of $H$.*

The following result of the same paper completely determines the minimax search strategy $\bar{S}$ for the continuous high–low search game.

**Theorem 5.7** *There is a unique strategy $\bar{S}$ satisfying*

$$\sup_{0<H<1} c(\bar{S}, H) = \inf_S \sup_{0<H<1} c(S, H) \equiv w_0 \doteq 0.62453572.$$

*The strategy $\bar{S}$ has a simple description in terms of a sequence $\{\lambda_k\}_{k=-\infty}^{\infty}$ of constants, with $1/2 = \lambda_0 < \lambda_1 < \cdots < 1$ and $\lambda_{-k} = 1 - \lambda_k$. Suppose that $(a, b)$ is the interval*

[1] A mapping $f$ such that both $f$ and $f^{-1}$ are continuous.

*of uncertainty after the guesses* $g_1, \ldots, g_n$ *have been made. Suppose that of these n numbers, k more of them have been to the left of* $(a, b)$ *than to the right. Then make the next guess* $g_{n+1} = (1 - \lambda_k)a + \lambda_k b$. *The* $\lambda_k$ *are given by the formula*

$$\lambda_k = \frac{w_{k+1}}{w_{k-1} + w_{k+1} + 1 - k}, \qquad \textit{where the } w_k \textit{ satisfy}$$

$$w_k = \frac{w_{k+1}(1 + w_{k-1})}{w_{k-1} + w_{k+1} + 1 - k}, \qquad k \geq 1 \quad \textit{and} \quad w_0 = w_1/2.$$

Suppose that an initial guess $\hat{g}_1$ has already been made and that the searcher re-evaluates his strategy. Assume without loss of generality that $\hat{g}_1$ was too low ($g_1 < H$). The new uncertainty interval for $H$ is $(a, b)$, with the original $\hat{g}_1 = a$ and $b = 1$. Suppose the searcher resets the time to 1 (as if starting a new game). If his subsequent guesses are labeled $g_1, g_2, \ldots$, his total cost is given by

$$(H - a) + \sum_{i=1}^{\infty} |g_i - H|, \quad H \in (a, b),$$

so this is what he must minimax.

Next consider another case in which the first guess $\hat{g}_1 = a$ was too low and the second guess $\hat{g}_2 = b$ was too high. Then $H \in (a, b)$ and if the next guesses are relabeled as $g_1, g_2, \ldots$, the total cost function will be

$$(H - a) + (b - H) + \sum_{i=1}^{\infty} |g_i - H| = (b - a) + \sum_{i=1}^{\infty} |g_i - H|,$$

so the variable part of the cost function,

$$\sum_{i=1}^{\infty} |g_i - H|, \quad H \in (a, b)$$

must be minimaxed.

Finally, suppose that $\hat{g}_1 < \hat{g}_2$ have both been too low and $\hat{g}_3$ has been too high. Then $H \in (a, b)$, where $a = \hat{g}_2$ and $b = \hat{g}_3$. Then calling the fourth guess $g_1$, the total cost is given by

$$(H - \hat{g}_1) + (H - a) + (b - H) + \sum_{i=1}^{\infty} |g_i - H|$$

$$= (b - a) + (a - \hat{g}_1) + (H - a) + \sum_{i=1}^{\infty} |g_i - H|.$$

$$= (b - \hat{g}_1) + (H - a) + \sum_{i=1}^{\infty} |g_i - H|.$$

So the searcher must minimax the variable part (depending on $H$) of the cost function,

$$(H - a) + \sum_{i=1}^{\infty} |g_i - H|.$$

In general, if at some point in time exactly $k$ of the guesses have been too low than too high, and the uncertainty interval of $H$ is $(a, b)$, the searcher must minimax the function

$$c_k = k(H - a) + \sum_{i=1}^{\infty} |g_i - H|, \quad H \in (a, b).$$

With this in mind, we define auxiliary games $\Gamma_k(a, b)$, $k = 0, 1, \ldots$, where $\Gamma_0(0, 1)$ is the original game. The idea is that $\Gamma_k$ is the game $\Gamma_0$ subject to the restriction that the first $k$ guesses must be 0 or more formally that the cost function $c_k$ is given by

$$c_k = k(H - a) + \sum_{i=1}^{\infty} |g_i - H|, \quad H \in (a, b).$$

The idea of the proof is to recursively decompose $\Gamma_0$ into the games $\Gamma_k(a, b)$ and to determine the minimax first move $\lambda_k$ in $\Gamma_k$. After the first guess, the game $\Gamma_k$ decomposes into $\Gamma_{k-1}$ and $\Gamma_{k+1}$, depending, respectively, on whether the guess is too high or too low. The existence of a (minimax) strategy in $\Gamma_k(0, 1)$, which guarantees the minimax value $w_k = \inf_S \sup_{0<H<1} c_k(S, H)$ follows as described above for the case $w_0$. By a simple scaling argument it is easy to see that the minimax value of $\Gamma_k(a, b)$ is given by $w_k \times (b - a)$. If we know the optimal first guess $\lambda_k$ in all games $\Gamma_k(0, 1)$ then we know the optimal $k$-th guess in the original game $\Gamma = \Gamma_0$ as a function of the high–low feedback.

In the case that there have been $k$ more guesses that were too high than too low, we obtain the reversed subgame $\Gamma'_k(a, b)$ with cost function

$$c'_k = k(b - H) + \sum_{i=1}^{\infty} |g_i - H|, \quad H \in (a, b).$$

It is clear that by left-right symmetry the minimax value of $\Gamma'_k(a, b)$ is the same as that of $\Gamma_k(a, b)$ – namely, $w_k \times (b - a)$.

**Lemma 5.8** *Any minimax strategy for* $\Gamma_0$ *must begin with an initial guess* $g_1 = \lambda_1 = 1/2$ *and furthermore* $w_0 = w_1/2$.

**Proof.** After the first guess $g_1$, the maximizing hider can either obtain the subgame $\Gamma_1(g_1, 1)$ or the subgame $\Gamma'_1(0, g_1)$, by respectively deciding that $H > g_1$ or $H < g_1$. In the former case the minimax value is $w_1 \times (1 - g_1)$, and in the latter it is $w_1 \times (g_1 - 0)$. Consequently the minimax value $w_0$ is given by

$$\begin{aligned} w_0 &= \min_{0 \le \hat{g}_1 \le 1} [w_1 \times (1 - g_1), w_1 \times (g_1 - 0)] \\ &= w_1 \min_{0 \le \hat{g}_1 \le 1} [(1 - g_1), g_1] \\ &= w_1/2, \text{ taking } g_1 = \lambda_1 = 1/2. \end{aligned}$$

■

**Lemma 5.9** *For $k \geq 1$, any minimax pure strategy for the searcher in $\Gamma_k$ must begin with a first guess of $g_1 = \lambda_k = w_{k+1}/(w_{k-1} + w_{k+1} + 1 - k)$. Furthermore, the minimax values $w_k$ satisfy*

$$w_k = \frac{w_{k+1}(1 + w_{k-1})}{w_{k-1} + w_{k+1} + 1 - k}.$$

**Proof.** The idea of the proof is that after the first guess $g_1 = x$ is made in $\Gamma_k(0, 1)$, the remaining game resembles $\Gamma_{k-1}$ or $\Gamma_{k+1}$, respectively, depending on whether $H < x$ or $H > x$. (Note that $H = x$ is never a best reply.) Observe that for a strategy $S$ in $\mathcal{S}$ with $g_1 = x$ and $H < x$ we have

$$\begin{aligned} c_k(S, H) &= k(H - 0) + (x - H) + \sum_{i=2}^{\infty} |g_i - H| \\ &= x + (k - 1)H + \sum_{i=2}^{\infty} |g_i - H| \\ &= x + c_{k-1}(S', H), \quad H \in (0, x) \end{aligned}$$

after renumbering the $g_i$'s and calling the resulting strategy $S'$. Thus the minimax value in this case is given by $x$ plus the minimax value of $\Gamma_{k-1}(0, x)$, which is

$$x + w_{k-1}(x - 0) = x + w_{k-1}x.$$

Similarly, if $x < H$, we have

$$\begin{aligned} c_k(S, H) &= k(H - 0) + (H - x) + \sum_{i=2}^{\infty} |g_i - H| \\ &= k(H - x) + kx + (H - x) + \sum_{i=2}^{\infty} |g_i - H| \\ &= kx + (k + 1)(H - x) + \sum_{i=2}^{\infty} |g_i - H| \\ &= kx + c_{k+1}(S', H), \quad H \in (x, 1), \end{aligned}$$

again after renumbering the $g_i$ and calling the resulting strategy $(x, 1)$. Since the minimax value of $\Gamma_{k+1}(x, 1)$ is $(1 - x)w_{k+1}$, the minimax value of $\Gamma_k(0, 1)$ in this case is given by

$$kx + (1 - x)w_k.$$

Combining these two estimates, we obtain

$$w_k = \inf_{0 \leq x \leq 1} \max[x + x\, w_{k-1}, kx + (1 - x)w_{k+1}].$$

It can be shown that these two lines intersect at $x$ between 0 and 1 at a height of

$$= \frac{w_{k+1}(1 + w_{k-1})}{w_{k-1} + w_{k+1} + 1 - k}.$$

The minimum is attained for $x = \lambda_k$ given by the formula stated in the theorem. Together, Lemmas 5.8 and 5.9 establish the main result, Theorem 5.7. ■

An algorithm is given in Alpern (1985) that recursively calculates the $w_k$ and hence also the $\lambda_k$. To within an error of $\pm 10^{-9}$, the first five values are as follows:

$$\lambda_1 = 0.88443951, \lambda_2 = 0.99119299, \lambda_3 = 0.99983241, \lambda_4 = 0.99999936, \lambda_5 = 1.$$

Thus using only five stored constants, a simple computer program can be written (using the algorithm of the above result) that carries out a pure high–low search that is nearly pure minimax search. The first four guesses for the minimax strategy $\bar{S}$ are given in the following table, as functions of the previous high–low feedback.

| | $g_1$ | $g_2$ | | $g_3$ | | | | $g_4$ | | | | | | | |
|---|---|---|---|---|---|---|---|---|---|---|---|---|---|---|---|
| $a_1$ | | 0 | 1 | 0 | 0 | 1 | 1 | 0 | 0 | 0 | 0 | 1 | 1 | 1 | 1 |
| $a_2$ | | | | 0 | 1 | 0 | 1 | 0 | 0 | 1 | 1 | 0 | 0 | 1 | 1 |
| $a_3$ | | | | | | | | 0 | 1 | 0 | 1 | 0 | 1 | 0 | 1 |
| $g_i$ | 0.5 | 0.058 | 0.942 | 0.001 | 0.279 | 0.721 | 0.999 | 0.000 | 0.007 | 0.083 | 0.474 | 0.525 | 0.916 | 0.992 | 0.999 |

### 5.2.2 Economic applications

The continuous high–low search game of Baston and Bostock has been adapted to suit economic models of production and of wage bargaining by Alpern and Snower (1987, 1988a,b,c, 1989, 1991), and both these models have been extended by Reyniers (1988, 1989, 1990, 1992).

The first model considers the problem faced by a firm that supplies a perishable good for which there is a fixed but uncertain demand. The full analysis is given in Alpern and Snower (1987). We will call this unknown demand level $H$ to make the links with the Baston–Bostock game clear. We normalize the maximum value of $H$ to 1, so that $H$ is assumed to lie in an interval $(H_1, 1)$. In each period $k = 1, 2, \ldots$, the firm supplies a quantity $g_k$ of the good. For example, in a so-called "newsboy" setting, the boy has to decide how many newspapers to bring to his stand. If the $g_k \leq H$, a "stockout" occurs, in which all the newspapers that are supplied are sold. In this case $g_k$ becomes the new lower bound on $H$, so that in period $k$ the uncertainty interval on $H$ is $(H_{k+1} = g_k, 1)$. On the other hand, if the amount supplied was larger than the demand, the exact demand $H$ can be calculated from the surplus $g_k - H$ that remains, i.e., $H = g_k - (g_k - H)$. In this case we may assume that a rational newsboy (or firm) will supply the exact amount demanded, $g_{k+t} = H$, in all future periods ($t \geq 1$). Note that higher supply levels $g_k$ yield more information about $H$, so that the firm may wish to sacrifice current profit to increase information for the future.

If the good (newspaper) costs the newsboy $C$ and is sold at a price $P$, then the single period opportunity cost of supplying a quantity $g$ when the actual demand is $H$ is given as follows. This cost is by definition zero when $g = H$. If too much is supplied, the cost is the total price paid by the newsboy for the unsold papers, $C(g - H)$. On the other hand if too little was supplied, the cost is the lost profit on the quantity shortfall, $(P - C)(H - g)$. Thus we have an asymmetric per period cost function

$$c(g, H) = \begin{cases} C(g - H), & \text{if } H \leq g, \\ (P - C)(H - g), & \text{if } H \geq g. \end{cases}$$

Note that in the case $P = 2C$, this normalizes to the same symmetric cost function $|g - H|$ proposed by Baston and Bostock. It is easy to see that the only undominated pure search (supplier) strategies $S$ are determined by an increasing sequence $H_1 < g_1 < g_2 < \cdots \leq 1$, with the understanding that $g_k$ will be supplied until the first period $N = N(S, H)$ when $g_{N-1} \geq H$, and thereafter $g_N = g_{N+1} = \cdots = H$. In the case that $H > \sup_k g_k$, we take $N = \infty$. Hence the total cost to the supplier-newsboy is given by

$$c^*(S, H) = \sum_{k=1}^{N-2} [(P - C)(H - g_k)]\delta^{k-1} + C(g_{N-1} - H)\delta^{N-2},$$

where the second term is eliminated if $N = \infty$. A discount rate $\delta \leq 1$ is included to make the model more realistic. Alpern and Snower (1987) derive the minimax (or minimax regret) strategy $\bar{S}$ for this problem, the one satisfying

$$\max_{H_1 \leq H \leq 1} c^*(\bar{S}, H) = \min_S \max_{H_1 \leq H \leq 1} c^*(S, H).$$

The case where some unsold inventory remains to be used in the next period was also analyzed.

Subsequently, many variations on this model have been studied. Among these include: a Bayesian approach to $H$ (Alpern and Snower, 1987), a selling price $P$ (which determines demand) controlled by the supplier (Alpern and Snower, 1988b, 1989), a delay in which the sign of $g_k - H$ is not known to the supplier until period $k + 2$ (Reyniers, 1988, 1990), an effect whereby stockouts decrease future demand (Reyniers, 1989).

The second model adapts the Baston–Bostock high–low search game to the problem of wage bargaining. In this model, the unknown quantity $H$ represents the marginal value to a firm of employing a specific worker (the searcher). We assume that the firm has an informational advantage over the worker in terms of knowing $H$; perhaps the firm gives the worker a test, without revealing his score. All the worker knows is that his value $H$ to the firm lies in an interval $[a_1, b_1]$. In each period $k$, the worker makes a *wage demand* $g_k$. In the basic model, the firm acts myopically and simply accepts any demand that gives it a profit in that period – namely, any $g_k \leq H$. Unlike the newsboy problem stated above (but like the Baston–Bostock model), the information feedback is symmetric: the worker learns whether his wage demand (guess) was too high or too low. At the beginning of each period, the worker knows that his true value $H$ is at least the highest accepted bid (called $a_k$) and no more than the lowest rejected bid (called $b_k$). A pure search strategy $S$ in this model is the same as in the Baston–Bostock model discussed earlier (a map $S : D \rightarrow [a_1, b_1]$). However in this model the cost function $c'(S, H)$ is not symmetric and not even continuous. The worker's optimal guess (demand) of his actual value $H$ obviously gives him $H$ (or $H - L$, where $L$ is the cost to him of laboring – but we ignore this in the basic model). So we measure his cost relative to $H$. If his demand of $g \leq H$ is accepted, his cost is $H - g$. If it is rejected because $g > H$, his cost is $H$ (as he earns nothing). Thus

$$c'(g, H) = \begin{cases} H - g, & \text{if } g \leq H, \\ H, & \text{if } g > H. \end{cases}$$

The total cost to the worker will be

$$\sum_{k=1}^{\infty} \delta^{k-1} c'(g_k, H).$$

When the uncertainty interval of $H$ is very small, the relative cost of a rejected demand becomes large. The problem is somewhat akin to the problem faced by a longjumper in gauging his takeoff point. Every inch before the legal takeoff location wastes an inch at the landing point, but going past that location results in a nullified jump. The basic model of Alpern and Snower (1988) considers the minimization of $c'(S, H)$, assuming a Bayesian or minimax approach to $H$.

Many variations of this basic model have been analyzed. For example, the firm may act strategically to distort the worker's learning process by sometimes not hiring a worker who demands less than his value. If the worker is not aware of this, he may lower his subsequent demands (Reyniers, 1992; see also Alpern and Snower, 1991).

### 5.2.3 Discrete high–low search

In the problems described in this section the search space $Q$ is a known set $n$ of points (locations) that are linearly ordered. Gilbert (1962) and Johnson (1964) consider the discrete search game, with an immobile hider under the additional assumption that if the searcher looks at the $K$-th location and does not detect the hider at that location, he still receives the information as to whether the number of the location he has just searched is greater or less than the number of the location that contains the hider. Solutions of this game were given only for $n \leq 11$.

Gal (1974b) considers a discrete search game in which the immobile hider chooses a number $H \in Q$ and at each step, $i$, the searcher gains the binary information $H \leq g_i$ or $H > g_i$. The game proceeds until the searcher locates $H$, with the cost being the number of steps made by the searcher during the game. A general solution for this game is presented. The author shows that the optimal strategies of both players are generally mixed strategies that are not unique and that the natural bisection strategy of the searcher is usually not optimal. A continuous problem of this type in which $H$ and $g_i$ are real numbers in a certain interval and the information $H \leq g_i$ or $H > g_i$ has a certain probability of being erroneous is presented and solved by Gal (1978). Another continuous version in which the searcher receives the binary information $H \leq g_i$ or $H > g_i$ and there is a travel cost of $b(g_{i+1} - g_i)$, where $b$ is a constant, in addition to the fixed cost of observation is solved by Murakami (1976).

**Remark 5.10** *An interesting generalization of high–low search in which $H$ is a node of a tree and $g_i$ is a chosen subtree has been considered by Ben-Asher, Farchi and Newman (1999). In each step the searcher obtains the information whether $H$ is in the subtree $g_i$ or not. The authors present an algorithm of complexity $O(n^4 \lg^3 n)$, $n$ being the number of nodes in the tree, for obtaining a minimax search strategy. Many works on high low search, including problems in which the evader can lie, are briefly described in Section 5.5 of Benkoski et al. (1991).*

## 5.3 Infiltration Games

Assume that at time $t = 0$ an infiltrator enters a set $Q$ through a known point $O$ on the boundary of $Q$, and for all $t \geq 0$ moves inside $Q$ (Figure 5.2). Suppose that the searcher has to defend a "sensitive zone" $B \subset Q$, so that he wishes to maximize the probability of capturing the infiltrator before he reaches the boundary of $B$. (The infiltrator has the opposite goal.) What are the optimal strategies of both players in this game?

Such a problem with a relatively simple structure is the case in which $Q$ is a narrow rectangle, as depicted in Figure 5.3.

A simpler, discrete problem with a similar flavor is the following: The search set is an array of $n$ ordered cells (vertices) with distance 1 between any two consecutive cells and cell $n + 1$ is the target. At time 0 the infiltrator and the searcher are both located at cell number 1. Thereafter, at the end of each time unit, the searcher can move to any cell with distance not exceeding a certain integer $w \geq 1$, while the infiltrator can only move a distance of 1. Both players then stay at the chosen cell for the next time unit and the searcher searches in his cell. The probability of capture is $1 - \lambda$, $0 < \lambda < 1$, for each time unit they stay in the same cell (independently of previous history) and zero otherwise. (We make the usual assumption in this book that the players do not receive any information concerning the movement of the other.) The infiltrator wins

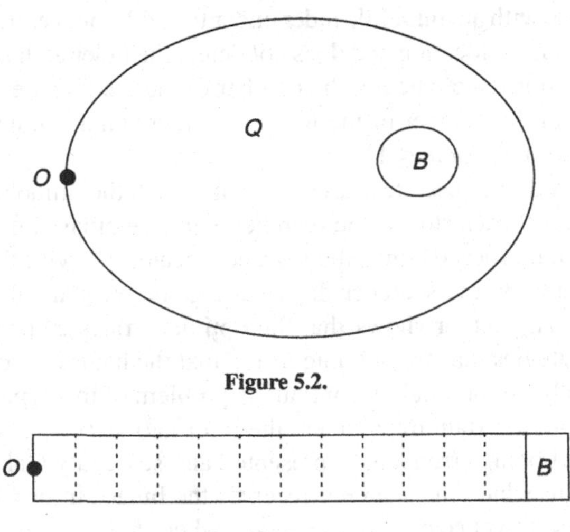


**Figure 5.2.**



**Figure 5.3.**

if he reaches the target (in finite time) without getting captured and loses otherwise. What are the optimal strategies of the players?

This apparently simple problem, proposed in Gal (1980), is still open. However, it attracted research in this area leading to interesting results by Lalley (1988), Auger (1991a, 1991b), Alpern (1992), Pavlovic (1995a, 2002), Baston and Garnaev (1996), Garnaev and Garnaeva (1996), and Garnaev et al. (1997). (A similar game with $n = 2$ and a more general detection function was considered by Stewart, 1981.)

Lalley considered the game under the simplifying assumptions that there is a safe zone (say at cell number 0) at which the infiltrator is located at time 0 and can stay for as long as he wants and remain safe from capture. He also made the assumption that the infiltrator wins only if he arrives at the target by time $t$ (where $t$ is known to both players).

Under the above assumptions Lalley presented a complete solution for $w = 1$ (denoted as the case "slow searcher") as follows: The optimal infiltration strategy is the following "random wait and run" strategy (called "*Admiral Farragut*" by Lalley): wait in the safe zone until time $\tau$, where $\tau$ is uniformly distributed over the integers $1 \le \tau \le t - n$, and then go full speed ahead toward the target reaching it (if not caught) at time $\tau + n + 1$. Denote by

$$m = t - n$$

the number of wait and run strategies and let

$$q = \left\lfloor \frac{t-1}{m} \right\rfloor \left( = 1 + \left\lfloor \frac{n-1}{m} \right\rfloor \right) \tag{5.6}$$

and

$$r = t - 1 - qm. \tag{5.7}$$

The optimal search strategy (called "*orderly fallback*" by Lalley) is to stay at cell number 1 for the first time unit, then withdraw with speed 1 for $\xi_1$ time units; rest for one time unit, then withdraw with speed 1 for $\xi_2$ time units.... There are $m$ such segments and the random variables $\xi_1, \xi_2, \ldots, \xi_m$ satisfy the following relations:

$$q - 1 \le \xi_j \le q, \quad 0 \le j \le m$$

$$\sum \xi_j = n - 1.$$

Each $\xi_j$ is obtained by sampling without replacement from an urn containing $r$ balls marked $q$ and $m - r$ balls marked $q - 1$. (Note that if $r = 0$ then all $\xi_j$ are equal to $q - 1$ and the searcher's strategy is pure.)

The value of the game (the probability of successfully infiltrating) is given by

$$v = \lambda^q \left[ \lambda \left( \frac{r}{m} \right) + \left( 1 - \frac{r}{m} \right) \right]. \tag{5.8}$$

Since $\left[ \lambda \left( \frac{r}{m} \right) + \left( 1 - \frac{r}{m} \right) \right]$ lies between $\lambda$ and 1 it follows that the value of the game is between $\lambda^{q+1}$ and $\lambda^q$. If $t \gg n$ then $q = 1$ and $v = \lambda \left[ \lambda \left( \frac{r}{m} \right) + \left( 1 - \frac{r}{m} \right) \right]$ (obviously approaching $\lambda$ as $t \to \infty$).

It is easy to see why the orderly fallback strategy is the best response to the randomized wait and run strategy. Denote by $k_\tau$ the number of times the searcher meets wait and run strategy number $\tau$, $\tau = 1, \ldots, m$. Since the number of time instances for possible meeting is $t - 1$, it follows that

$$\sum_{\tau=1}^{m} k_\tau = t - 1. \tag{5.9}$$

Then the probability of successful infiltration is

$$\frac{1}{m}\sum_{\tau=1}^{m} \lambda^{k_\tau}. \tag{5.10}$$

Since the function $\lambda^x$ is convex it follows that minimizing (5.10) under the constraint (5.9) leads to integers $k_\tau$ as close as possible to $(t-1)/m$, i.e., $q+1$ ($r$ times) and $q$ ($m-r$ times) with the minimal value given by (5.8). Such $k_\tau$, with the appropriate randomization, are obtained in the orderly fallback strategy. Note that such an optimal solution can be obtained for $w = 1$. Thus, $w > 1$ should not improve the chance of the searcher to win the game. This *a priori* surprising result, was proved by Auger (1991b).

To see why the wait and run strategies are the best response to the orderly fallback strategy, consider the simple case in which $t - 1 = q \times m$ so that the orderly fallback strategy is pure and meets each wait and run strategy $q$ times. We have to show that in this case any infiltration trajectory meets the orderly fallback trajectory at least $q$ times. This follows from considering the location of the infiltrator versus the location of a searcher who follows the orderly fallback trajectory: at time $t = 0$ the infiltrator is below the searcher and by time $t$ the infiltrator has to arrive at cell number $n+1$ so that at time $t$ his location is above the searcher's location. Thus the infiltrator has to "pass thorough" the searcher's trajectory, which means that their location has to coincide at least $q$ times.

We have just shown that for the simple case $t - 1 = q \times m$ the randomized wait and run strategy and the orderly fallback trajectory are both best response to each other. Thus, they have to be optimal strategies. The general case is more complicated by the fact that the orderly fallback strategy includes some segments of length $q-1$ and some others of length $q$ in a randomized fashion.

Auger (1991b) has demonstrated that, while the searcher does not benefit from speed larger than 1 in Lalley's problem, $w > 1$ does give a definite advantage over $w = 1$ for the original infiltration problem without the safe zone. In this paper Auger also considered the simplest non-trivial case $n = 3$ (for $n = 1, 2$ the game is trivial, yielding values $\lambda$ and $\lambda^2$, respectively). For $n = 3$ and $w > 1$ he showed that $v = \lambda^2((\lambda/2) + (1/2))$. He also suggested an approach for finding the solution for the infinite time period by solving the game for a finite $t$ and letting $t \to \infty$.

An asymptotic solution for the original game with $\lambda$ close to 1 (and $w = 1$) was obtained by Pavlovic (1995a) for $n \le 9$ and later (2002) for all $n$.

An interesting extension of the game with a safe zone is the infiltration on $k$ arcs introduced by Auger (1991a). Denote the safe zone (cell 0) by $O$ and the target (cell $n+1$) by $A$. Now, assume that there are $k$ non-intersecting arcs (possibly with different number of vertices on each arc) that connect $O$ and $A$, rather than the single arc in the original problem. Denote the total number of vertices on the arcs, excluding $O$ and $A$ by $N$. Then the number of wait and run strategies on the $k$ arcs is $kt - N$. It turns out that expression (5.8) for the value remains true with $m$ (= the number of wait and run strategies for the one arc) replaced by $kt - N$, where $q$ and $r$ in (5.6) and (5.7) are obtained by replacing $m$ by $kt - N$. A simplified proof of the result for a more general case of limited resources is presented in Baston and Garnaev (1996) and Garnaev et al. (1997). (The special case of the circle is presented in Garnaev and Garnaeva, 1996.) They solved the game on the $k$ arcs in which the searcher has limited resources and can look for the infiltrator only $L$ times, for any $L \leq t-1$. (If $L = t - 1$, then the game is the infiltration on $k$ arcs considered by Auger.) It is interesting to note that also in this case the value is still given by (5.8) with $t - 1$ replaced by $L$ and $m$ replaced by $kt - N$ in (5.6), (5.7) and (5.8).

Consider the infiltration on $k > 1$ arcs with no time bound for the infiltrator ($t \to \infty$). Then $q = \lfloor t - 1/kt - N \rfloor = 0$ and $r = t - 1$. Thus,

$$v = \lambda \frac{t-1}{kt-N} + 1 - \frac{t-1}{kt-N} \sim 1 - \frac{1-\lambda}{k} \text{ as } t \to \infty.$$

Alpern (1992) has extended this result for a general graph. He showed that if $O$ and $A$ are vertices of any graph and $k$ is the minimal number of vertices separating them (minimal cut), then the value of the infiltration game with no time bound is $1 - (1 - \lambda/k)$.

## 5.4 Searching in Discrete Locations

Assume that a stationary object is hidden in one of $n$ (labeled) locations. For each location $i$, $i = 1, \ldots, n$ there are three known parameters: $c_i$, the cost of a single search in location $i$; $\alpha_i$, the probability of finding the object by a single search in location $i$ if it is in this location; and $p_i$, the probability that the object is in location $i$. The parameters $c_i$ and $\alpha_i$ remain fixed while $p_i$, which is equal to a known *a priori* probability at the beginning, is updated by Bayes' law after each (unsuccessful) attempt. The problem of finding the object in minimal expected cost is a classical problem solved by Blackwell (see Matula, 1964). Although the functional equation obtained by a dynamic programming formulation of this problem is complicated, the optimal policy is easy to describe: at each stage look into the location which maximizes $p_i\alpha_i/c_i$.

Ross (1969) and (1983) extends the classical model to an optimal search or stop problem. Using dynamic programming he characterizes the optimal search strategy for the case that a (possibly location dependent) reward is earned if the object is found and the searcher may decide to stop searching at any time.

Gittins (1989) obtains the classical result as a special case of his well known index for multi-armed bandit processes. In his book he also presents the solution of the location search within the context of a search game. (The results he presents there are based on Roberts and Gittins, 1978, Gittins and Roberts, 1979, and Gittins, 1979.) The search game is based on the assumptions of the classical model but the object is hidden in one of the locations by an adversary who knows all the relevant parameters. Thus, at the beginning, the location of the object is not given by a known *a priori* distribution but is determined by the hider as the worst probability distribution (for searching). For simplicity the problem is solved under the assumption that $c_i = 1, i = 1, \ldots, n$. The general idea is for the hider to make all the locations equally "attractive", with the same $p_i \alpha_i$ for all $i = 1, \ldots, n$ such that

$$p_i = \bar{p}_i = \frac{1/\alpha_i}{\sum_{j=1}^{n} 1/\alpha_j}. \tag{5.11}$$

This aim is totally achievable if the hider is allowed to move the object after each (unsuccessful) search. In this case the optimal strategy for the hider is to choose the location of the object according to $p_i = \bar{p}_i, i = 1, \ldots, n$, independently of previous search history, for each time. (The optimal strategy of the searcher is exactly the same: each time to search location $i$ with probability $\bar{p}_i$.) Such a model, with a cost for moving the object, was considered by Norris (1962). He obtained the above solution in the special case that the cost of moving is zero.

However, in the original problem, in which the hider chooses the probability vector once for the whole game, if he chooses $p_i = \bar{p}_i, i = 1, \ldots, n$ in the beginning, then after an unsuccessful search in location $i$ the *a posteriori* probability of location $i$ will drop below $\bar{p}_i$ and will increase above $\bar{p}_j$ for all $j \neq i$. Thus, all the other locations will be more attractive after the first search. Still, the optimal Hiding strategy is to choose the probabilities near $\bar{p}_i$, and the optimal strategy of the *searcher* is, for each time period to choose the probabilities $\bar{p}_i$ (exactly) for searching location $i$. The exact expression of the Hiding strategy and the value are complicated because of the discreteness in time.

A neat solution that can be explicitly presented is obtained for the continuous version of the search game (with a stationary target). Here, the time is continuous and the detection of the object in location $i$ is given by a Poisson process with rate $\alpha_i$. The optimal hiding probabilities are exactly given by (5.11) and, for each infinitesimal time interval $\Delta t$, it is optimal to search during $\bar{p}_i \times \Delta t$ time period in location $i$. Thus, the *a posteriori* probability distribution for the location of the object remains the same all the time, being equal to the *a priori* distribution given by (5.11).

It is easy to extend Gittins' continuous time results to the case of general $c_i$. For an infinitesimal time interval, spending cost $\Delta t$ at location $i$, searching it during $1/c_i \Delta t$, leads to detection probability $\alpha_i/c_i \Delta t$ rather than $\alpha_i \Delta t$. Thus (5.11) is changed into:

$$\hat{p}_i = \frac{c_i/\alpha_i}{\sum_{j=1}^{n} c_j/\alpha_j}, \quad i = 1, \ldots, n.$$

The numbers $\hat{p}_i, i = 1, \ldots, n$, are the optimal strategy for the hider as well as the for the searcher for a single period. Note, however, that the length of such a period at location $i$ is proportional to $1/c_i$ so that the minimax search intensity, $\hat{q}_i$, has to be proportional to $(1/c_i) \times (c_i/\alpha_i) = 1/\alpha_i$. Thus,

$$\hat{q}_i = \frac{1/\alpha_i}{\sum_{j=1}^{n} 1/\alpha_j}, \quad i = 1, \ldots, n.$$

(Thus, somewhat surprisingly, the search intensity is independent of the cost coefficients $c_i$.) The probability of detection in each period is $\left(\sum_{i=1}^{n} (\alpha_i/c_i) \hat{p}_i^2\right) \Delta t$ with a search cost of $\Delta t$. Thus the expected (minimax) search cost is

$$\frac{1}{\sum_{i=1}^{n} \alpha_i / c_i \hat{p}_i^2} = \frac{\left(\sum_{j=1}^{n} c_j / \alpha_j\right)^2}{\sum_{i=1}^{n} (\alpha_i / c_i)(c_i^2 / \alpha_i^2)} = \sum_{j=1}^{n} \frac{c_j}{\alpha_j}.$$

In another interesting model the hider chooses a node in a graph and the search cost is composed of the travel cost plus $c_i$, the cost of inspecting the chosen node $i$. The case of linear ordered nodes is solved in Kikuta (1990, 1991). The more general case of a tree is solved in Kikuta and Ruckle (1994) and Kikuta (1995).

It should be noted that if $c_i = c$ for all $i$, then the game can be transformed into the search on an equivalent tree with no inspection cost, solved in Section 3.3 as follows. To each node add a new arc of length $c/2$ with a terminal node at its end. The optimal strategies for the equivalent tree (with no inspection cost) are optimal for the problem with inspection cost, $c$, but the expected (optimal) cost is reduced by $c/2$. (If $c_i$ are not all equal, then adding similar arcs, with length $c_i/2$ to each node, leads to a close but not identical problem.)

Many other discrete search games have been considered by Ruckle in his book "Geometric games and their applications" (1983). In particular, he describes a search and pursuit game in a cyclic graph. This game is not just a discrete version of the princess and monster on the circle because here the princess cannot be caught while moving between vertices. Thus, the discrete analog of the Sweeping Lemma 4.6 for $k = 2$ does not hold.

The problem of two competing players searching for the same object located in one of a set of labeled boxes is considered by Nakai (1986), Garnaev and Sedykh (1990), and by Fershtman and Rubinstein (1997). The problem of searching for different objects, with each player trying to find his object before his opponent does, is discussed by Nakai (1990).

Baston and Garnaev (2000) investigated another game in the framework of discrete location search with a known hiding probability for each location. In this game the searcher is opposed not by the hider but by a protector who can allocate resources to locations making it more difficult for the searcher to find the object.

**Remark 5.11** *Kikuta and Ruckle (1997) and Ruckle and Kikuta (2000) considered "Accumulation Games" described as follows. The hider hides material at various locations and has to accumulate a given amount by a fixed time in order to win. The searcher examines these locations and can confiscate any material he finds. The 1997 paper discusses discrete material and location versions of this game and the 2000 paper studies accumulation of continuous material over two types of continuous regions: the interval and the circle.*

# Part Two

# Search Games in Unbounded Domains

# Chapter 6

# General Framework

## 6.1 One-Dimensional Search Games

The work on search problems in unbounded domains was initiated by Bellman (1963) (and independently by Beck, 1964) who introduced the following problem. "An immobile hider is located on the real line according to a known probability distribution. A searcher, whose maximal velocity is one, starts from the origin $O$ and wishes to discover the hider in minimal expected time. It is assumed that the searcher can change the direction of his motion without any loss of time. It is also assumed that the searcher cannot see the hider until he actually reaches the point at which the hider is located and the time elapsed until this moment is the duration of the game." This problem is usually called the *linear search problem* (LSP). It has attracted much research, some of it quite recent. A detailed discussion is presented in Chapter 8.

Beck and Newman (1970) presented a solution for the search on the real line considered as a game. As we have already pointed out in the introduction, if the capture time is used as the cost function and no restrictions are imposed on the hider, then the value of the game is infinite. Thus, in order to have a reasonable problem, one should either assume that the hider is in some way restricted or that a normalized cost function is used. Beck and Newman used the first method and allowed the (immobile) hider to choose his location by using a probability distribution function $h$, which has to belong to the class $\mathcal{H}_\lambda$ defined as follows.

**Definition 6.1** *The class $\mathcal{H}_\lambda$ is the set of all hiding strategies in $Q$, which satisfy the condition that the expected distance of the hiding point $H$ from the origin is less than or equal to the constant $\lambda$. In other words, any $h \in \mathcal{H}_\lambda$ satisfies*

$$\int_Q |H|\, dh \leq \lambda, \tag{6.1}$$

*where $|H|$ is the distance of $H$ from the origin.*

Such a restriction is the "natural" one for one-dimensional problems because $C(S, H) \geq |H|$ for all $S$ so that condition (6.1) is necessary in order to get a finite value for the game and, as we shall see in Chapter 8, condition (6.1) is also sufficient. We shall also see in Chapter 8 that the optimal search strategy does not depend on the constant $\lambda$, so that there is no need for the searcher to know the value of $\lambda$.

Another method of handling search games in unbounded domains, which is used by Gal (1974a) and Gal and Chazan (1976), is to impose no restriction on the hiding strategies but to normalize the cost function. Such a "natural" normalized cost function is the function $\hat{C}$ defined as

$$\hat{C}(S, H) = C(S, H)/|H|, \tag{6.2}$$

where $C$ is the capture time. Since $C(S, H) \geq |H|$, it readily follows that the only normalization of the type $C/|H|^{\gamma}$ which yields a finite value is the normalization with $\gamma = 1$. This approach will be referred to as using *normalized cost.*

Using a minimax criterion for a normalized cost is common in computer science literature for searching for a target under incomplete information (see, e.g., Papadimitriou and Yannakis (1991); Koutsopias et al. (1996); Burley (1996); or Blum et al. (1997). It is assumed that the goal of the searcher is to devise a strategy, so that the total distance traversed has the best possible ratio to the shortest path. This type of ratio is used not only for search problems but is generally used in the so-called online algorithms. The analysis of such online algorithms is a very active research area in computer science (see, for example, the book by Borodin and El-Yaniv (1998); an application for load balancing by Berman et al. (2000); and some other applications, including the star search, described in Section 9.2, by Schuierer (2001)). The quality of an online algorithm is measured in such an analysis by comparing its performance with an optimal offline algorithm. The ratio guaranteed by a specific strategy is called the *competitive ratio*. Our minimax approach is equivalent to finding the minimal competitive ratio.

Actually, the two approaches just discussed (either restricting the hider by (6.1) or using normalized cost) lead to equivalent results under the following *scaling assumption* which holds for all the games to be considered in Part II.

**Scaling Assumption**. For any positive constant $\beta$, there exists a mapping $\Phi_\beta$ of the search space Q onto itself such that $\Phi_\beta(O) = O$ and for all $Z_1, Z_2 \in Q$

$$d(\Phi_\beta(Z_1), \Phi_\beta(Z_2)) = \beta d(Z_1, Z_2).$$

Under the scaling assumption, any game with the restriction (6.1), with $\lambda$ being any positive constant, can be easily transformed into the game having the restriction (6.1) with $\lambda = 1$. This can be done by a construction that is similar to the one used in the scaling lemma of Chapter 2 (Proposition 2.5), with $\Phi_{1/\lambda}$.

We now show that, under the scaling assumption, if the game with the normalized cost function (6.2) has a finite value then the game with the restriction (6.1) with $\lambda = 1$ is equivalent to it. Assume that the game with restriction (6.1), has a value $v$. Let $\bar{h}(H)$ be an $\varepsilon$-optimal hiding strategy that satisfies (6.1) with $\lambda = 1$. Then

$$\int_Q C(S, H)\, d\bar{h}(H) \geq (1 - \varepsilon)v \quad \text{for all } S. \tag{6.3}$$

Since any hiding strategy with $E|H| < \lambda$ is dominated by a hiding strategy with $E|H| = \lambda$, we may assume that under $\bar{h}$, $E|H| = 1$. If we define the hiding strategy $\hat{h}(H)$ by

$$d\hat{h} = |H|\, d\bar{h}(H),$$

then the strategy $\hat{h}$ satisfies for all $S$

$$\int_Q \hat{C}(S, H)\, d\hat{h}(H) = \int_Q \frac{C(S, H)}{|H|}\, d\hat{h}(H) = \int_Q C(S, H)\, d\bar{h}(H) \geq (1 - \varepsilon)v.$$

On the other hand, assume that $\hat{v} < \infty$ and let $\hat{h}$ be an $\varepsilon$-optimal for the normalized cost approach, that is,

$$\int_Q \hat{C}(S, H)\, d\hat{h}(H) \geq (1 - \varepsilon)\hat{v},$$

where $\hat{v}$ is the value of the search game with the normalized cost.

Let $\delta > 0$ be sufficiently small so that

$$\int_{|H| \geq \delta} \hat{C}(S, H)\, d\hat{h}(H) > (1 - 2\varepsilon)\hat{v}. \tag{6.4}$$

Denote

$$\int_{|H| \geq \delta} \frac{d\hat{h}(H)}{|H|} = b,$$

and define a hiding strategy $h_b(H)$ by

$$dh_b(H) = \begin{cases} \dfrac{1}{b|H|}\, d\hat{h}(H), & \text{for } |H| \geq \delta, \\ 0, & \text{for } |H| < \delta. \end{cases}$$

Then $h_b$ satisfies

$$\int_Q |H|\, dh_b(H) = \frac{1}{b} \int_{|H| \geq \delta} d\hat{h}(H) \leq \frac{1}{b} \tag{6.5}$$

and for all $S$,

$$\int_Q C(S, H)\, dh_b(H) = \frac{1}{b} \int_{|H| \geq \delta} \frac{C(S, H)}{|H|}\, d\hat{h}(H) > \frac{(1 - 2\varepsilon)\hat{v}}{b}, \quad \text{by (6.4).}$$

Under the scaling assumption, we can use the construction of the scaling lemma with the mapping $\Phi_b$ and obtain a hiding strategy $h_1$ with $\int_Q |H|\, dh_1(H) \leq 1$, which makes sure that the expected capture time exceeds $(1 - \varepsilon)\hat{v}$.

We have thus shown that $v = \hat{v}$, so that the two approaches lead to equivalent results.

**Remark 6.2** *If Assumption 1 does not hold, then the two approaches may lead to different results. Such an example, in which $Q$ is the half line $-\varepsilon \leq x \leq \infty$, is discussed in Section 8.2.1.*

The normalizing approach is more convenient for us to work with than using the "restrictive approach" (i.e., using (6.1)), and we shall generally use the cost function $\hat{C}$

given by (6.2). Working with the normalizing approach also has the following advantage: If $\hat{s}$ is an $\varepsilon$-optimal search strategy for the normalizing approach, then for any hiding strategy $h$ that satisfies $\int_Q |H|\, dh \le 1$,

$$c(\hat{s}, h) = \int_Q c(\hat{s}, H)\, dh = \int_Q \frac{c(\hat{s}, H)}{|H|} |H|\, dh$$
$$\le \sup_H \frac{c(\hat{s}, H)}{|H|} \le (1+\varepsilon)v.$$

Thus, $\hat{s}$ is also $\varepsilon$-optimal for the restrictive approach. On the other hand, it can be easily seen that an $\varepsilon$-optimal search strategy for the restrictive approach need not be an $\varepsilon$-optimal strategy for the normalizing approach.

It should be noted that if the hider is mobile (see Section 8.5), then the appropriate restriction is $\int_Q |H(0)|\, dh \le 1$, and the appropriate normalization is $\hat{C} = C/H(0)$, where $H(0)$ is the location of the hider at time $t = 0$.

We shall deal mainly with one-dimensional search games and present only a brief discussion on multidimensional problems in Section 6.2. In Chapter 7, we develop the main tools for obtaining minimax trajectories for a rather general family of search games. These tools will enable us to solve the search games on the infinite line presented in Chapter 8. We shall also use these tools in some of the games presented in Chapter 9, including searching on a finite set of rays and in the plane.

## 6.2 Multidimensional Search Games

In this section, we discuss the framework of search games in the case that the search space is the entire $N$-dimensional Euclidean space with $N \ge 2$. The rules of the game are similar to those used in Part I. The searcher has to start from the origin $O$ and move along a continuous trajectory, with maximal velocity not exceeding 1. The hider is immobile and stays at a point $H$. The game terminates when the hider is within a distance of $r(>0)$ from the searcher. We shall be concerned with the appropriate restriction, which has to be imposed on the hider's strategies, in order to obtain a finite value for the game. In Section 6.1 we noted that condition (6.1) is necessary and sufficient to obtain a finite value for one-dimensional search games. Such a condition is not appropriate for multidimensional search games. In order to find the adequate moment restriction, we derive a necessary and sufficient condition for the finiteness of the value of the search game in the entire plane. Such a condition is

$$\int_Q |H|^2\, dh \le \lambda. \tag{6.6}$$

First we show that (6.6) is indeed a necessary restriction. If we allow hiding strategies with $\int_Q |H|^2\, dh = \infty$, then the hider can use the strategy $\bar{h}$ with the following probability density $f$ in the plane. Let $H = (|H|, \theta), 0 \le \theta \le 2\pi$ and

$$f(H) = \begin{cases} \dfrac{1}{\pi |H|^3}, & \text{if } |H| \ge 1, \\ 0, & \text{otherwise.} \end{cases}$$

Since any trajectory $S$ covers in each time interval $[0, t]$ an area which does not exceed $2rt$, it follows that, under $\bar{h}$, the probability that the capture time exceeds $t$ satisfies

$$\Pr(C(S, H) > t) \geq \int_{\sqrt{(1+2rt)/\pi}}^{\infty} \frac{2}{x^3}\, dx = \frac{\pi}{1+2rt}.$$

Thus, for any $S$

$$c(S, \bar{h}) = \int_Q C(S, H)\, d\bar{h} = \int_0^{\infty} Pr(C(S, H) > t)\, dt$$
$$\geq \int_0^{\infty} \frac{\pi}{1+2rt}\, dt = \infty$$

which implies that $v = \infty$. On the other hand, if the hider's strategy has to satisfy condition (6.6), then the searcher can guarantee a finite capture time using the trajectory $\bar{S}$ defined as follows. Start by moving in a straight line to the circle with radius $2r$ around the origin and encircle it, then move to the circle with radius $4r$ around the origin and encircle it, then to the circle with radius $6r$, etc. The trajectory $\bar{S}$ satisfies, for all $H$,

$$C(\bar{S}, H) \leq |H| + r + 2\pi \sum_{i=1}^{\lceil |H|/2r \rceil} 2ri$$
$$< \frac{\pi}{2r}|H|^2 + (3\pi + 1)|H| + (4\pi + 1)r.$$

Since $E|H|^2 \leq \lambda$ and consequently $E|H| \leq \sqrt{E|H|^2} \leq \sqrt{\lambda}$, it follows that for any admissible $h$

$$C(\bar{S}, h) < \frac{\pi}{2r}\lambda + (3\pi + 1)\sqrt{\lambda} + (4\pi + 1)r,$$

so that the expected capture time guaranteed by $\bar{S}$ is finite.

From the foregoing discussion, it follows that restriction (6.6) is the moment condition that has to be used in the case of an unbounded two-dimensional search. By using a similar argument, it is easy to show that the appropriate moment condition for $N$-dimensional unbounded search is $E|H|^N \leq \lambda$. In the case that we use the approach of normalizing the capture time, a similar argument leads to the conclusion that the appropriate normalizing function is $|H|^N$. Thus, the normalized cost function should be $\hat{C}(S, H) = C(S, H)/|H|^N$ for $N$-dimensional search games.

# Chapter 7

# On Minimax Properties of Geometric Trajectories

## 7.1 Introduction

In this chapter we develop an important tool for finding minimax trajectories for searching unbounded search spaces. The main results, which are extensively used in Chapters 8 and 9, are Corollary 7.11 and Theorem 7.18. The proofs, as well as most other mathematical details, are mainly for experts and can be skipped at first reading.

In many search games, including the ones discussed in Chapters 8 and 9, finding minimax search trajectories is a special case of the following game.

The searcher chooses a sequence of positive numbers $X = \{x_j\}$. We consider both infinite sequences in which $0 \le j < \infty$ and doubly infinite sequences in which $-\infty < j < \infty$. The hider chooses an integer $i$. Let $X^{+i}$ be the sequence $X$ with indices shifted $i$ to the right:

$$X^{+i} \equiv \{x_{i+j}\}, \quad 0 \le j < \infty \text{ (or } -\infty < j < \infty \text{ for doubly infinite sequences).} \tag{7.1}$$

Let the loss $\hat{C}(X, i)$ of the searcher be given as

$$\hat{C}(X, i) = F(X^{+i}), \tag{7.2}$$

where $F$ is a functional (i.e., a mapping from the space of all positive sequences into the real numbers) that satisfies several requirements to be specified in Section 7.2.

Finding minimax search trajectories for games of this type requires solving

$$\inf_X \sup_i \hat{C}(X, i).$$

A typical example, extensively used in the linear search games of Chapter 8, amounts to solving

$$\inf_X \sup_i \frac{\sum_{j=-\infty}^{\infty} \alpha_j x_{i+j}}{x_i}$$

for a given sequence $\alpha_j \geq 0$, $-\infty < j < \infty$.

The positive geometric sequences $A_a$

$$A_a = \{a^j\} \tag{7.3}$$

play a special role in searching for minimax solutions. Theorem 7.1, below, states that for any positive sequence $X$, the cone spanned by $\{X^{+i}, 0 \leq i < \infty\}$ contains a sequence, which is "close enough" to a sequence $A_a$.

**Theorem 7.1** *Let $x_i$, $0 \leq i < \infty$, be a positive sequence and let $k$ be any positive integer. Let $W_k$ be the $(k+1)$-dimensional convex cone spanned by the set $\{(x_i, x_{i+1}, \ldots, x_{i+k}),$ $0 \leq i < \infty\}$. Thus,*

$$W_k = \left\{ Y\colon Y = \sum_{i=0}^{n} \theta_i (x_i, x_{i+1}, \ldots, x_{i+k});\ \theta_i \geq 0, 0 \leq i \leq n, 0 < n < \infty \right\}.$$

*$\bar{W}_k$ will denote the closure of $W_k$. Define*

$$a = \limsup_{n\to\infty} (x_n)^{1/n}.$$

*Then*

$$\frac{1}{\sum_{i=0}^{k} a^i}(1, a, a^2, \ldots, a^k) \in \bar{W}_k. \tag{7.4}$$

*(The case where $a = \infty$ is included. In this case (7.4) means $(0, 0, \ldots, 0, 1) \in \bar{W}_k$.)*

Using Theorem 7.1 we shall show, in Section 7.2, that a minimax search strategy can be found in the family of the geometric sequences $\{A_a, a > 0\}$. This result will extensively be used in Chapters 8 and 9.

The continuous version of the theorem follows.

**Theorem 7.2** *Let $X(\tau)$, $0 \leq \tau < \infty$, be any measurable positive function that is bounded on any finite interval. Define*

$$b = \ln\left( \limsup_{n\to\infty} \left( \int_n^{n+1} X(\tau)\, d\tau \right)^{1/n} \right).$$

*Suppose $-\infty < b < \infty$. For any $\varepsilon > 0$, $\varepsilon_1 > 0$, and $D > 0$, there exists a positive number $L$ and a nonnegative function $\theta(t)$, $0 \leq t < \infty$, such that $\theta'(t)$ is continuous and the function $f(\tau)$ defined by*

$$f(\tau) = \int_0^L \theta(t) X(t+\tau)\, dt$$

*satisfies*

$$(1-\varepsilon)e^{b\tau} \leq f(\tau) \leq (1+\varepsilon)e^{b\tau}$$

*and*

$$(1-\varepsilon_1)e^{b\tau} \leq f'(\tau) \leq (1+\varepsilon_1)e^{b\tau}$$

*for all* $0 \leq \tau \leq D$.

The proofs of Theorems 7.1 and 7.2 can be found in Appendix 2 of Gal (1980) or in Gal and Chazan (1976).

Theorem 7.2 is needed for the continuous versions of the game, described in Section 7.2.1, with the analogous property that in this case, the minimax solution is an exponential function. This result will be used in Section 9.2. Finally, in Section 7.3, we establish the uniqueness of the solution for functionals belonging to a class that contains problems in Chapters 8 and 9.

The results of this chapter are mostly based on Gal and Chazan (1976) and Gal (1972).

## 7.2 Minimax Theorems

In essence our theorems are valid for any homogenous unimodal functional $F$. We now formally present the required properties for $F$. We shall present in full detail the theorems for the case in which $X$ is a sequence and, at the end of the chapter, we briefly present some of the results for the continuous functions.

We now consider the case in which $X$ is positive infinite sequence $\{x_j\}$, $0 \leq j < \infty$. The case in which $X$ is a doubly infinite sequence will be discussed separately.

At the first stage, we shall assume that $F$ depends only on a finite number of terms of $X$. Thus, in this case, there is a positive integer $k$ such that

$$F(X) = F(x_0, x_1, \ldots, x_k)$$

i.e.,

$$F(X^{+i}) = F(x_i, x_{i+1}, \ldots, x_{i+k}).$$

We require that $F$ satisfy the following conditions for all $x_0 > 0$, $x_1 > 0, \ldots,$ $x_k > 0$:

$$F(X) \text{ is continuous,} \tag{7.5}$$

$$F(\alpha X) = F(X) \quad \text{for all } \alpha > 0, \quad \text{and} \tag{7.6}$$

$$F(\alpha X + (1-\alpha)Y) \leq \max[F(X), F(Y)] \quad \text{for all } X, Y \text{ and } 0 \leq \alpha \leq 1.$$

For functionals, which satisfy the homogeneity condition (7.6), the above unimodality condition is equivalent to the more convenient condition:

$$F(X+Y) \leq \max[F(X), F(Y)] \quad \text{for all } X \text{ and } Y. \tag{7.7}$$

In addition, we assume that $F$ satisfies the following two requirements, which actually hold for all "reasonable" functionals that satisfy (7.5)–(7.7).

$$F(A_\infty) \equiv \liminf_{a\to\infty} F\left(\frac{1}{a^k}, \frac{1}{a^{k-1}}, \dots, \frac{1}{a}, 1\right) = \liminf_{\varepsilon_k, \varepsilon_{k-1}, \cdots, \varepsilon_1 \to 0} F(\varepsilon_k, \varepsilon_{k-1}, \dots, \varepsilon_1, 1) \tag{7.8}$$

and

$$F(A_0) \equiv \liminf_{a\to 0} F(1, a, \dots, a^k) = \liminf_{\varepsilon_1, \varepsilon_2, \dots, \varepsilon_k \to 0} F(1, \varepsilon_1, \varepsilon_2, \dots, \varepsilon_k). \tag{7.9}$$

We now present two examples of $F$, which satisfy the above required conditions.

**Example 7.3** *Let $\alpha_0, \alpha_1, \dots, \alpha_k$ be any nonnegative numbers and $\beta_0, \beta_1, \dots, \beta_k$ be nonnegative numbers, not all zero. Define*

$$F(X) = \frac{\sum_{j=0}^{k} \alpha_j x_j}{\sum_{j=0}^{k} \beta_j x_j}. \tag{7.10}$$

*It can be easily verified that conditions (7.5)–(7.9) hold.*

**Example 7.4** *Let $0 \le \beta \le 1$ and*

$$F(X) = \frac{x_0 + \sum_{j=1}^{k} \sqrt{x_{j-1}^2 + x_j^2 - 2\beta x_{j-1} x_j}}{x_1}.$$

*This type of functional is used in Section 9.3. We show there that $F$ satisfies (7.5)–(7.9).*

The unimodality condition of the homogenous functional $F$ immediately leads to the following result.

**Lemma 7.5** *Let $F$ satisfy (7.6) and (7.7), $X$ be a positive sequence and $\beta_0, \dots, \beta_l$ a sequence of nonnegative numbers. If*

$$Y = \sum_{i=0}^{l} \beta_i X^{+i} \tag{7.11}$$

*(see (7.1)), then*

$$F(Y) \le \max_{0\le i\le l} F(X^{+i}). \tag{7.12}$$

Note that condition (7.6) implies that for any geometric sequence $A_a$ and any positive integer $i$,

$$F(A_a^{+i}) = F(a^i, a^{i+1}, \dots) = F(1, a, \dots) = F(A_a). \tag{7.13}$$

It follows from Lemma 7.5 that for any $X$, any linear combination of $X^{+i}$ with nonnegative coefficients is at least as good as $X$. However, by Theorem 7.1 this linear combination can be made as close as desired to a geometric sequence. This is the idea behind the following minimax theorem.

**Theorem 7.6** *If $F$ satisfies conditions* (7.5)–(7.9), *then for any positive sequence* $X = \{x_i\}_{i=0}^{\infty}$ *we have*

$$\limsup_{i\to\infty} F(X^{+i}) \geq \inf_{0<a<\infty} F(A_a). \tag{7.14}$$

Theorem 7.6 implies that an optimal (or an $\varepsilon$-optimal) search strategy for the class of games defined in Section 7.1 can be found among the geometric sequences $A_a$. This follows from the fact that (7.13) and Theorem 7.6 imply that for any positive sequence $X$

$$\begin{aligned}\sup_{0\leq i<\infty} F(X^{+i}) &\geq \limsup_{i\to\infty} F(X^{+i})\\ &\geq \inf_{0<a<\infty} F(A_a) = \inf_{0<a<\infty} \sup_{0\leq i<\infty} F(A_a^{+i}).\end{aligned}$$

Thus

$$\inf_X \sup_i F(X^{+i}) \geq \inf_{0<a<\infty} \sup_{0\leq i<\infty} F(A_a^{+i}) = \inf_{0<a<\infty} F(A_a),$$

which implies

$$\inf_X \sup_i F(X^{+i}) = \inf_{0<a<\infty} F(A_a)$$

(since the inf is obtained within the subset of geometric series).

**Proof.** Let us denote

$$q = \inf_{0<a<\infty} F(A_a). \tag{7.15}$$

If (7.14) does not hold, then there exists a positive sequence $X = \{x_i\}_{i=0}^{\infty}$, an integer $i_0$, and a positive number $\varepsilon$ such that for all $i \geq i_0$

$$F(X^{+i}) < q - \varepsilon. \tag{7.16}$$

Without loss of generality, we may assume that $i_0 = 0$ so that (7.16) holds for all $i \geq 0$.

Let $a = \limsup_{n\to\infty} x_n^{1/n}$. If $0 < a < \infty$, then Theorem 7.1 implies that for any $\delta > 0$ there exist nonnegative numbers $\beta_0, \beta_1, \ldots, \beta_l$, such that the sequence

$$Y = \{y_j\}_{j=0}^{\infty} = \sum_{i=0}^{l} \beta_i X^{+i} \tag{7.17}$$

satisfies

$$|y_j - a^j| < \delta \quad \text{for } 0 \leq j \leq k. \tag{7.18}$$

It follows from Lemma 7.5 and (7.16) that

$$F(y_0, y_1, \ldots, y_k) = F(Y) \leq \sup_{0\leq i<\infty} F(X^{+i}) < q - \varepsilon.$$

Using (7.5) and (7.18), we obtain

$$F(A_a) = F(1, a, \ldots, a^k) \leq q - \varepsilon$$

which contradicts (7.15). Thus Theorem 7.6 has been proved for $0 < a < \infty$. If $a = \infty$ and (7.16) hold, then Theorem 7.1 implies that for any $\delta > 0$, we can find a sequence $Y$, given by (7.17), satisfying $y_k = 1$ and $y_j < \delta$ for all $0 \le j \le k-1$. Thus, by Lemma 7.5 and (7.8) we would have

$$\liminf_{a\to\infty} F\left(\frac{1}{a^k}, \frac{1}{a^{k-1}}, \ldots, \frac{1}{a}, 1\right) \le q - \varepsilon, \text{ which contradicts (7.15).}$$

A similar argument can be used for the case $a = 0$. ■

We now illustrate the use of Theorem 7.6 by two simple examples.

**Example 7.7** *Let F be defined as*

$$F = \frac{9x_0 + x_2}{x_1}.$$

*This is a special case of Example 7.3 and so satisfies the required conditions. It follows from Theorem 7.6 that for any positive sequence* $X = \{x_i\}_{i=0}^{\infty}$,

$$\begin{aligned}\sup_{0\le i<\infty} F(X^{+i}) &= \sup_{0\le i<\infty} \frac{9x_i + x_{i+2}}{x_{i+1}} = \inf_{0<a<\infty}\left(\frac{9a^i + a^{i+2}}{a^{i+1}}\right)\\ &= \inf_{0<a<\infty}\left(\frac{9}{a} + a\right) = 6\end{aligned}$$

*with an optimal solution* $X = \{1, 3, \ldots, 3^i, \ldots\}$.

**Example 7.8** *Let*

$$F(X) = \frac{x_0 + (2+\beta)x_2 + x_4}{x_1 + x_3},$$

*where* $\beta \ge 0$. *This is also a special case of Example 7.3. Thus, it follows from Theorem 7.6 that for any positive sequence* $X = \{x_i\}_{i=0}^{\infty}$,

$$\begin{aligned}\sup_{0\le i<\infty} F(X^{+i}) &= \sup_{0\le i<\infty} \frac{x_i + (2+\beta)x_{i+2} + x_{i+4}}{x_{i+1} + x_{i+3}}\\ &= \inf_{0<a<\infty} \frac{1 + (2+\beta)a^2 + a^4}{a + a^3}\\ &= \inf_{0<a<\infty} \left(a + \frac{1}{a}\right) + \frac{\beta}{a + 1/a}. \qquad (7.19)\end{aligned}$$

*It can be easily verified that for* $\beta \le 4$ *the minimum of* (7.19) *is* $2 + \beta/2$ *with the optimal* $a = 1$, *while for* $\beta > 4$ *the minimum of* 7.19 is $2\sqrt{\beta}$ with the optimal $a$ satisfying $(a + 1/a) = \sqrt{\beta}$.

We now consider the case when $F(X)$ may depend on an infinite number of terms of $X$.

**Theorem 7.9** *Let $\{F_k(X)\}_{k_0}^{\infty}$, $k_0 > 0$ be a sequence of functionals and assume that each $F_k$ satisfies conditions (7.5)–(7.9) and that for any positive sequence $X = \{x_i\}_{i=0}^{\infty}$, we have*

$$F_{k+1}(x_0, x_1, \ldots, x_k, x_{k+1}) \geq F_k(x_0, x_1, \ldots, x_k). \tag{7.20}$$

*For any positive sequence $X$, define*

$$F(X) \equiv \lim_{k\to\infty} F_k(X) \tag{7.21}$$

$$F(A_\infty) \equiv \lim_{k\to\infty} F_k(A_\infty) \quad \text{and} \quad F(A_0) \equiv \lim_{k\to\infty} F_k(A_0) \tag{7.22}$$

*(see (7.8) and (7.9)). Then*

$$\limsup_{i\to\infty} F(X^{+i}) \geq \inf_{0\leq a\leq\infty} F(A_a). \tag{7.23}$$

It should be noted that the inf in the right-hand side of (7.23) contains the end points $a = 0$ and $a = \infty$ because it may happen that

$$F(A_\infty) < \inf_{0<a<\infty} F(A_a)$$

(see Gal, 1972).

**Proof.** Let

$$a = \limsup_{n\to\infty} x_n^{1/n}.$$

It has been established in the proof of Theorem 7.6 that for any integer $k$

$$\limsup_{i\to\infty} F_k(X^{+i}) \geq F_k(A_a).$$

Thus, it follows from (7.20) and (7.21) that for all $k$

$$\limsup_{i\to\infty} F(X^{+i}) \geq F_k(A_a). \tag{7.24}$$

Since $F(A_a)$ is defined by (7.21) (or (7.22)) it follows from (7.24) that

$$\limsup_{i\to\infty} F(X^{+i}) \geq F(A_a) \geq \inf_{0\leq a\leq\infty} F(A_a).$$

■

Schuierer (2001) has shown that (7.20) can be replaced by

$$F_{k+1}(x_0, x_1, \ldots, x_k, x_{k+1}) \geq F_k(x_1, x_2, \ldots, x_k, x_{k+1}),$$

and the result corresponding to (7.23) is then

$$\sup_k F_k(X) \geq \inf_a \sup_k F_k(A_a).$$

We now illustrate the use of Theorem 7.9. Let $L$ be a positive integer and let $\beta_j$, $0 \le j \le L$, and $\alpha_j$, $0 \le j < \infty$ be nonnegative sequences (not all zero). Let

$$F(X) = \frac{\sum_{j=0}^{\infty} \alpha_j x_j}{\sum_{j=0}^{L} \beta_j x_j}. \tag{7.25}$$

Then

$$\limsup_{i \to \infty} F(X^{+i}) \ge \inf_{0 \le a \le \infty} \frac{\sum_{j=0}^{\infty} \alpha_j a^j}{\sum_{j=0}^{L} \beta_j a^j}.$$

This result is established by using Theorem 7.9 with

$$F_k(X) = \frac{\sum_{j=0}^{k} \alpha_j x_j}{\sum_{j=0}^{L} \beta_j x_j}.$$

**Remark 7.10** *Condition* (7.20) *in Theorem 7.9 can be omitted, and then instead of (7.23), we obtain*

$$\limsup_{k \to \infty} \limsup_{i \to \infty} F_k(X^{+i}) \ge \limsup_{k \to \infty} \inf_{0 \le a \le \infty} F_k(A_a). \tag{7.26}$$

*The proof of* (7.26) *is similar to the proof of Theorem 7.9.*

Theorem 7.9 can be extended to the case where $X$ is a doubly infinite positive sequence, i.e.,

$$X = \{x_i, -\infty < i < \infty\}.$$

In this case, if $F_k(X) = F(x_{-k}, \ldots, x_0, \ldots, x_k)$ satisfies (7.5)–(7.9) and a monotonicity condition as in (7.20), then

$$\limsup_{i \to \infty} F(X^{+i}) \ge \inf_{0 \le a \le \infty} F(A_a), \quad \text{and}$$

$$\limsup_{i \to -\infty} F(X^{+i}) \ge \inf_{0 \le a \le \infty} F(A_a). \tag{7.27}$$

The proof of this result is almost identical to the proof of Theorem 7.9, and we do not present it here. Inequalities (7.27) can be used to establish the following important result.

**Corollary 7.11** *Let $L$ be a positive integer, and let $\beta_j, -L < j < L$, and $\alpha_j, -\infty < j < \infty$ be nonnegative sequences (not all zero). Then*

$$\limsup_{i \to \pm\infty} \frac{\sum_{j=-\infty}^{\infty} \alpha_j x_{i+j}}{\sum_{j=-L}^{L} \beta_j x_{i+j}} \ge \inf_{0<a<\infty} \frac{\sum_{j=-\infty}^{\infty} \alpha_j a^j}{\sum_{j=-L}^{L} \beta_j a^j}.$$

*Thus,*

$$\inf_X \sup_i \frac{\sum_{j=-\infty}^{\infty} \alpha_j x_{i+j}}{\sum_{j=-L}^{L} \beta_j x_{i+j}} = \inf_{0<a<\infty} \frac{\sum_{j=-\infty}^{\infty} \alpha_j a^j}{\sum_{j=-L}^{L} \beta_j a^j}$$

*for any positive sequence $X$.*

We shall use this corollary extensively in the linear search problem (Chapter 8), the Star search (Section 9.2), and in the rendezvous on the line (Section 17.4).

The next example will be useful for the linear or star search under the assumption that the hider's distance from the origin is at least 1 (see Remarks 8.2 and 9.5).

**Example 7.12** *Consider*

$$v = \inf_X \sup_{i \geq 0} \frac{\sum_{j=0}^{i+M-1} x_j}{x_i}, \tag{7.28}$$

*where $M \geq 2$ is an integer and $x_i \geq 1$ for all $i \geq 0$.*

*Obviously, any $X$ with a finite $v$ has to satisfy $x_i \to \infty$ as $i \to \infty$. Thus, if we let $x_i$, $-\infty < i < 0$ be any converging positive sequence, then*

$$\inf_X \sup_{i \geq 0} \frac{\sum_{j=0}^{i+M-1} x_j}{x_i} \geq \inf_X \limsup_{i \to \infty} \frac{\sum_{j=0}^{i+M-1} x_j}{x_i} = \inf_X \limsup_{i \to \infty} \frac{\sum_{j=-\infty}^{i+M-1} x_j}{x_i}.$$

*Now, for any $k \geq M$, let*

$$F_k(X) = \frac{\sum_{j=-k}^{M-1} x_j}{x_1} \quad \text{and} \quad F(X) = \frac{\sum_{j=-\infty}^{M-1} x_j}{x_1}.$$

*Then we can use Theorem 7.9 and obtain*

$$\begin{aligned} \limsup_{i \to \infty} F(X^{+i}) &\geq \inf_{0 \leq a \leq \infty} F(A_a) \\ &= \inf_{a > 1} \sum_{j=-\infty}^{M-1} a^j \\ &= \inf_{a > 1} \frac{a^M}{a-1} = \frac{M^M}{(M-1)^{M-1}}, \end{aligned} \tag{7.29}$$

*with $M/(M-1)$ as the optimal value for $a$. It now easily follows that*

$$\inf_X \sup_{i \geq 0} \frac{\sum_{j=0}^{i+M-1} x_j}{x_i} = \frac{M^M}{(M-1)^{M-1}},$$

*with $x_i = (M/(M-1))^i$, $i = 0, 1, \ldots$ being a minimax solution.*

**Remark 7.13** *Instead of considering the set of all positive sequences, we may consider only a subset $B$ of it. Let $a_X = \limsup_{n \to \infty} x_n^{1/n}$. Then the modification of Theorem 7.6 (and Theorem 7.9) states that for any $X \in B$*

$$\limsup_{i \to \infty} F(X^{+i}) \geq \inf_{X \in B} F(A_{a_X}).$$

*The proof of this extension of Theorem 7.6 (and Theorem 7.9) is based on the following argument. If $0 < a_X < \infty$, then Theorem 7.1 implies that there exist nonnegative* *constants* $\beta_0, \beta_1, \ldots,$ *such that* $Y = \sum_{i=0}^{\infty} \beta_i X^{+i}$ *"close" to the geometric sequence* $1, a_X, a_X^2, \ldots$. *On the other hand, Lemma 7.5 implies that if* $X \in B$, *then*

$$\sup_{0 \le i < \infty} F(X^{+i}) \ge F(Y) \approx F(1, a_X, a_X^2, \ldots)$$

$$\ge \inf_{a_X : X \in B} F(1, a_X, a_X^2, \ldots).$$

*The cases* $a_X = 0$ *and* $a_X = \infty$ *can be handled as in the proof of Theorem 7.6.*

**Example 7.14** *Let* $I$ *be either* $0$ *or* $-\infty$. *Consider the expression*

$$q = \inf_X \sup_{i > I} \frac{x_i + 100x_{i+2}}{x_{i+1}}$$

*where* $X = \{x_i, I < i < \infty\}$ *belongs to the set* $B$ *given by the condition*

$$0 \le R_1 < x_i < R_2, \quad I < i < \infty.$$

*If* $R_1 > 0$, *then* $a_X = 1$ *for all* $X \in B$. *Thus, for both cases* $I = -\infty$ *and* $I = 0$, $q = 101$ *and* $x_i =$ *constant is a minimax solution.*

*If* $R_1 = 0$, *then we have to distinguish between two cases. If* $I = -\infty$, *then we still have* $\limsup_{n \to -\infty} x_n^{1/n} \ge a_X = \limsup_{n \to -\infty} R_2^{1/n} = 1$ *for all* $X \in B$, *so that*

$$q \ge \inf_{a \ge 1} \frac{1 + 100a^2}{a} \ge 101$$

*and the foregoing result still holds. On the other hand, if* $I = 0$, *then* $\{a_X \colon X \in B) = \{a, 0 \le a \le 1\}$. *Thus,*

$$q = \inf_{0 < a \le 1} \frac{1 + 100a^2}{a} = 20$$

*with the optimal* $a$ *is* $1/10$, *and* $x_i = \alpha/10^i$, $i > 0$, *where* $0 < \alpha < 10R_2$ *is a minimax solution.*

### 7.2.1 Minimax theorems for the continuous case

Some search games have the following structure: The searcher chooses a positive continuous function $X(\theta)$, $-\infty < \theta < \infty$, and the hider chooses a real number $t$. The payoff (to the hider) is $F(X^{+t})$, where $X^{+t}(\theta) \equiv X(t + \theta)$ and $F$ is a functional, which satisfies several continuity and unimodality conditions analogous to (7.5)–(7.9). The continuous analog of Theorem 7.6 is the following equality

$$\inf_X \sup_t F(X^{+t}) = \inf_{-\infty < b < \infty} F(e^{b\theta}). \tag{7.30}$$

The full details are presented in Gal (1980, chapter 6.4). The general idea of the proof is that for any positive $X$ and any positive kernel $\beta$, the positive function $Y$ defined by the convolution

$$Y(\theta) = \int \beta(\gamma) X(\theta - \gamma)\, d\gamma$$

satisfies, for the relevant functionals,

$$\sup_t F(Y^{+t}) \le \sup_t F(X^{+t}).$$

On the other hand, it follows from Theorem 7.2 that, for any positive function $X$ there exists a kernel $\beta(\cdot)$ such that $Y$ is as close as desired to an exponential function $e^{b\theta}$. This observation leads to (7.30).

In the present book we will use only two special cases of (7.30):

- **Case 1.**

$$F(X) = \frac{\int_{-\infty}^{\infty} X(\theta)\, dA(\theta)}{X(0)},$$

where $A(\theta)$ is any measure, is extensively analyzed in the next section.

- **Case 2.**

$$F(X) = \frac{\int_{-\infty}^{2\pi} \sqrt{X^2(\theta) + X'^2(\theta)}\, d\theta}{X(0)}.$$

In this case (7.30) implies that

$$\begin{aligned} &\inf_X \sup_t \frac{\int_{-\infty}^{t+2\pi} \sqrt{X^2(t+\theta) + X'^2(t+\theta)}\, d\theta}{X(t)} \\ &= \inf_{-\infty<b<\infty} \int_{-\infty}^{2\pi} \sqrt{e^{2b\theta} + b^2 e^{2b\theta}}\, d\theta \\ &= \min_{b>0} e^{2\pi b} \sqrt{1 + \frac{1}{b^2}}. \end{aligned} \tag{7.31}$$

## 7.3 Uniqueness of the Minimax Strategy

In this section $F(X)$ has the form

$$F(X) = \frac{\int_{-\infty}^{\infty} X(\theta)\, dA(\theta)}{X(0)} \tag{7.32}$$

where $A(\theta)$ is a measure and $X(\theta) > 0$ for all $-\infty < \theta < \infty$. This case will be encountered in Chapters 8 and 9.

If $F$ is given by (7.32), then (7.30) implies that

$$\inf_X \sup_t \frac{\int_{-\infty}^{\infty} X(t+\theta)\, dA(\theta)}{X(t)} = \inf_{-\infty<b<\infty} \int_{-\infty}^{\infty} e^{b\theta}\, dA(\theta). \tag{7.33}$$

Thus, if the minimum of the right-hand side of (7.33) is obtained at $b = \bar{b}$, then

$$X(\theta) = \alpha e^{\bar{b}\theta}, \tag{7.34}$$

where $\alpha$ is a positive constant, is a minimax solution for the left-hand side of (7.33). We now investigate the conditions under which (7.34) is the unique solution of that problem. Such conditions are supplied by the following theorem.

**Theorem 7.15** *Let $A(\theta)$ be a measure, which is not concentrated at $\theta = 0$, and let $X(t)$, $-\infty < t < \infty$ be a positive function, which is integrable on any finite interval. Let $f(b)$ be the bilateral Laplace transform of $A(\theta)$,*

$$f(b) = \int_{-\infty}^{\infty} e^{b\theta}\, dA(\theta). \tag{7.35}$$

*Assume that $f(b)$ attains its minimum at a point $-\infty < \bar{b} < \infty$ and that $f'(\bar{b})$ satisfies*

$$\int_{-\infty}^{\infty} \theta e^{\bar{b}\theta}\, dA(\theta) = 0. \tag{7.36}$$

*If*

$$\frac{\int_{-\infty}^{\infty} X(t+\theta)\, dA(\theta)}{X(t)} \le f(\bar{b}) \quad \text{for all } -\infty < t < \infty, \tag{7.37}$$

*then*

(a) *If $A(\theta)$ is not arithmetic,*[1] *then $X(t) = \alpha e^{\bar{b}t}$, where $\alpha$ is any positive constant.*

(b) *If $A(\theta)$ is arithmetic with span $\lambda$, then $X(t) = \alpha(t)e^{\bar{b}t}$, where $\alpha(t)$ is a positive periodic function having period $\lambda$.*

It should be noted that since $f(b)$ is a smooth convex function, then condition (7.36) is always satisfied if $\bar{b}$ is an interior point of the interval of convergence of $f(b)$, as occurs in all our applications. (But condition (7.36) may sometimes hold even when $\bar{b}$ is an end point of the interval of convergence of $f(b)$, or when $f(b)$ converges at a single point.)

Also note that the proof of Theorem 7.15 actually provides a direct proof for (7.33) under the condition (7.36).

In order to establish Theorem 7.15 we first prove the following lemma.

**Lemma 7.16** *Let $P(\theta)$, $-\infty < \theta < \infty$, be the cumulative distribution function of a probability measure which is not concentrated at $\theta = 0$, satisfying*

$$\int_{-\infty}^{\infty} \theta\, dP(\theta) = 0 \tag{7.38}$$

*and let $W(t)$ be a positive function, which is integrable on each finite interval. Assume that for each $t$, $-\infty < t < \infty$, the integral*

$$\int_{-\infty}^{\infty} W(t+\theta)\, dP(\theta)$$

[1] We use *arithmetic* in the sense used by Feller (1971), namely, a measure $A$ is *arithmetic* if it is concentrated on a set of points of the form $0, \pm\lambda, \pm 2\lambda, \ldots$. The largest $\lambda$ with this property is called the *span* of $A$.

*is defined and satisfies*

$$\int_{-\infty}^{\infty} W(t+\theta)\,dP(\theta) \le W(t). \tag{7.39}$$

*Then*

(a) *If $P(\theta)$ is not arithmetic, $W(t)$ is a constant.*

(b) *If $P(\theta)$ is arithmetic with span $\lambda$, $W(t)$ is periodic having period $\lambda$.*

**Proof.** Let $u_i, i = 1, 2, \ldots,$ be a sequence of independent random variables, each having the distribution $P$. Denote

$$e_n = \sum_{i=1}^{n} u_i \tag{7.40}$$

and

$$w_n = -W(e_n). \tag{7.41}$$

Condition (7.39) implies that $w_n$ is a negative submartingale.[2] Hence, there exists a random variable $w^*$ such that

$$w_n \to w^* \text{ with probability } 1 \tag{7.42}$$

(see Breiman, 1968, chapter 5.4). We distinguish between two cases.

(a) *$P(\theta)$ is not arithmetic.*
In this case, (7.38) implies that the random walk defined by (7.40) visits every interval infinitely often, with probability 1. This, together with (7.42), implies that if $W$ is a continuous function, then it has to be a constant. If $W$ is not continuous, define

$$W_\varepsilon(t) = \frac{1}{2\varepsilon}\int_{-\varepsilon}^{\varepsilon} W(t+\theta)\,d\theta \tag{7.43}$$

and note that $W_\varepsilon$ is a continuous function satisfying the conditions of the lemma and so must be a constant $\alpha_\varepsilon$. It is easily verified that $\alpha_\varepsilon$ has the same value, $\alpha$, for each $\varepsilon$, so that for all real $t$

$$\lim_{\varepsilon\to 0} W_\varepsilon(t) = \alpha.$$

On the other hand,

$$\lim_{\varepsilon\to 0} W_\varepsilon(t) = W(t)\ a.s. \text{ (almost surely)}$$

which implies that $W_\varepsilon(t) = \alpha\ a.s.$ This proves Lemma 7.16($a$).

(b) *$P(\theta)$ is arithmetic with span $\lambda$.*
In this case, (7.38) implies that the random walk defined by (7.40) visits every point $j\lambda$ (where $j$ is any integer) infinitely often, with probability 1. Hence, (7.42) implies that $W(j\lambda)$ has the same value for each integer $j$. In the same manner, we define

[2] A sequence of ramdom variables $y_1, y_2, \ldots$ is called a *submartingale* if $E|y_n| < \infty$, $n = 1, 2, \ldots,$ and $E(y_{n+1} \mid y_n, \ldots, y_1) \ge y_n$ *a.s.*, $n = 1, 2, \ldots$.

$w_n = -W(\beta + e_n)$, where $\beta$ is any real number, and deduce that $W(\beta + j\lambda)$ has the same value for every integer $j$. ■

We can now prove Theorem 7.15.

**Proof.** Define a probability distribution function $P(\theta)$ and a positive function $W(t)$ by

$$dP(\theta) = \frac{e^{\bar{b}\theta}}{f(\bar{b})} dA(\theta) \quad \text{and} \quad W(t) = X(t)e^{-\bar{b}t}.$$

Since $W(t)$ and $P(\theta)$ satisfy the conditions of Lemma 7.16 ((7.38) follows from (7.36) and (7.39) from (7.37)) it follows that $W(t) = \alpha(t)$, where $\alpha(t)$ is $a.s.$ a constant in case (a), and a periodic function in case (b). ■

**Example 7.17** *Assume that the equation*

$$\int_{R-\beta}^{\infty} X(R+\theta)\, d\theta = \gamma X(R), \tag{7.44}$$

*with $\beta > 0$, holds for all $-\infty < R < \infty$. The function*

$$f(b) = \int_{-\beta}^{\infty} e^{b\theta}\, d\theta = -\frac{e^{-\beta b}}{b}$$

*converges for all $b < 0$, attaining its minimum at $\bar{b} = -1/\beta$ with $f(\bar{b}) = \beta e$, and $f'(\bar{b}) = 0$. Thus, it follows from* (7.33) *that if $\gamma < \beta e$, then equation* (7.44) *has no positive solution.*

*If $\gamma = \beta e$, then equation* (7.44) *has the unique solution*

$$X(R) = \alpha e^{-R/\beta},$$

*where $\alpha$ is any positive constant.*

*If $\gamma > \beta e$, then it is easily verified that equation* (7.44) *has two solutions of the type $e^{\xi R}$ (as well as any linear combination of them).*

In the special case that the measure $A$ has only atoms at the integers we obtain the discrete version of Theorem 7.15.

**Theorem 7.18** *Let $A_j$, $-\infty < j < \infty$, be a nonnegative sequence with the following (bilateral) generating function*

$$\phi(a) = \sum_{j=-\infty}^{\infty} A_j a^j, \quad a > 0.$$

*Assume that $\phi(a)$ attains its minimum*[3] *at a positive number $\bar{a}$, $0 < \bar{a} < \infty$ and that $\phi'(\bar{a})$ satisfies*

$$\sum_{j=-\infty}^{\infty} jA_j \bar{a}^{j-1} = 0. \tag{7.45}$$

[3]Note that $\phi(a)$ is convex for $a > 0$. Thus, a local minimum is also a global minimum.

*Then any positive sequence* $X = \{x_j, -\infty < j < \infty\}$ *that satisfies, for all* $-\infty < i < \infty$, *the inequality*

$$\frac{\sum_{j=-\infty}^{\infty} A_j x_{i+j}}{x_i} \leq \phi(\bar{a}), \tag{7.46}$$

*has the following form:*

(i) *If the span of* $\{A_j\}$ *is* 1, *then* $x_i = \alpha \bar{a}^i$, *where* $\alpha$ *is a positive constant.*

(ii) *If the span of* $\{A_j\}$ *is* $\lambda > 1$ *(i.e.,* $A_j = 0$ *for every* $j$ *that is not a multiple of* $\lambda$*), then* $x_i = \alpha_i \bar{a}^i$, *where* $\alpha_i$ *is a positive periodic sequence with period* $\lambda$.

**Proof.** Define a discrete measure $A(\theta)$ such that the measure of the point $\theta = j$ is $A_j$ and the measure of any interval not containing integral points is zero. Define $X(t)$ as the step function $X(t) = x_i$ for $i - 1 < t \leq i$, $-\infty < i < \infty$. Thus defined, $A(\theta)$ and $X(t)$ satisfy the conditions of Theorem 7.15 with $\bar{b} = \ln \bar{a}$. Since $A(\theta)$ is arithmetic, it follows that $X(t) = \alpha(t)\bar{a}^t$, where $\alpha(t)$ is a periodic function with the period length equal to $\lambda$, the span of $A(\theta)$. Thus, if $\lambda = 1$, then $x_i = X(i) = \alpha \bar{a}^i$, and if $\lambda$ is an integer greater than 1, then $x_i = \alpha_i \bar{a}^i$, where $\alpha_i$ is a periodic sequence with period $\lambda$. ■

Note that condition (7.45) automatically holds if $\bar{a}$ is an interior point of the interval of convergence of $\phi(a)$.

**Remark 7.19** *If condition* (7.45) *(resp.* (7.36) *for the continuous case) is not satisfied, then it is proven by Gal (1972) that the minimax strategy is no longer unique up to a multiplicative constant. Such a situation may occur only if the minimum of the generating function* $\phi(a)$ *is attained at a point* $a$, *which is an endpoint of the interval of convergence of* $\phi(a)$ *(or if* $\phi(a)$ *converges only at a single point).*

Theorem 7.18 will be extensively used in Chapters 8 and 9.

**Example 7.20** *Let* $A_1 = A_{-1} = 1/2$ *and* $A_j = 0$ *for all other* $j$. *Then* $\bar{a} = 1$ *and* $\phi(\bar{a}) = 1$. *Thus, Theorem 7.18 implies that for any positive sequence* $X = \{x_j, -\infty < j < \infty\}$, *if*

$$\frac{x_{j-1} + x_{j+1}}{2} \leq x_j, \quad \text{for all } -\infty < j < \infty,$$

*then* $x_j$ *is a constant. This is a well-known result: If a positive sequence* $\{x_j, -\infty < j < \infty\}$ *is concave then* $x_j =$ *constant.*

*It should be noted that, usually, the problem*

$$\inf_X \sup_{i > i_0} \frac{\sum_{j=-\infty}^{\infty} A_j x_{i+j}}{x_i},$$

*where* $x_i > 0$, $A_j \geq 0$, *and* $i_0 > -\infty$, *does not have a unique (geometric) solution. For example, any sequence* $Y = \{y_i\}$, *which is strictly concave for* $i_0 \leq i < \infty$, *satisfies*

$$\frac{y_{i-1} + y_{i+1}}{2} < y_i, \quad \text{for all } i > i_0$$

*and thus is a non-constant solution to the problem*

$$\inf_X \sup_{i>i_0} \frac{x_{i-1}+x_{i+1}}{2x_i}.$$

*Still, Corollary 7.11 implies that any such sequence satisfies*

$$\limsup_{i\to\infty} \frac{y_{j-1}+y_{j+1}}{2y_i} = 1.$$

*Some other examples and a more detailed discussion are given by Gal (1972).*

**Remark 7.21** *If we remove the assumption that the sequence X is positive, then the equation*

$$x_{i-1}+x_{i+1} = \beta x_i, \quad -\infty < i < \infty \tag{7.47}$$

*has an infinite number of solutions for all real $\beta$, including $\beta < 2$. This is so because starting with any $x_0$ and $x_1$, if we recursively define*

$$x_{i+1} = \beta x_i - x_{i-1} \quad i \geq 1, \text{ and}$$

$$x_{i-1} = \beta x_i - x_{i+1}, \quad i \leq 0,$$

*then such a sequence would satisfy (7.47).*

*A similar statement holds for all the other examples of Section 7.2. Thus, the requirement that X be positive is indispensable for all the results of Chapter 7.*

# Chapter 8

# Search on the Infinite Line

## 8.1 Introduction

The problem of searching for a target hidden on the infinite line has already been introduced in Section 6.1. We first briefly present a short, informal, discussion on the linear search problem (LSP), which was suggested by Bellman (1963) and, independently, by Beck (1964).

A target is located on the real line, at a point $z$, with a known probability distribution $F$. A searcher, whose maximal velocity is 1, starts from the origin $O$ and wishes to discover the target in minimal expected time. (Changing the direction of motion can be done instantaneously, except in Section 8.4.)

A search plan is equivalent to a list of turning points. There need not be a first turning point. (Actually, we will show that "effective" trajectories start with infinite "oscillations"). Thus, a search trajectory $S$ is equivalent to the set of the turning points $\ldots x_{-n} \ldots x_0 \ldots x_n \ldots$. At the $i$-th stage of his trajectory $S$ the searcher goes from the origin $O$ to point $x_i$, then to $-x_{i+1}$, and back to $O$.

Let $I$ be the first integer that satisfies $x_I < z \leq x_{I+2}$ for $z > 0$ and $x_{I+2} \leq -z < x_I$ for $z < 0$. Then the time required for $S$ to reach $z$ is given by

$$C(S, z) = |z| + 2 \sum_{j=-\infty}^{I+1} x_j \tag{8.1}$$

It is obvious that the condition

$$E(|\, z\,|) = \int_0^\infty |z|\, dF(z) < \infty \tag{8.2}$$

is necessary for the finiteness of the expected search time because for any hiding point $z$, the time to reach $z$ is at least $|z|$ for any trajectory. On the other hand, if (8.2) is satisfied, then there is a trajectory with a finite expected search time. Such a trajectory is the "geometric" trajectory $\bar{S}$ with $x_i = (-2)^i$, since it reaches any point $z$ by time $9|z|$. (If $2^i < |z| \leq 2^{i+1}, -\infty < i < \infty$, then the time to reach $z$ by $\bar{S}$

is bounded by $|z| + 2\sum_{-\infty}^{i+1} 2^j < 9|z|$ (because $2^i < |z|$). In fact, Beck (1965) has proved that condition (8.2) is both necessary and sufficient for the problem to have an optimal search trajectory.

Beck and others (1964–1995) have done much work on finding the characteristic properties of the optimal solution and presented optimal solutions for the following distributions:

- rectangular (go to one end point and then return to the other),
- normal ($x_n \sim \sqrt{2n \times \ln n}$) (Beck and Beck, 1986), and
- triangular (infinite number of turning points!).

A necessary and sufficient condition for a distribution on a compact interval to have a nonterminating minimizing search strategy was presented by Baston and Beck (1995). (Their paper actually proves this result for a more general cost function $t^\alpha$, $\alpha \geq 1$).

The LSP with bounded resources (maximizing the probability of discovery) was considered by Foley, Hill, and Spruill (1991). Recently, Hipke et al. (1999) have considered the minimax search in an interval. A comprehensive survey with some useful insight is presented by Bruss and Robertson (1988).

As Bruss and Robertson note, although it is immediately clear what the problem is, it turns out to be much harder than one's intuition would indicate. In general, an effective general solution, from the computational point of view was not presented. In fact, although the problem is interesting and has an "applicational appeal," no algorithm for solving the problem for a general probability distribution function has been found during about 37 years since the LSP was first presented. We do have, however, a *dynamic programming* algorithm for finding an approximate solution with any desired accuracy. This algorithm is presented in detail in Section 8.7. The description of the topics presented in this chapter is given in the end of the next section.

## 8.2 The Linear Search Problem as a Search Game

We now describe the LSP within the framework of a search game. Beck and Newman (1970) have used the restrictive approach, but we will use the normalization approach, (see Section 6.1). We shall usually assume that the hider is immobile and that capture occurs the first time that the searcher passes the point occupied by the hider, but in Sections 8.5 and 8.6 we shall consider some variations of the game in which these assumptions do not hold.

As usual, we assume that the searcher can use any continuous trajectory $S(S(t)$ being the point visited by $S$ at time $t$) that satisfies

$$S(0) = 0, \quad \text{and for all } t_2 > t_1 \geq 0: \quad |S(t_2) - S(t_1)| \leq t_2 - t_1.$$

The hider can choose any real number $H$, and the cost function $\hat{C}(S, H)$ is given by

$$\hat{C}(S, H) = \frac{C(S, H)}{|H|} \tag{8.3}$$

where $C(S, H)$ is the capture time. In the case that mixed strategies are used, we shall denote the expected value of the normalized cost by $\hat{c}$.

At first we shall identify some properties that a search trajectory must have in order to be "efficient." Such a property is obviously

$$\sup_{t>0} S(t) = \infty \quad \text{and} \quad \inf_{t>0} S(t) = -\infty \tag{8.4}$$

because otherwise there exist hiding points $H$ that are never discovered by $S$, i.e., $v(S) = \infty$. Moreover, any mixed search strategy $s$ with $v(s) < \infty$ has to satisfy condition (8.4) with probability 1. Thus, it can be assumed that all the relevant mixed strategies make their probabilistic choice among the trajectories that satisfy (8.4).

We have already mentioned in Section 3.1 that, when searching for an immobile hider, it can be assumed that the velocity along the search trajectory is 1. Using a dominance argument of a similar type, it can be shown that it is sufficient to consider search trajectories $S(t)$ with $S'(t) = 1$ or $S'(t) = -1$ for all $t > 0$ except a denumerable set of *turning times* defined as follows.

**Definition 8.1** *The trajectory S(t) has a left turning time (LT) at the time $t_0 > 0$ if there exists an $\varepsilon > 0$ such that*

$$S'(t) = 1 \ \text{for}\ t_0 - \varepsilon < t < t_0 \quad \text{and} \quad S'(t) = -1 \quad \text{for}\ t_0 < t < t_0 + \varepsilon.$$

*Similarly, the trajectory $S(t)$ has a right turning time (RT) at $t_0$ if*

$$S'(t) = -1 \quad \text{for}\ t_0 - \varepsilon < t < t_0 \quad \text{and} \quad S'(t) = 1 \quad \text{for}\ t_0 < t < t_0 + \varepsilon.$$

Obviously, between any two LT's there exists at least one RT and between any two RT's there exists at least one LT.

We also have the following property:

*Let S be any search trajectory with $v(S) < \infty$. Then for any $t_0 > 0$*
*(a) The number of the turning points before $t_0$ is infinite, and*
*(b) The number of the turning points after $t_0$ is infinite.*

Property *(b)* readily follows from (8.4), while property *(a)* follows from the fact that if there is a first left turning point, then the hider can obtain an arbitrarily large payoff by choosing $H = -\varepsilon$ with $\varepsilon > 0$ being very small, while if there is a first right turning point, then the hider can obtain any large payoff by choosing $H = \varepsilon$. Thus, the set of turning points is a doubly infinite sequence.

**Remark 8.2** *In the approach that we use, every admissible search trajectory has to start by making an infinite number of small "oscillations" near 0. This phenomenon follows from the normalization we use to define the cost function. (Note, however, that the discussion in Section 8.2.1 implies that, even if we use the nonnormalized cost function with the restriction on the absolute moment of H, the optimal trajectory still has an infinite number of oscillations near 0). In order to obtain a "practical" solution, which does have a first turning point, one could assume that there is a positive (very small) discovery radius r, so that all the points H with $|H| < r$ are discovered at $t = 0$. From the optimal search strategy that we shall present, one can immediately obtain an* *optimal strategy for the above-mentioned versions of the game by simply ignoring all the turning points before a certain instant $t_0$ where $t_0$ is very small. Such a solution will use trajectories which have a first turning point and do not have small "oscillations", (see Example 7.12 and Remark 9.5).*

From our previous conclusions it follows that any search trajectory S can be represented by a doubly infinite sequence of positive numbers $\{x_i\}_{i=-\infty}^{\infty}$, with the convention that each $x_i$ with an even $i$ represents the location of a LT, while if $i$ is an odd integer, then $-x_i$ represents the location of a RT. For any even $i$, the searcher moves from the point $x_i$ to $-x_{i+1}$ and then to $x_{i+2}$, etc. For any time $t$, if the interval covered by that time is $[a(t), b(t)]$ then we can assume that the next turning point has to be outside of that interval (because a turning point inside that interval just wastes time without discovering any new points and can be skipped). Thus, we can consider only search trajectories that satisfy

$$-x_{2i+1} < -x_{2i-1} < \cdots 0 \cdots < x_{2i} < x_{2i+2}. \tag{8.5}$$

Such trajectories will be referred to as *periodic and monotonic*.

A minimax trajectory for the LSP will be derived in Section 8.2.1 and optimal strategies will be derived in Section 8.3. In Section 8.4 we analyze the linear search when changing the direction of motion requires some time and cannot be done instantaneously (as originally assumed in the LSP). In Section 8.5 we present a minimax trajectory for capturing a hider who moves away from the origin. In Section 8.6 we consider linear search when there is some probability of not finding the target even when the searcher visits it's location. Finally, we describe in Section 8.7 a *dynamic programming* algorithm for computing, with any desired accuracy, the optimal search trajectory of the LSP for known hiding distribution.

### 8.2.1 The minimax search trajectory

In this section, we present the minimax pure strategy of the searcher. In other words, we find a trajectory $\bar{S}$ that minimizes $v(S)$, where

$$v(S) = \sup_H \hat{C}(S, H) = \sup_H \frac{C(S, H)}{|H|}.$$

First we note that by property (8.5) of Section 8.2, any hiding point $H$ has to satisfy $x_i < H \le x_{i+2}$ for some even $i$, or $-x_{i+2} \le H < -x_i$ for some odd $i$. Since

$$C(S, H) = |H| + 2 \sum_{j=-\infty}^{i+1} x_j,$$

where $x_i < |H| \le x_{i+2}$, we have

$$v(S) = \sup_H \left(1 + 2\frac{\sum_{j=-\infty}^{i+1} x_j}{|H|}\right) = 1 + 2 \sup_{-\infty<i<\infty} \frac{\sum_{j=-\infty}^{i+1} x_j}{x_i} \tag{8.6}$$

with the hider choosing the best response among $\{x_i + \varepsilon\}$, choosing $H = x_i + \varepsilon$ if on the $i$-th stage $S$ goes right and $H = -(x_i + \varepsilon)$ if on that stage $S$ goes left.

We now use the results obtained in Chapter 7. It follows from Corollary 7.11 of Section 7.2 that

$$\inf_S \sup_{-\infty<i<\infty} \frac{\sum_{j=-\infty}^{i+1} x_j}{x_i} = \min_{a>0} \frac{\sum_{j=-\infty}^{i+1} a^j}{a^i} = \min_{a>1} \frac{a^2}{a-1} = 4. \tag{8.7}$$

Moreover, the minimax sequence $x_i = \beta 2^i$ is unique, up to a positive constant $\beta$. This follows from Theorem 7.18 of Section 7.3 because the minimum $\bar{a} = 2$ of the (bilateral) generating function, $\phi(a) = a^2/(a-1)$, satisfies $\phi'(\bar{a}) = 0$. (Note that $\beta = \beta_0$ and $\beta = \beta_0 4^j$, where $j$ is any integer, represent the same trajectory.) Thus, the minimal cost assured by using a pure strategy (fixed trajectory) is, by (8.6) and (8.7),

$$\bar{V} = \inf_S v(S) = 1 + 2 \times 4 = 9.$$

We will use a simplified notation choosing $\beta = 1$. Thus writing $\bar{S} = \{2^i\}_{i=-\infty}^{\infty}$ will actually mean $\bar{S} = \{\beta 2^i\}_{i=-\infty}^{\infty}$ with $\beta > 0$.

**Remark 8.3** *Consider the case in which the hider is restricted to symmetric hiding distributions. Then, for a given search trajectory S, the hider's best response is to choose $H = x_i\ (+\varepsilon)$ with probability 1/2 and $H = -x_i(-\varepsilon)$ with probability 1/2, for some i. Thus, the hider is captured either on the $i+1$ stage or on the $i+2$ stage (each with probability 1/2). This leads to an average capture time of*

$$x_i + \frac{1}{2}\left(2\sum_{j=-\infty}^{i} x_j + 2\sum_{j=-\infty}^{i+1} x_j\right) = x_i + 2\sum_{j=-\infty}^{i} x_j + x_{i+1}.$$

*In this case*

$$v(S) = 1 + \sup_{-\infty<i<\infty} \frac{2\sum_{j=-\infty}^{i} x_j + x_{i+1}}{x_i} \quad \text{(by Corollary 7.11)}$$

$$= 1 + \min_{a>0} \frac{2\sum_{j=-\infty}^{i} a^j + a^{i+1}}{a^i} = 3 + \min_{a>1}\left(\frac{2}{a-1} + a\right) = 4 + 2\sqrt{2} \doteq 6.83$$

*with $\bar{a} = 1 + \sqrt{2}$. Thus, in this case the minimal cost assured by using a pure strategy (fixed trajectory) is about 6.83. This result will be used for the rendezvous search on the line with an unknown distribution (see Section 17.4 Book II). This solution was presented by Baston and Gal (1998).*

It should be noted that the trajectory $\bar{S} = \{2^i\}_{i=-\infty}^{\infty}$ is still minimax for the case when the capture time is used as the cost function and the hider's strategy has to satisfy $E|H| \leq \lambda$. In order to show it we first note that

$$\sup_{h:E|H|\leq\lambda} c(\bar{S}, h) = \sup_{h:E|H|\leq\lambda} \int \hat{C}(\bar{S}, H) \times |H| dh$$

$$\leq \lambda \sup_H \hat{C}(\bar{S}, H) = 9\lambda.$$

On the other hand, any hiding strategy, which has atoms of probability masses $\lambda/|H|$ and $1-\lambda/|H|$ at the points $H$, ($|H| \geq \lambda$) and 0, respectively, is an admissible strategy. Thus

$$\begin{aligned}
\inf_S \sup_{h:E|H|\leq\lambda} c(S,h) &\geq \inf_S \sup_{H:|H|\geq\lambda} \frac{\lambda}{|H|} C(S,H) \\
&= \lambda \inf_S \sup_{H:|H|\geq\lambda} \hat{C}(S,H) \\
&\geq \lambda \left(1 + 2 \sup_{i:x_i\geq\lambda} \frac{\sum_{j=-\infty}^{i+1} x_j}{x_i}\right), \text{ and since } x_i \to \infty \text{ as } i \to \infty \\
&\geq \lambda \left(1 + 2 \limsup_{i\to\infty} \frac{\sum_{j=-\infty}^{i+1} x_j}{x_i}\right) \geq 9\lambda
\end{aligned}$$

by Corollary 7.11 of Section 7.2.

The discussion presented in Section 6.1 implies that the minimax search trajectory, just presented, and the optimal mixed strategies, to be presented in the next section, are still optimal if one uses the capture time as the cost function, with the restriction $E|H| \leq 1$ (rather than using the normalized cost function $\hat{C}$). In order to obtain the optimal search strategy in the case that the constraint is $E|H| \leq \lambda$, one has to use the scaling lemma (Proposition 2.5) with the mapping $\Phi(x) = \lambda x$, but it is readily seen that $\Phi$ maps the optimal search strategy into itself. Thus, the minimax trajectory and the optimal mixed strategy are uniformly optimal for all $\lambda > 0$, so that it is not necessary to have any knowledge about the value of $\lambda$.

### 8.2.2 Minimax trajectory on a half-line

We have already noted, in Section 6.1, that the equivalence between the two approaches holds whenever the scaling assumption (of Section 6.1) holds. We now show that if this assumption does not hold, then the two approaches may lead to different optimal strategies.

Consider the same game described in Section 8.1, except that the search space $Q$ is the half line $-\varepsilon \leq z < \infty$ where $\varepsilon > 0$. If one uses the normalized cost function, then the minimax pure value $\bar{V}$ would be 9, the same as in Section 8.2 because by Corollary 7.11, any search trajectory S satisfies

$$\limsup_{i\to-\infty} \frac{\sum_{j=-\infty}^{i+1} x_j}{x_i} \geq \min_{a>0} \sum_{j=-\infty}^{1} a^j = 4.$$

Now consider the case of using the capture time as the cost and restricting the hider to strategies that satisfy $E|H| \leq 1$. If the searcher uses the trajectory $\tilde{S}$ defined by

$$\tilde{S}(t) = \begin{cases} -t \text{ for } 0 \leq t \leq \varepsilon, & \text{and} \\ t - 2\varepsilon \text{ for } \varepsilon < t \end{cases}$$

then, for any $H$

$$C(\tilde{S}, H) \leq |H| + 2\varepsilon.$$

Thus, for any admissible hiding strategy $h$, we have

$$c(\hat{S}, h) \leq 1 + 2\varepsilon.$$

Consequently, the pure value is much less than 9.

## 8.3 Optimal Strategies

In the previous section, we showed that if the searcher wishes to use a fixed trajectory that ensures him a minimal worst-case cost, then he should use the "geometric" trajectory $\{2^i\}_{i=-\infty}^{\infty}$. We now assume that the searcher wishes to minimize the expected cost, looking for an optimal mixed strategy $\bar{s}$ that achieves this goal. We shall use the same method adopted by Beck and Newman (1970) to demonstrate that $\bar{s}$ belongs to the family $A = \{s_a\}$ of mixed strategies defined as follows.

**Definition 8.4** *For any $a > 1$, the strategy $s_a$ chooses a trajectory $S = \{x_i\}$ with $x_i = a^{i+u}$, where $u$ is a random variable uniformly distributed in the interval $[0, 2)$.*

The strategy $s_a$ is a random choice among the geometric trajectories with rate of increase $a$. (The random variable $u$ is restricted to $[0, 2)$ because $S$ is a periodic function of $u$ with period 2.) Each strategy $s_a$ has the following property.

**Lemma 8.5** *For any hiding point $H$,*

$$\hat{c}(s_a, H) = 1 + \frac{a+1}{\ln a}. \tag{8.8}$$

**Proof.** Assume for convenience that $H > 0$. Then for any trajectory $S = \{x_i\}_{i=-\infty}^{\infty}$ the capture time $C(S, H)$ satisfies

$$C(S, H) = H + 2\sum_{j=-\infty}^{i+1} x_j,$$

where $x_i < H \leq x_{i+2}$ (and $i$ is even). Thus,

$$\begin{aligned}
c(s_a, H) &= H + 2E\left(\sum_{j=-\infty}^{i+1} x_j \mid x_i < H \leq x_{i+2}\right), \\
&= H + 2E\left(\frac{a}{1-\frac{1}{a}} x_i \mid x_i < H \leq a^2 x_i\right) \quad \text{(by Definition 8.4)} \\
&= H + \frac{2a^2}{a-1} E\left(x_i \mid \frac{H}{a^2} \leq x_i < H\right) \\
&= H + \frac{2a^2}{a-1}\int_0^2 \frac{1}{2} H a^{u-2} du = H\left(1 + \frac{a+1}{\ln a}\right)
\end{aligned}$$

which establishes (8.8). ■

Let

$$q = \min_{a>1}\left(1 + \frac{a+1}{\ln a}\right) \equiv 1 + \frac{\bar{a}+1}{\ln \bar{a}}. \tag{8.9}$$

The function $(a+1)/(\ln a)$ is unimodal for $a > 1$.

$$\left(\text{If } 0 \leq \theta \leq 1, \text{ then } \frac{\theta a_1 + (1-\theta)a_2 + 1}{\ln(\theta a_1 + (1-\theta)a_2)} \leq \frac{\theta(a_1+1) + (1-\theta)(a_2+1)}{\theta \ln(a_1) + (1-\theta)\ln(a_2)} \leq \max\left(\frac{a_1+1}{\ln a_1}, \frac{a_2+1}{\ln a_2}\right).\right)$$

It thus follows that its derivative vanishes at $\bar{a}$. Thus

$$\frac{\bar{a}+1}{\ln \bar{a}} = \bar{a} \tag{8.10}$$

and

$$q = \bar{a} + 1.$$

A numerical calculation gives: $\bar{a} \doteq 3.6$ so that $q \doteq 4.6$. By using $s_{\bar{a}}$, the searcher can make sure that the expected cost does not exceed $q$, which is about half of the value guaranteed by the *fixed* trajectory presented in the previous section. We now show that $s_{\bar{a}}$ is indeed an optimal strategy by proving that the value of the game is $q$. The proof constructs an $\varepsilon$-optimal strategy of the hider.

**Theorem 8.6** *For any $\varepsilon > 0$, there exists a hiding strategy $h_\varepsilon$ such that for all $S$, we have*

$$\hat{c}(S, h_\varepsilon) \geq (1-\varepsilon)q,$$

*where $q$ is given by (8.9).*

**Proof.** Denote

$$R = \exp\left(\frac{1-\varepsilon}{\varepsilon}\right), \tag{8.11}$$

and let $h_\varepsilon$ be the mixed strategy that chooses the hiding point $H$ using the probability distribution function $G$ defined as follows

$$\begin{aligned}
G(H) &= 0, \quad H < -R \\
G(-R) &= \varepsilon/2 \\
G(H) &= (1 + \ln R - \ln|H|)\varepsilon/2, \quad -R \leq H \leq -1 \\
G(H) &= 1/2, \quad -1 \leq H \leq 1 \\
G(H) &= 1/2 + (\ln H)\varepsilon/2, \quad 1 \leq H < R \\
G(H) &= 1, \quad H \geq R
\end{aligned} \tag{8.12}$$

In other words $h_\varepsilon$ has density $\varepsilon/(2|H|)$ for $1 \leq |H| \leq R$ and has two probability atoms, of mass $\varepsilon/2$ each, at $H = -R$ and $H = R$. Since the hiding point $H$ satisfies

$1 \le |H| \le R$ and $|H| = R$ with a positive probability, we need consider (against $h_\varepsilon$) only search trajectories which have a first turning point (bigger than 1) and a last turning point (at $R$). Let $S$ be such a trajectory. Then we can assume that $S$ starts by going from 0 to $x_0$ and back to 0, then to $-x_1$ and back to 0, etc. The last two steps of $S$ are of length $R$ each. Thus $S$ can be represented by a set of positive numbers $\{x_i\}_{i=0}^n$ with $x_{i+2} > x_i$, $0 \le i \le n-2$. We denote for convenience $x_{-1} = x_{-2} = 1$. The trajectory $S$ satisfies

$$
\begin{aligned}
\hat{c}(S, h_\varepsilon) &= \sum_{i=0}^{n} \int_{x_{i-2}}^{x_i} \left( 2\sum_{j=0}^{i-1} x_j + H \right) \frac{1}{H}\, dG(H) \\
&= 1 + 2\sum_{i=0}^{n-1} x_i \left( \int_{x_{i-1}}^{R} \frac{1}{H}\, dG(H) + \int_{x_i}^{R} \frac{1}{H}\, dG(H) \right) \\
&= 1 + \varepsilon \sum_{i=0}^{n-1} x_i \left( \frac{1}{x_i} + \frac{1}{x_{i-1}} \right) \quad \text{(by (8.12))} \\
&\ge 1 + \varepsilon n \left( 1 + \left( \prod_{i=0}^{n-1} \frac{x_i}{x_{i-1}} \right)^{1/n} \right) \\
&= 1 + \varepsilon n \left( 1 + R^{1/n} \right) \quad \text{(by (8.11))} \\
&= 1 + (1-\varepsilon) \frac{1 + R^{1/n}}{\ln(R^{1/n})} > (1-\varepsilon) \left( 1 + \frac{1 + R^{1/n}}{\ln(R^{1/n})} \right) \\
&\ge (1-\varepsilon) q \quad \text{(by (8.9))}.
\end{aligned}
$$

■

Note that for search games in general the searcher has an optimal strategy and that the game has a value, which implies that the hider has $\varepsilon$-optimal strategies. (See Appendix A). However, the hider need not have an optimal strategy. An example demonstrating this possibility is the search game with a mobile hider presented in Section 4.2. The same phenomenon occurs in the LSP.

The intuitive reason for this fact is that if an optimal hiding strategy would have existed, then it would have to use a probability density which is proportional to $1/|H|$ for all $0 < |H| < \infty$. However, such a probability density, obviously, does not exist.

**Theorem 8.7** *Any hiding strategy $h$ satisfies*

$$\inf_s c(s, h) < q$$

*where $q$ is given by* (8.9).

The details of the proof are given in Gal (1980, section 7.3).

It is worth noting that the worst possible outcome of using the search strategy $s_{\bar{a}}$ ($\bar{a} \doteq 3.6$) is a loss of

$$1 + 2 \sum_{j=-\infty}^{1} \bar{a}^j \doteq 10.9,$$

while the expected cost of the strategy $s_2$, which uses only minimax trajectories ($a = 2$), is $1 + 3/\ln 2 \doteq 5.3$. Thus, use of $s_{\bar{a}}$ yields (the minimal) expected cost of 4.6 but risks a maximal cost of 10.9, while use of $s_2$, which yields an expected cost of 5.3, minimizes the maximal cost (which in this case is equal to 9). The expected cost of any search strategy $s_a$ with $2 < a < \bar{a}$ lies between 4.6 and 5.3, while the maximal cost lies between 9 and 10.9. All the strategies $s_a$ with the parameter $a$ lying outside the segment $[2, \bar{a}]$ are dominated by the family $\{s_{\bar{a}}: 2 \leq a \leq \bar{a}\}$ with respect to the expected and the maximal cost.

## 8.4 Search with a Turning Cost

In this section we consider a more realistic version of the LSP, which has not been considered before in the literature. In this model the time spent in changing the direction of moving is not 0, as is usually assumed in the LSP, but a constant $d > 0$. Here, any search trajectory with a finite expected search time must have a first step because starting with an infinite number of oscillations takes infinite time. Therefore, assume for convenience that the search trajectory starts by going to $x_0 > 0$, then turning and going to $-x_1$, then turning and going to $x_2$, etc. (We can obviously assume that the searcher always goes with his maximal speed, 1, as is always the case with an immobile hider.) Thus

$$S = \{x_i\}_{i=0}^{\infty},$$

and denote

$$y_i = x_i + \frac{d}{2}, \quad i = 1, 2, \ldots .$$

In this case the normalized cost function (in the worst case) is not bounded near 0. Thus the reasonable cost function is the time to reach the target, $C(S, H)$, under the restriction $E|H| \leq \lambda$. For convenience we assume $\lambda = 1$. Thus we are interested in

$$\bar{V} = \inf_S \sup_{h: E|H| \leq 1} c(S, h).$$

We shall show that

$$9 + d \leq \bar{V} \leq 9 + 2d. \tag{8.13}$$

The left inequality follows from equality (8.7), which implies that for any $S$ and any $\delta$, there always exist an $x_i$, as large as desired, with

$$\frac{2\sum_{j=0}^{i+1} y_j + x_i}{x_i} \sim \frac{2\sum_{j=0}^{i+1} y_j + y_i}{y_i} > 9 - \delta.$$

Thus, if the hider chooses $h$ as

$$H = \begin{cases} -\varepsilon \text{ with probability } 1 - \dfrac{1}{x_i} & \text{and} \\ x_i + \varepsilon \text{ with probability } \dfrac{1}{x_i}, \end{cases}$$

then $E|H| \approx 1$ and, for a large enough $x_i$

$$c(S, h) \approx (2x_0 + d + \varepsilon)\left(1 - \frac{1}{x_i}\right) + \left(2\sum_{j=0}^{i+1} y_j + x_i\right)\frac{1}{x_i} \geq 9 - \delta + d$$

with $\delta > 0$ arbitrarily small.

In order to prove the right inequality of (8.13) we present a trajectory $S$ that satisfies for all $x_I < |H| \leq x_{I+2}$:

$$C(S, H) \leq 9x_I + 2d \leq 9|H| + 2d$$

so that for any $h$ with $E|H| \leq 1$

$$c(S, h) \leq 9 + 2d.$$

We use the following approach. For any real $\gamma$, a sufficient condition for $v(S) \leq 9+\gamma$ is the condition

$$\text{for all } |H| = x_i(+\varepsilon): \quad C(S, H) \leq 9x_i + \gamma(+\varepsilon),$$

which will hold if the following conditions hold:

$$2\sum_{0}^{i+1} y_j = 8\left(y_i - \frac{d}{2}\right) + \gamma, \quad i = 0, 1, \ldots \tag{8.14}$$

$$2y_0 = \gamma, \quad (\gamma > d/2)$$

$$y_i \geq d/2, \quad i = 0, 1, \ldots$$

Equality (8.14) is equivalent to (denoting $\frac{\gamma}{2} = b + 2d$)

$$y_{i+1} = 3y_i - \sum_{j=0}^{i-1} y_i + b, \quad i = 0, 1, \ldots \tag{8.15}$$

$$y_0 = b + 2d \left(= \frac{\gamma}{2}\right)$$

$$y_i > d/2, \quad i = 0, 1 \ldots.$$

We now look for the minimal $b$ which satisfies (8.15). It turns out that the general solution of (8.15) is

$$y_i = (y_0 + i\beta)2^i, \tag{8.16}$$

where $\beta \geq 0$ is a nonnegative parameter. (Because by (8.15) $y_{i+1} - y_i = 3y_i - 4y_{i-1}$, denoting $y_i = 2^i \alpha_i$ it easily follows that $\alpha_{i+1} - \alpha_i = \alpha_i - \alpha_{i-1}$, which leads to (8.16).)

Using (8.16) for $i = 0, 1$ in (8.15) it follows that $\beta = y_0 - d$. Since $\beta \geq 0$ and $\gamma = 2y_0$, it easily follows that $\gamma \geq 2d$. On the other hand, the value $9 + 2d$ can be achieved by the following trajectory

$$y_i = d2^i, \qquad x_i = d2^i - d/2, \quad i = 0, 1, \ldots$$

with the time to reach $x_i + \varepsilon$ being (neglecting $O(\varepsilon)$)

$$2 \sum_0^{i+1} y_i + x_i = 2d(2^{i+2} - 1) + d2^i - d/2 = 9x_i + 2d.$$

Since $E|H| \leq 1$, the last equation guarantees expected time not exceeding $9 + 2d$.

Is $9 + 2d$ the best possible constant? This is still an open problem. (Note that (8.14) is a sufficient but not a necessary condition.)

## 8.5 Search for a Moving Hider

This problem is an extension of the search on the real line to the case of a moving hider. At time $t = 0$, the hider is located at a point $-\infty < H(0) < \infty$ on the real line, and at any $t \geq 0$, the hider moves away from the origin with velocity $w < 1$. All the other rules of the game are the same as the LSP.

For any search trajectory $S = \{x_i\}$ the loss of the searcher $\hat{C}(S, H)$ is equal to the distance $C(S, H)$ traveled by the searcher (until he meets the hider) divided by $|H(0)|$.

We are interested in the minimax pure strategy of the searcher; thus, we want to find

$$\inf_S \sup_H \frac{C(S, H)}{|H(0)|}.$$

Note that we need only consider turning points $x_i$ satisfying

$$x_i > w \left( 2 \sum_{j=-\infty}^{i-1} x_j + x_i \right) \tag{8.17}$$

because by the time the searcher reaches $x_i$ the hider has already traveled the distance on the RHS of (8.17). Thus, any stage that does not satisfy (8.17) is inefficient.

Now, if the hider is captured during stage $i + 2$, then it is easily seen that the worst case for the searcher occurs if at the end of stage $i$ (when the searcher is at distance $x_i$ from $O$) he "just" misses the hider who is at that time at distance $x_i + \varepsilon$ from $O$, where $\varepsilon > 0$ is small. This means that $|H(0)| = y_i + \varepsilon$, where (see (8.17))

$$y_i = x_i - w \left( 2 \sum_{j=-\infty}^{i-1} x_j + x_i \right) = (1 - w)x_i - 2w \sum_{j=-\infty}^{i-1} x_j, \tag{8.18}$$

and that the hider is captured at distance $d$ from $O$, where $d$ satisfies

$$d = y_i + \varepsilon + wC(S, H).$$

Since

$$C(S, H) = 2 \sum_{j=-\infty}^{i+1} x_j + d = 2 \sum_{j=-\infty}^{i+1} x_j + y_i + \varepsilon + wC(S, H),$$

it follows that

$$C(S, H) = \frac{1}{1-w} \left( y_i + \varepsilon + 2 \sum_{j=-\infty}^{i+1} x_j \right).$$

So in this case

$$\hat{C}(S, H) = \frac{C(S, H)}{|H(0)|} = \frac{1}{1-w} \left( 1 + \frac{2 \sum_{j=-\infty}^{i+1} x_j}{y_i + \varepsilon} \right), \tag{8.19}$$

where $y_i$ satisfies (8.18). It follows that for trajectories $S$ that always discover the hider,

$$\inf_S \sup_H \hat{C}(S, H) = \frac{1}{1-w} \left( 1 + 2 \inf_S \sup_i \frac{\sum_{j=-\infty}^{i+1} x_j}{y_i} \right). \tag{8.20}$$

It will turn out that the minimax trajectory of (8.20) always discovers the hider, no matter how large $H(0)$ is, so that it is indeed the minimax search trajectory. We now find the optimal solution of (8.20). First note that, by (8.18)

$$y_i = (1-w) \left( \sum_{j=-\infty}^{i} x_j - \sum_{j=-\infty}^{i-1} x_j \right) - 2w \sum_{j=-\infty}^{i-1} x_j$$

$$= (1-w) \sum_{j=-\infty}^{i} x_j - (1+w) \sum_{j=-\infty}^{i-1} x_j,$$

which readily implies that

$$\sum_{j=-\infty}^{i} x_j = \frac{1}{1-w} \sum_{j=-\infty}^{0} \left( \frac{1+w}{1-w} \right)^{-j} y_{i+j}. \tag{8.21}$$

Thus

$$\inf_S \sup_i \frac{\sum_{j=-\infty}^{i+1} x_j}{y_i} = \frac{1}{1-w} \inf_S \sup_i \frac{\sum_{j=-\infty}^{0} ((1+w)/(1-w))^{-j} y_{i+j+1}}{y_i}, \tag{8.22}$$

where $y_i$, $-\infty < i < \infty$ is any positive sequence.

The solution of (8.22) is given by Theorem 7.18 of Section 7.3 by

$$\frac{1}{1-w}\inf_{a>0}\sum_{j=-\infty}^{0}\left(\frac{1+w}{1-w}\right)^{-j}a^{j+1}=\frac{1}{1-w}\inf_{a>0}\frac{a}{1-(1+w)/(a(1-w))},$$

which is equal to

$$4\frac{1+w}{(1-w)^2}. \tag{8.23}$$

This is obtained by the unique (up to a multiplicative constant) solution

$$y_i=\bar{a}^i,\quad -\infty<i<\infty, \tag{8.24}$$

the optimal $a$ being

$$\bar{a}=2\frac{1+w}{1-w}.$$

It easily follows from (8.24) and (8.21) that $x_i=\beta\bar{a}^i$ where $\beta>0$ is a constant. This sequence gives the value of (8.20) as

$$\frac{1}{1-w}\left(1+8\frac{1+w}{(1-w)^2}\right). \tag{8.25}$$

In order to show that (8.25) is the value of the game, it remains to show that the trajectory

$$x_i=\beta\left(\frac{2(1+w)}{1-w}\right)^i,\quad -\infty<i<\infty \tag{8.26}$$

always captures the hider. A sufficient condition for that is that for any $H(0)$

$$x_i>|H(0)|+w\left(2\sum_{j=-\infty}^{i-1}x_j+x_i\right)$$

(see (8.17)) for any large enough $i$. This condition follows from the fact that, by (8.18), we have

$$x_i-w\left(2\sum_{j=-\infty}^{i-1}x_j+x_i\right)=y_i\sim\left(\frac{2(1+w)}{1-w}\right)^i\to\infty.$$

Note that if $w=0$, then $x_i=\beta 2^i$, which is the minimax search trajectory obtained in Section 8.2.

**Remark 8.8** *The problem of searching for a target that moves randomly on the real line was considered by McCabe (1974).*

## 8.6 Search with Uncertain Detection

The setting of this problem is the same as in Section 8.2 except that when the searcher reaches the hiding point he is not sure to find the target. Instead we assume that the probability of finding the hider on his $k$-th visit of the hiding point H is $P_k$, where $\sum_{k=1}^{\infty} P_k = 1$. (For convenience we assume that when we search an interval, its end points are visited twice.)

We shall now find the minimax periodic monotonic trajectory (see (8.5)). In this case, if $I$ satisfies

$$x_I < |H| \leq x_{I+2}, \tag{8.27}$$

then (noting that the searcher reaches $H$ from below on his odd number visit and from above in the even number visit)

$$\hat{C}(S, H) = \frac{1}{|H|}\left(\sum_{k=1}^{\infty} P_{2k-1}\left(|H| + 2\sum_{j=-\infty}^{I+2k-1} x_j\right) + \sum_{k=1}^{\infty} P_{2k}\left(-|H| + 2\sum_{j=-\infty}^{I+2k} x_j\right)\right)$$

$$= \gamma + \frac{2}{|H|}\left(\sum_{j=-\infty}^{I+1} x_j + \sum_{j=2}^{\infty} q_j x_{I+j}\right), \tag{8.28}$$

where

$$q_j = \sum_{k=j}^{\infty} P_k \tag{8.29}$$

and

$$\gamma = \sum_{k=1}^{\infty} P_{2k-1} - \sum_{k=1}^{\infty} P_{2k}. \tag{8.30}$$

Inequality (8.27) implies that the maximum, over $H$, of (8.28) satisfies

$$v(S) = \gamma + 2 \sup_{-\infty < I < \infty} \frac{1}{x_I}\left(\sum_{j=-\infty}^{I+1} x_j + \sum_{j=2}^{\infty} q_j x_{I+j}\right) \tag{8.31}$$

$$\geq \gamma + 2 \inf_{1 < a < \infty}\left(\sum_{j=-\infty}^{1} a^j + \sum_{j=2}^{\infty} q_j a^j\right) \tag{8.32}$$

by Corollary 7.11. Thus, expression (8.32) is the value that can be obtained by using a periodic monotonic trajectory.

In order to obtain this value, one would have to use the trajectory

$$x_i = \beta \bar{a}^i, \quad -\infty < i < \infty,$$

where $\beta > 0$ is a constant and $\bar{a}$ is the value of the $a$ that minimizes (8.32). In this trajectory, the positive and the negative rays are visited alternatively.

On the other hand, contrary to the linear search problems considered before, the optimal trajectory need not be periodic and monotonic, as will be shown by the following example.

### 8.6.1 Search with a delay

Assume that we find the target only on our *third* encounter, i.e., $P_3 = 1$ and $P_j = 0$ for $j \neq 3$. It follows from (8.32) that the value obtained by using periodic monotonic trajectories is equal to

$$1 + 2 \inf_{0<a<\infty} \left( \sum_{j=-\infty}^{1} a^j + a^2 + a^3 \right) \doteq 20. \tag{8.33}$$

On the other hand, a smaller value can be obtained by the following trajectory $\tilde{S}$. We use a nested set of turning points

$$-x_{2i+1} < -x_{2i-1} < \cdots 0 \cdots < x_{2i} < x_{2i+2}$$

(as in the LSP) but during stage $j$, instead of moving from 0 to $(-1)^j x_j$ and back to 0, we move along the following four segments:

$$0 \to (-1)^j x_j \to (-1)^{j-2} x_{j-2} \to (-1)^j x_j \to 0.$$

Then, the distance traveled at stage $j$ is equal to

$$2x_j + 2(x_j - x_{j-2}) = 4x_j - 2x_{j-2}.$$

Now, for

$$(-1)^I x_I < H \leq (-1)^{I+2} x_{I+2}$$

or

$$(-1)^I x_I > H \geq (-1)^{I+2} x_{I+2}$$

the target is found at stage $I + 2$, so that

$$\hat{C}(\tilde{S}, H) = \frac{1}{|H|} \left( \sum_{j=-\infty}^{I+1} (4x_j - 2x_{j-2}) + x_{I+2} + (x_{I+2} - x_I) + (|H| - x_I) \right)$$

$$= 1 + \frac{2}{|H|} \left( \sum_{j=-\infty}^{I+2} x_j + x_{I+1} \right).$$

Thus,

$$v(\tilde{S}) = 1 + 2 \sup_{-\infty<I<\infty} \frac{1}{x_I} \left( \sum_{j=-\infty}^{I+2} x_j + x_{I+1} \right).$$

Let $\bar{a}$ be the $a$ that minimizes

$$\sum_{j=-\theta}^{2} a^j + a.$$

It turns out that $\bar{a} \doteq 1.45$. If we choose $x_j = \beta \bar{a}^j$, $-\infty < j < \infty$, then $v(\tilde{S}) \doteq 17.2$, which is less than the value obtained in (8.33).

It follows from the previous example that in some of the problems with probability of detection less than 1, the optimal trajectory does not have the simple properties of monotonicity and periodicity.

### 8.6.2 Geometric detection probability

The following simple model was presented as an open problem in Gal (1980) and is still open. Assume that the searcher detects the hider with probability $p$ each time he passes the target point. In other words, the probability of capture in the $i$-th time the searcher visits the point $H$ is

$$P_i = (1-p)^{i-1} p.$$

Thus (see (8.29) and (8.30))

$$q_j = (1-p)^{j-1}, \quad 2 \le j < \infty$$

and

$$\gamma = \frac{p}{2-p}$$

so that (8.32) is equal to

$$\frac{p}{2-p} + 2 \inf_{1<a<\infty} \left( \sum_{j=-\infty}^{1} a^j + \sum_{j=2}^{\infty} (1-p)^{j-1} a^j \right) \tag{8.34}$$

$$= \frac{p}{2-p} + 2 \min_{1<a<\infty} \left( 1 + \frac{1}{a-1} + \frac{a}{1-a(1-p)} \right)$$

$$= \frac{p}{2-p} + \frac{8}{p} = \hat{V} \quad \left( \text{taking } \bar{a} = \frac{2}{2-p} \right). \tag{8.35}$$

Expression (8.35) represents the minimax value, obtained by monotonic and periodic trajectories.

Thus, if the minimax trajectory is monotonic and periodic, then $\hat{V}$, given by (8.35), is the pure value of the game. The conjecture that the above geometric trajectory is the minimax solution seems quite reasonable, but it is still not known whether the minimax trajectory is indeed monotonic and periodic. A similar difficulty is associated with the optimal mixed strategy.

## 8.7 A Dynamic Programming Algorithm for the LSP

In this section we return to the original formulation of the linear search problem (LSP) as an optimization problem in which the probability distribution function (PDF) of the target is known and we look for the trajectory S with the minimal expected search time.

We now present a dynamic programming (DP) algorithm for solving the LSP, for any PDF, with any desired accuracy. This approach was suggested by Bruce and

Robertson (1988) for numerically solving the LSP for a probability distribution that has a finite number of atoms. (It has also been used by Washburn, 1995, to find the best starting point for such a distribution.) We will generalize the algorithm and obtain $\varepsilon$-optimal solutions for the LSP for a general distribution function. We shall also show how to use DP to solve the double linear search problem (to be described in the sequel), which is equivalent to solving the rendezvous search on the line (see Section 16.3 Book II).

We now show how to use DP to produce an approximation, with any desired accuracy, for any PDF. We first find an $\frac{\varepsilon}{2}$-optimal trajectory for a distribution with a finite support. Then we show how to extend the trajectory to get an $\varepsilon$-optimal trajectory for a distribution with an unbounded support.

Assume that the target point $Z$ has a PDF $F$, with a finite support. For any given positive number $e$, let $F^e$ be an *e-grid* obtained from $F$ by moving all the probability mass of the interval $((i-1)e, ie)$, $1 \le i < \infty$ into the grid point $ie$ and similarly for $(-ie, -(i-1)e)$ into $-ie$, and let $Z^e$ be a random variable with PDF $F^e$. Also, let $F_e$ be an *e-grid* obtained from $F$ by moving all the probability mass of $((i-1)e, ie)$ into the grid point $(i-1)e$ and similarly for $(-ie, -(i-1)e)$ into $-(i-1)e$, and denote the corresponding random variable by $Z_e$.

Let $\tilde{T}$, $\tilde{T}^e$, $\tilde{T}_e$ denote the minimal expected discovery time for the PDF's $F$, $F^e$, $F_e$, respectively. Then $\tilde{T}_e \le \tilde{T} \le \tilde{T}^e$ because any optimal trajectory for $F^e$ when used for the original $F$ yields a lower expected discovery time and any optimal trajectory for the original $F$, used for $F_e$ yields a lower expected discovery time.

We now show that for a small $e$, $\tilde{T}^e - \tilde{T}_e$ is small. Let $\delta > 0$. Put all the probability mass of $-\delta \le Z \le \delta$ into 0. Then the distribution function obtained by this operation, $F(\delta)$, has an expected discovery time $\tilde{T}(\delta)$ satisfying

$$\tilde{T} - 4\delta \le \tilde{T}(\delta) \le \tilde{T}$$

because the original problem can be reduced to search with PDF $F(\delta)$ (after time $4\delta$) by moving to $\delta$, then to $-\delta$, and then returning to 0.

For $F(\delta)$ choose $e = \delta/k$, where $k$ is large. Then

$$\tilde{T}^e(\delta) \ge \tilde{T}_e(\delta) \ge \frac{k}{k+1}\tilde{T}^e(\delta)$$

(because $Z_e(\delta)$ dominates $k/(k+1)Z^e(\delta)$).

Thus,

$$\tilde{T}^e(\delta) \le \left(1 + \frac{1}{k}\right)\tilde{T}_e(\delta) \le \left(1 + \frac{1}{k}\right)\tilde{T}(\delta).$$

Recall that $F^e$ can be reduced to $F^e(\delta)$ by adding $4\delta$ time units at the beginning. Thus, we get

$$\tilde{T}^e \le 4\delta + \tilde{T}^e(\delta) \le 4\delta + \left(1 + \frac{1}{k}\right)\tilde{T}(\delta) \le 4\delta + \left(1 + \frac{1}{k}\right)\tilde{T}.$$

Now recall that

$$E(|z|) = \int_0^\infty |z|\, dF(z) \le \tilde{T}.$$

Choosing $\delta = (\varepsilon/16)E(|z|)$ and $k = 4/\varepsilon$, we obtain

$$\tilde{T} \le \tilde{T}^e \le \frac{\varepsilon}{4}E(|z|) + \left(1+\frac{\varepsilon}{4}\right)\tilde{T} \le \left(1+\frac{\varepsilon}{2}\right)\tilde{T}. \tag{8.36}$$

Thus, solving the LSP with the PDF $F^e$ gives the desired accuracy for a bounded $F$.

For a PDF $F$ with an unbounded support, we can get an approximation with any desired accuracy by truncating $F$ at $R$, where $R$ is large enough. Specifically, consider the truncated distribution $F_R$ in which the probability mass of $F$ that lies above $R$ is put at the point $z = R$ and the probability mass of $F$ that lies below $-R$ is put at the point $z = -R$. Since we assume a finite first absolute moment we can choose a large enough $R$ such that

$$\int_{|z|\ge R} |z| dF(z) \le \frac{\varepsilon}{24}E(|z|) \tag{8.37}$$

(Note that (8.37) implies also that $R\int_{|z|\ge R} dF(z) \le \frac{\varepsilon}{24}E(|z|)$.)

Let $\tilde{T}_R$ be the optimal search time for $F_R$ and let $S_R^e$ denote a trajectory that guarantees (8.36), for $F_R$. Extend $S_R^e$ as follows. After searching in the interval $[-R, R]$, if the target has not been found, then go back to $z = 0$, reaching it at time $\tau$, and use the following turning points: $2R, -4R, 8R, \ldots$. Denote the extended trajectory by $S_\infty$. Now, for any $z$ with

$$R2^i < |z| \le R2^{i+1}, \quad i = 0, 1, 2, \ldots, \tag{8.38}$$

the time to reach $z$ by $S_\infty$ is bounded by

$$\tau + |z| + 2\sum_{1}^{i+1} R2^j < \tau + |z| + 2R\sum_{-\infty}^{i+1} 2^j < 9|z| + \tau \tag{8.39}$$

(by (8.38)).

Now compare the expected search time of $S_R^e$ against $F_R$ denoted by $\tilde{T}_R$ to the expected time of $S_\infty$ against $F$, denoted by $\tilde{T}_\infty$. For $z \in [-R, R]$ the discovery time is the same while for the tails the discovery time under $F_R$ is at least $\tau - 3R$ (because the end points of $[-R, R]$ are searched at time $\tau - 3R$ or $\tau - R$). Thus (by (8.37) and (8.39)),

$$\begin{aligned}\tilde{T}_\infty - \tilde{T}_R &\le \int_{|z|\ge R} (9|z| + 3R)\, dF(z) \\ &\le (9+3)\frac{\varepsilon}{24}E\left(|z|\right) = \frac{\varepsilon}{2}E\left(|z|\right) \le \frac{\varepsilon}{2}\tilde{T}.\end{aligned} \tag{8.40}$$

Combining (8.36) and (8.40) it follows that if $e$, $k$, and $R$ are chosen as described, then finding an optimal trajectory $S_R^e$ for $F_R^e$ and then extending it as described above to $S_\infty$ would yield, for the original $F$, an expected time

$$\tilde{T}_\infty \le \tilde{T}_R^e + \frac{\varepsilon}{2}\tilde{T} \le (1+\varepsilon)\tilde{T},$$

which shows that $S_\infty$ is an $\varepsilon$-optimal search trajectory as desired.

Thus, an algorithm for solving the LSP for an *e-grid* would enable us to solve any LSP, with a general PDF, within any desired accuracy.

For convenience we take the measuring unit as $e$ so that the point $i$ has distance $i$ from the origin. Note that if there are $n$ point masses on each side of the origin, then the number of possible trajectories for $F$ is at least $2^n$ because it exceeds the number of possible sets of right turning points i.e., $\sum_{i=0}^{n}\binom{n}{i}$. Thus we have to deal with an exponential number of trajectories. However, our following dynamic programming (DP) algorithm has a complexity of $O(n^2)$.

Assume that there are (at most) $n$ point masses on each side of the origin with probabilities $p(l)$, $-n \leq l \leq n$ and denote

$Q(i, j)-$ the probability that the target is above $i$ or below $-j$ :

$$Q(i, j) = \sum_{l=i+1}^{n} p(l) + \sum_{l=j+1}^{n} p(-l). \tag{8.41}$$

We shall recursively construct the following two functions:

$VR(i, j)$ : $Q(i, j)\times$ the expected remaining time given that the interval covered is $(-j, i)$ and we are now at $i$, and

$VL(i, j)$ : $Q(i, j)\times$ the expected remaining time given that the interval covered is $(-j, i)$ and we are now at $-j$.

The recursive formula for $VR(i, j)$, $-n < i, j < n$ is

$$VR(i, j) = min\{(Q(i, j) + VR(i + 1, j); (i + j + 1) \times Q(i, j) + VL(i, j + 1))\}. \tag{8.42}$$

To see why (8.42) holds, note that if the searcher decides to go from $i$ to $i + 1$, then the expected conditional remaining time satisfies

$$\frac{VR(i, j)}{Q(i, j)} = 1 \times \frac{p(i + 1)}{Q(i, j)} + \left(1 + \frac{VR(i + 1, j)}{Q(i + 1, j)}\right) \times \left(1 - \frac{p(i + 1)}{Q(i, j)}\right),$$

which is equal to the first term of the right side of (8.42), while if the searcher decides to go from $i$ to $-j - 1$, then the expected conditional remaining time satisfies

$$\frac{VR(i, j)}{Q(i, j)} = (i + j + 1) \times \frac{p(-j - 1)}{Q(i, j)} + \left(i + j + 1 + \frac{VL(i, j + 1)}{Q(i, j + 1)}\right) \times \left(1 - \frac{p(-j - 1)}{Q(i, j)}\right)$$

which is equal to the second term of the right side of (8.42). Similarly,

$$VL(i, j) = min\{Q(i, j) + VL(i, j + 1); (i + j + 1) \times Q(i, j) + VR(i + 1, j)\}.$$

In order to complete the DP scheme we use the following boundary conditions:

$$VR(n,j) = \sum_{k=j+1}^{n} p(-k) \times (n+k), \quad VR(i,n) = \sum_{k=i+1}^{n} p(k) \times (k-i),$$

$$VL(n,j) = \sum_{k=j+1}^{n} p(-k) \times (k-j), \quad VL(i,n) = \sum_{k=i+1}^{n} p(k) \times (n+i).$$

The minimal expected discovery time $\tilde{T}$ is obtained at the last stage when $i = j = 0$:

$$\tilde{T} = VR(0,0) = VL(0,0) = \min\{Q(0,0) + VR(1,0),\, Q(0,0) + VL(0,1)\}.$$

The number of operations of the DP algorithm can be calculated as follows. We first have a preprocessing stage of calculating $Q(i,j)$ for $-n \leq i, j \leq n$ requiring $O(n^2)$ number of operations. Then, the $VR$ and $VL$ calculations require $O(n^2)$ operations, each of constant time (assuming constant table lookup time for the pre-calculated function $Q(i,j)$). Thus, the overall complexity of the DP algorithm is $O(n^2)$.

**Remark 8.9** *A (discrete) two dimensional extension of the LSP aims to minimize the expected time to reach n given points in the plane. It is called the "minimum latency problem" (Blum et al., 1994). This problem is shown to be at least as hard as the Travelling Salesman Problem and therefore is NP hard. (For approximation algorithms see Goemans and Kleinberg, 1998.)*

A similar approach can be used to solve the double linear search problem (DLSP) in which two cooperating agents 1 and 2 are searching for a target that is hidden on one of two infinite lines, line 1 or line 2. The probability of the target being on each line is 1/2 with the same conditional probability distribution function (PDF) $F(z)$. As the search begins, each agent is placed at the origin of the corresponding line and starts to move, in a coordinated way, with maximal combined velocity normalized to 1. In Book II we will show that DLSP is equivalent to rendezvous search on the line. Thus, the DP algorithm yields (approximated) optimal strategies for two agents to meet on the line.

As for the LSP, the DP approach finds the minimal expected discovery time for the discretized problem, which amounts to using an *e-grid* with PDF $F^e$ (choosing $e = 1$) and assuming that in the optimal solution for $F^e$ the following assumption holds: for any time interval $ie < t \leq (i+1)e$ only one player moves.

We now present the DP scheme: Let $E(i,j,k,l)$ be the event that the target is above $i$ or below $-j$ on the first line or above $k$ or below $-l$ on the second line. $-n \leq i, j, k, l \leq n$. Denote

$$P = P(E(i,j,k,l)) = Q(i,j)/2 + Q(k,l)/2$$

(see (8.41)).

In order to use the DP algorithm for DSLP we need to calculate four value functions: $VRR(i,j,k,l)$, $VRL(i,j,k,l)$, $VLR(i,j,k,l)$, $VLL(i,j,k,l)$.

For example, $VRL(i, j, k, l) = P\times$ the expected remaining time given that the first player covered $(-j, i)$ and is now at $i$ and the second player covered $(-l, k)$ and is now at $-l$.

The other three functions are defined in a similar way.

We now present one of the recursion formulas. (Obtaining the other three recursion formulas is straightforward):

$$VRL(i, j, k, l) = \min \begin{cases} P + VRL(i+1, j, k, l), \\ (i + j + 1)P + VLL(i, j+1, k, l), \\ P + VRL(i, j, k, l+1), \\ (k + l + 1)P + VRR(i, j, k+1, l). \end{cases}$$

If one of the arguments is equal to $n$ then we obtain a similar (simpler) recursion formula with three possibilities (instead of four).

In the case that two arguments are equal to $n$, if these arguments are of the same player, then the problem reduces to LSP; if the argument $n$ appears (once) for both players, then we obtain similar (simpler) recursion formulas with two possibilities (instead of four).

If three arguments are equal to $n$, then we obtain the boundary condition of the LSP of the corresponding player.

The complexity of our algorithm is $O(n^4)$ since the $V$ functions have to be calculated for $O(n^4)$ argument values.

# Chapter 9

# Star and Plan Search

## 9.1 Introduction

In this chapter we first present several search problems, which can be solved using the tools developed in Chapter 7. The first search space, presented in Section 9.2, is a finite set of rays radiating from the origin (a *star*), and the second, in Section 9.3, is the boundary of an unbounded plane sector. Minimax search trajectories are found, for each of these problems, using the results presented in Section 7.2. The continuous minimax result, presented in Section 7.2.1, is used for obtaining the required search trajectory for the search in the plane presented in Section 9.4. We then present, in Section 9.5, several interesting problems of "swimming in a fog," i.e., attempting to minimize the time to reach the shore given its shape and some information about its location around the starting point. Finally, in Section 9.6 we present a problem of detecting a submarine with a known location (only) at $t = 0$. The optimal search strategy is still unknown.

## 9.2 Star Search

Suppose we are searching for an immobile hider in a star $Q$ that consists of $M$ ($M > 1$) unbounded rays radiating from the origin $O$. This problem is an extension of the search on the real line ($M = 2$), and the results that we obtain are indeed of a similar nature. This problem was first considered by Gal (1974a) as a minimax problem. The search problem with a known probability distribution of the target (on the star) is an extension of the linear search problem (LSP). Considering a probability distributions with bounded support, Kella (1993) found a sufficient condition under which an optimal policy has the property that it visits every direction only once (thereby exhausting it). A special case with this property, already mentioned in Section 8.1, is the LSP with a uniform target distribution (or any other convex PDF).

There has been a great deal of interest in computer science in search problems like the star search or the LSP. In these problems a searcher has to traverse a graph, $Q$, from a starting point to a target but must operate with incomplete information.

The searcher starts with a limited amount of information about the environment to be searched and learns additional information as the search proceeds (in our case about the part of $Q$ that contains the target). A model of this type was used by Papadimitriou and Yannakis (1991) and by Burley (1996). The star search has been considered by Baeza-Yates et al. (1993) within the context of searching sequentially for a record that is known to be on one of $M$ large tapes given that we have only one drive and that we must rewind the current tape before searching any other. Kao et al. (1996) call it the *cow-path problem* presenting it as an example of searching in an unknown environment. Lopez-Otiz and Schuierer (1997, 1998), Brocker and Schuierer (1999), Brocker and Lopez-Otiz (1999), and Schuierer (1999, 2001), used a model of this type for searching a target in a polygon and other problems in computational geometry. The optimal strategy for several searchers who start from the origin and search in parallel is presented in Hammar et al. (1999). Recently, Jallet and Stafford (2001) reconstructed the optimal solution obtained by Gal (1974a), using a different technique and also analyzed the solution under some additional information: (*a*) When the maximal distance of the object from the origin is given, and (*b*) When the probability that the object lies on each ray is given. For the LSP (i.e., $M = 2$) the analysis for case (*a*) above was carried out by Hipke et al. (1999). Somewhat similar ideas were used in finding competitive ratios (i.e., worst-case analysis) for on-line algorithms. (See, for example, the on-line load balancing algorithm for related machines, presented by Berman et al. (2000).)

The star search will be analyzed using the following framework. A pure strategy of the hider is given by

$$H = (|H|, m)$$

where $1 \leq m \leq M$ is the ray number and $|H| > 0$ is the distance from the origin (see Figure 9.1).

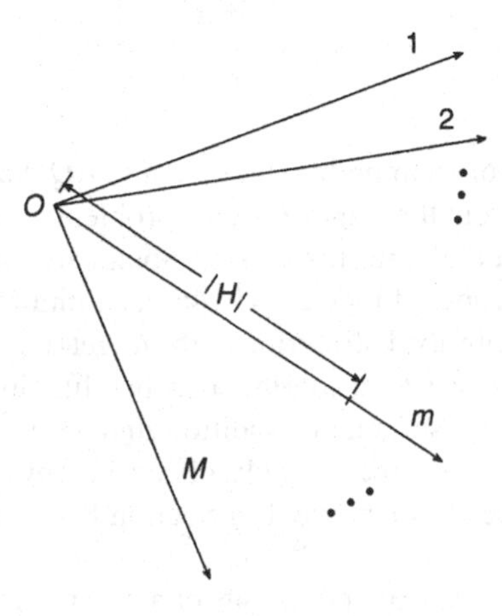


**Figure 9.1.**

Similar to the discussion presented in Section 8.1 for the search on the line ($M = 2$), any admissible trajectory $S$ of the searcher can be described by an infinite number of pairs

$$S = \{(x_i, N_i)\} \tag{9.1}$$

where $-\infty < i < \infty$, $x_i > 0$ and $1 \le N_i \le M$. At the $i$th stage of the trajectory $S$, the searcher starts from the origin, walks a distance $x_i$ (or $|H|$ if the hider is to be found at this stage) along ray $N_i$ and returns to the origin $O$. The capture time $C(S, H)$ is given by

$$C(S, H) = 2 \sum_{i=-\infty}^{i_H - 1} x_i + |H| \tag{9.2}$$

where $i_H$ is the number of the stage during which the hider is discovered.

We shall use the normalized cost function $\hat{C}(S, H)$, discussed in Section 6.1 and used in Chapter 8. Thus the cost is the capture time divided by the distance of the hiding point from the origin,

$$\hat{C}(S, H) = \frac{C(S, H)}{|H|}.$$

As usual, we denote the value of the search trajectory $S$ by

$$v(S) = \sup_H \hat{C}(S, H).$$

In order to find the minimax search trajectory, we first establish several properties of the search trajectories. At first, we introduce a new representation for the search trajectories, which can be handled more conveniently than (9.1). This representation, which will be used from now on, is

$$S = \{(x_i, J_i)\}_{i=-\infty}^{\infty}, \tag{9.3}$$

where $x_i$ is the distance traveled along the chosen ray at the $i$th stage, and $J_i$ is the minimal stage number greater than $i$ in which the same ray is visited again. The index $J_i$ should satisfy the following three requirements.

- For every $i$,

$$i < J_i < \infty. \tag{9.4}$$

- For every $i$, there exists an $l < i$ such that

$$J_l = i \quad (l \text{ will be denoted by } J_i^{-1}). \tag{9.5}$$

- For every integer $l$, the set

$$G_l = \{i \colon i \le l \text{ and } J_i > l\} \tag{9.6}$$

  contains $M$ members.

For example, if $M = 4$ and the trajectory $S$ satisfies (see (9.1))

$$\ldots, \quad N_{-3} = 3, \quad N_{-2} = 4, \quad N_{-1} = 2, \quad N_0 = 1,$$
$$N_1 = 4, \quad N_2 = 2, \quad N_3 = 3, \quad N_4 = 1, \quad \ldots,$$

then $J_{-3} = 3$ because the next visit to ray number 3 (which is visited when $i = -3$) occurs when $i = 3$. Similarly, $J_{-2} = 1$, $J_{-1} = 2$, $J_0 = 4, \ldots$, etc. Thus,

$$G_1 = \{-3, -1, 0, 1\}$$

because ray number 3 is visited when $i = \mathbf{-3} \leq 1$ and after that when $i = 3 > 1$, ray number 2 is visited when $i = \mathbf{-1} \leq 1$ and after that when $i = 2 > 1$, ray number 1 is visited when $i = \mathbf{0} \leq 1$ and after that when $i = 4 > 1$, and ray number 4 is visited when $i = \mathbf{1} \leq 1$. Similarly,

$$G_2 = \{-3, 0, 1, 2\}, \quad G_3 = \{0, 1, 2, 3\}, \quad \text{etc.}$$

It is obvious that (9.3) does not define a unique trajectory but a class of $M!$ ("equivalent") trajectories, which can be generated from one trajectory by changing the enumeration of the $M$ rays but for each of them, $v(S)$ is the same.

It is also obvious that if the function $J_i$ does not satisfy any one of the requirements (9.4)–(9.6), then either after or before a certain stage the searcher will not visit a certain ray, so that the value of such a trajectory is infinite.

For example, if $M = 3$ and the trajectory $S$ satisfies (see (9.1))

$$N_{2i} = 1, \qquad N_{2i+1} = 2, \quad -\infty < i < \infty$$

then ray number 3 is not visited and indeed, for all $l$, $G_l = \{l - 1, l\}$ contains only two members.

On the other hand, if the trajectory $S$ satisfies (9.4)–(9.6), then it follows from (9.4) and (9.5) that for any visit on a specific ray there corresponds a following visit and also a preceding visit on this ray. In addition, we note that if at least one ray is not visited, then $G_l$ contains less than $M$ members for all $l$. Thus, it follows from (9.6) that all the rays are visited and from (9.4) and (9.5) that for each of them, there is no first visit and no last visit.

It follows from dominance considerations that we need deal only with trajectories having the monotonicity property defined as follows

$$\text{for every } i\colon x_{J_i} > x_i. \tag{9.7}$$

Obviously, we may also assume that

$$v(S) < \infty. \tag{9.8}$$

A trajectory $S$ satisfying conditions (9.4)–(9.8) will be called an *admissible trajectory*. We shall show that for all $M$, there exist admissible trajectories.

For each admissible trajectory $S$

$$v(S) = \sup_H \frac{C(S, H)}{|H|} = 1 + 2 \sup_{-\infty<i<\infty} f_i(S), \tag{9.9}$$

where

$$f_i(S) \equiv \frac{\sum_{j=-\infty}^{J_i - 1} x_j}{x_i}. \tag{9.10}$$

This follows from the fact that against a known trajectory $S$, the best point for the hider belongs to the set $\{(x_i + 0, N_i)\}$ (see (9.1)).

A trajectory $X$ satisfying

$$\text{for each } i\text{: } J_i = i + M \tag{9.11}$$

will be called a periodic trajectory. If $S$ is a periodic (and monotonic) trajectory, then

$$f_i(S) = \frac{\sum_{j=-\infty}^{i+M-1} x_j}{x_i}. \tag{9.12}$$

It follows from Corollary 7.11 that

$$\sup_{-\infty<i<\infty} \frac{\sum_{j=-\infty}^{i+M-1} x_j}{x_i} \geq \inf_{a>0} \sum_{j=-\infty}^{M-1} a^j = \frac{M^M}{(M-1)^{M-1}}. \tag{9.13}$$

Furthermore, it follows from Theorem 7.18 that if

$$\sup_{-\infty<i<\infty} \frac{\sum_{j=-\infty}^{i+M-1} x_j}{x_i} = \frac{M^M}{(M-1)^{M-1}}$$

then

$$x_i = \alpha \left(\frac{M}{M-1}\right)^i, \tag{9.14}$$

where $\alpha$ is a positive constant.

Hence, the trajectory $\{(x_i, J_i)\}$, where $x_i$ is given by (9.14) and $J_i = i + M$, which is obviously an admissible trajectory, is optimal among the periodic trajectories. We shall show that it is the best (pure) strategy among all the possible trajectories.

To this end, we present the following lemmas.

**Lemma 9.1** *Every admissible trajectory satisfies*

$$\lim_{i\to-\infty} x_i = 0 \quad \textit{and} \quad \lim_{i\to\infty} x_i = \infty. \tag{9.15}$$

**Proof.** Any trajectory that does not satisfy (9.15) has an infinite value. ■

We now use an interchange argument to show that any non-monotonic trajectory is dominated by some monotonic trajectory.

**Lemma 9.2** *For each trajectory $S = \{(x_i, J_i)\}$ there exists a trajectory $S^1 = \{x_i^1, J_i^1\}$ satisfying for every $-\infty < i < \infty$*

$$x_{i+1}^1 \geq x_i^1 \tag{9.16}$$

*and*

$$\sup_{-\infty<i<\infty} f_i(S^1) \leq \sup_{-\infty<i<\infty} f_i(S) \tag{9.17}$$

*where $f_i$ is defined by* (9.10). *Furthermore, if there exists an $i_0$ satisfying $x_{i_0+1} < x_{i_0}$, then there exists a $j_0$ satisfying $f_{j_0}(S^1) < f_{j_0}(S)$.*

**Proof.** We present an outline of the proof. The full details appear in Gal (1980) (or Gal, 1974a). Assume that there exists an $i_0$ with

$$x_{i_0+1} < x_{i_0}. \tag{9.18}$$

Replace the trajectory $S$ by the trajectory $S'$, which is identical to $S$ before stage $i_0$ and from stage $i_0$ on, the role of the rays visited at stages number $i_0$ and $i_0 + 1$ is interchanged. It is easily seen that (9.18) implies

$$f_{i_0}(S') = f_{i_0+1}(S)$$
$$f_{i_0+1}(S') = f_{i_0}(S)$$
$$f_l(S') < f_l(S), \quad l = J_{i_0+1}^{-1}$$

and for any other $k$

$$f_k(S') = f_k(S).$$

Thus,

$$\sup_{-\infty<i<\infty} f_i(S') \leq \sup_{-\infty<i<\infty} f_i(S).$$

It is possible to repeatedly apply the same operation on $S'$ and obtain in the limit a trajectory $S^1$ that satisfies (9.16). ■

We will also use the following lemma.

**Lemma 9.3** *Let*

$$d_i = \frac{\sum_{j=-\infty}^{i+M-1} x_j}{x_i} \tag{9.19}$$

$$D_i = \left(\prod_{k=1}^{i+M-1} d_k\right)^{1/M}, \tag{9.20}$$

*and, as in* (9.13), *let*

$$q = \inf_{a>0} \sum_{j=-\infty}^{M-1} a^j = \frac{M^M}{(M-1)^{M-1}}. \tag{9.21}$$

*If $D_i \leq q$ for all $-\infty < i < \infty$, then $D_i = q$.*

**Proof.** Define

$$y_i = \left(\prod_{k=1}^{i+M-1} x_k\right)^{1/M}$$

and

$$E_i = \frac{\sum_{j=-\infty}^{i+M-1} y_j}{y_i}.$$

Using Holder's inequality (see Hardy et al., 1952), we obtain

$$D_i \geq \frac{\sum_{j=-\infty}^{i+M-1} \left(\prod_{k=j}^{j+M-1} x_k\right)^{1/M}}{\left(\prod_{k=i}^{i+M-1} x_k\right)^{1/M}} = E_i \tag{9.22}$$

so that the assumption of the lemma implies also that $E_i$ $(\leq D_i) \leq q$. Using Theorem 7.18 we obtain that $E_i = q$, and consequently, $D_i = q$ for all $i$. ■

It is now possible to show that the periodic trajectory defined by (9.14) is the minimax trajectory for the searcher, guaranteeing a value of

$$\bar{V} = 1 + 2\frac{M^M}{(M-1)^{M-1}}.$$

**Theorem 9.4** *Let $S$ be a search trajectory and let $f_i(S)$ be defined by (9.10) and $q = M^M/(M-1)^{M-1}$. If for all $-\infty < i < \infty$*

$$f_i(S) \leq q, \tag{9.23}$$

*then for all $i$*

*(a)* $f_i(S) = q$,

*(b)* $J_i = i + M$, *and*

*(c)* $x_i = \alpha(M/(M-1))^i$, *where $\alpha$ is any positive constant.*

**Proof.** The proof is given in two parts:

1. First assume that

$$x_i \leq x_{i+1} \quad \text{for every } -\infty < i < \infty. \tag{9.24}$$

Since $S$ is an admissible strategy, we can use condition (9.6) together with (9.23) and deduce that for every $-\infty < k < \infty$,

$$\begin{aligned} q &\geq \left(\prod_{i \in G_k} f_i(S)\right)^{1/M} \qquad \text{(by (9.24))} \\ &\geq \left(\prod_{i=k-M+1}^{k} \frac{\sum_{j=-\infty}^{i+M-1} x_j}{x_i}\right)^{1/M} = D_{k-M+1}, \end{aligned} \tag{9.25}$$

$D_{k-M+1}$ being defined by (9.20). Using Lemma 9.3, we deduce that for every $k$, $D_k = q$. Finally, we use (9.23) and (9.25) and establish $(a)$.

In order to prove $(b)$ note that assumption (9.24) together with $(a)$ imply that

$$i < k \rightarrow J_i < J_k. \tag{9.26}$$

This implies that the set $G_l$ defined by (9.6) consists of the integers $l-M+1, l-M+2, \ldots, l$ (because, otherwise, there would exist at least one integer $k < J_l^{-1}$ with $J_k > l = J_{J_l^{-1}}$ contradicting (9.26)). Similarly, $G_{l-1}$ consists of the integers $l-M, l-M+1, \ldots, l-1$. Thus for every $l$: $J_l^{-1} = l - M$ so that $J_{l-M} = l$ establishing ($b$).

In order to prove ($c$), it remains to note that proposition ($b$) implies that

$$f_i(S) = \frac{\sum_{j=-\infty}^{i+M-1} x_j}{x_i}$$

so that ($c$) follows from the uniqueness Theorem 7.18 under assumption (9.23). This completes the proof under assumption (9.24).

2. If there exists an integer $i_0$ satisfying

$$x_{i_0+1} < x_{i_0}, \tag{9.27}$$

then we can use Lemma 9.2 and obtain a trajectory $S^1$ satisfying, for all $-\infty < i < \infty$

$$x_i^1 \le x_{i+1}^1, \quad f_i(S^1) \le q \quad \text{and} \quad f_{i_0}(S^1) < q.$$

However, the existence of this trajectory contradicts part 1. Therefore, inequality (9.27) cannot hold. ■

We now present, as an example, the optimal search strategy for the case where $M = 3$. Using Theorem 9.4 we see that the optimal strategy has the form

$$S = \left\{ \alpha \left(\frac{3}{2}\right)^i, i+3 \right\},$$

where $\alpha$ is any positive constant.

If we choose $\alpha = 1$, then the stages $i = 0$ to $i = 4$ (for example) are carried out by going a distance of 1 along the first ray (and returning to the origin), then a distance of 3/2 along the second ray, then a distance of $(3/2)^2$ along the third ray and then a distance of $(3/2)^3$ along the first ray, etc. The value of this game is

$$1 + 2\frac{3^3}{2^2} = 14.5.$$

We have previously discussed the fact that the optimal trajectory has no first step. In order to apply it to a real world situation, one would have to assume that the object cannot be hidden closer than $\varepsilon$ from the origin. In this case we can modify the minimax trajectory and define it to be

$$x_i = \varepsilon \left(\frac{M}{M-1}\right)^i, \quad 0 \le i < \infty$$

and $J_i = i + M$ as before. Thus, in this trajectory there is a first step that is to travel a distance $\varepsilon$ on the first ray and return to the origin. The second step is to travel a distance of $\varepsilon(M/M-1)$ on the second ray and return to the origin, etc.

**Remark 9.5** *If we assume that the distance of the hider from O exceeds a known constant, (say 1), then the minimax search trajectory has a first step, corresponding to $i = 0$, and, for all $i \geq 0$ the distance, $x_i$, traveled along the chosen ray satisfies $x_i \geq 1$. In this case the minimax trajectory $\bar{X}$ (assuming that it is periodic) has to satisfy*

$$V(\bar{X}) = 1 + 2 \sup_{i \geq 0} \frac{\sum_{j=0}^{i+M-1} \bar{x}_j}{\bar{x}_i} = 1 + 2 \inf_X \sup_{i \geq 0} \frac{\sum_{j=0}^{i+M-1} x_j}{x_i}.$$

*This problem has been analyzed in Example 7.12. The minimax value obtained is equal to*

$$1 + 2\frac{M^M}{(M-1)^{M-1}}$$

*with the minimax trajectory satisfying: $\bar{x}_i = (M/M-1)^i$ , $i = 0, 1, \ldots$.*

Note that as $M$ becomes large, the minimax value $\bar{V}$ satisfies

$$\frac{\bar{V}}{M} = \frac{1}{M} + 2\left(\frac{M}{M-1}\right)^M \to 2e \doteq 5.4. \tag{9.28}$$

We now discuss optimal (mixed) strategies. It is very reasonable that the results presented in Section 8.3 can be extended to $M$ rays as was conjectured by Gal (1980, p. 172). Specifically, the optimal (mixed) strategy should be periodic and monotonic with $x_i = \bar{a}^{i+u}$, where $u$ is uniformly distributed in $[0, M)$ and $\bar{a}$ minimizes

$$\begin{aligned} \hat{c}(s_{\bar{a}}, H) &= 1 + \frac{2}{|H|} E\left(\sum_{j=-\infty}^{i+M-1} a^{j+u} \mid a^{i+u} < |H| \leq a^{i+M+u}\right) \\ &= 1 + \frac{2(a^M - 1)}{M(a-1)\ln a}. \end{aligned} \tag{9.29}$$

Kella (1993) proved this conjecture under the assumption that, against symmetric hiding distribution (i.e., the same probability distribution for all the $M$ rays) there exists an optimal search trajectory, which is *periodic and monotonic*. His proof is an extension of the approach used by Beck and Newman (1970) to $M > 2$.

Note that if $M = 2$ (the line search considered by Beck and Newman), then the optimality of periodic and monotonic trajectories is obvious. However, proving this property for $M > 2$, which is the step needed to complete the proof of Gal's conjecture, may involve many technical details.

Kella (1993) used the fact that $\bar{a}$ minimizes (9.29) to show that as $M \to \infty$, $\bar{a} \sim 1 - \theta/M$, where $\theta (\doteq 1.59)$ is the unique positive root of

$$2(1 - e^{-x}) - x$$

and that

$$\frac{v}{M} \sim \frac{2}{\theta(2-\theta)} \ (\doteq 3.09). \tag{9.30}$$

Note that if $M$ is large, then (9.28) and (9.30) imply that $v/\bar{V} \doteq 0.57$, while if $M = 2$ then the results of Sections 8.2.1 and 8.3 show that $v/\bar{V} \doteq 0.51$. Thus, the expected relative gain of using mixed strategies is about the same (albeit a little smaller for large $M$).

## 9.3 Search on the Boundary of a Region

In the paper "*Searching in the plane*," Baeza-Yates et al. (1993) discuss several search problems. Most of these problems were already considered by Gal (1980), but they also presented the following new search problem: We look for a target (point), which lies on a continuous curve that bisects the plane into two halfplanes. We assume that the searcher can shorten his path by moving off the curve (in the plane). The curve considered in the 1993 paper is two rays, radiating from the origin $O$, for which the following *bow–tie search* algorithm was presented: Walk along one ray some distance, walk in the plane, in a straight line, to the second ray (to the last visited point) walk along the second ray some distance, then return to the last point of departure on the first ray and repeat. Baeza-Yates et al. present the best bow–tie search with exponentially increasing distances on the rays and ask whether this search trajectory is optimal (among all possible trajectories). We shall show that it is indeed optimal.

The above described problem is a generalization of the search on the infinite line with the angle between the two rays, $\theta$, satisfying $0 < \theta < \pi$. (Note that $\theta = \pi$ is the LSP.) Let $[x_i, x_{i+2}]$, $-\infty < i < \infty$ be the segment walked along the first ray if $i$ is even and along the second ray if $i$ is odd (see Figure 9.2).

It is easy to see that if the target is discovered in the $I + 2$ stage at $H \in [x_I, x_{I+2}]$, then the time to reach $H$ is

$$C(S, H) = |H| + x_{I+1} + \sum_{j=-\infty}^{0} \sqrt{x_{I+j}^2 + x_{I+j+1}^2 - 2x_{I+j}x_{I+j+1}\cos\theta},$$

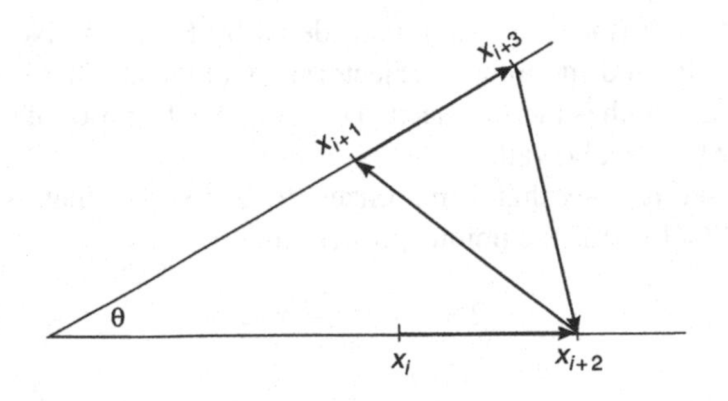


**Figure 9.2.**

and the normalized cost is

$$\hat{C}(S,H) = 1 + \frac{x_{I+1} + \sum_{j=-\infty}^{0} \sqrt{x_{I+j}^2 + x_{I+j+1}^2 - 2x_{I+j}x_{I+j+1}\cos\theta}}{|H|}.$$

Since the worst case occurs if the target is discovered in the $I+2$ stage with $H = x_I + \varepsilon$ for some $I$, it follows that

$$\inf_S \sup_H \hat{C}(S,H) = 1 + \inf_S \sup_I \frac{x_{I+1} + \sum_{j=-\infty}^{0} \sqrt{x_{I+j}^2 + x_{I+j+1}^2 - 2x_{I+j}x_{I+j+1}\cos\theta}}{x_I}.$$

This leads us to the functional

$$F(X) = \frac{x_1 + \sum_{j=-\infty}^{0} \sqrt{x_j^2 + x_{j+1}^2 - 2x_j x_{j+1}\cos\theta}}{x_0}. \tag{9.31}$$

In order to use the general minimax theorems of Chapter 7 (Theorem 7.9) we have to show that

$$F_k(X) = \frac{x_1 + \sum_{j=-k}^{0} \sqrt{x_j^2 + x_{j+1}^2 - 2x_j x_{j+1}\cos\theta}}{x_0}$$

is unimodal. ($F_k(X) = F_k(x_{-k}, \ldots, x_0, x_1)$ is obviously a homogenous continuous functional in the arguments $x_{-k}, \ldots, x_0, x_1$.)

We will need the following inequality

$$\begin{aligned}&\sqrt{(x_j+y_j)^2 + (x_{j+1}+y_{j+1})^2 - 2(x_j+y_j)(x_{j+1}+y_{j+1})\cos\theta}\\ &\quad \le \sqrt{x_j^2 + x_{j+1}^2 - 2x_j x_{j+1}\cos\theta} + \sqrt{y_j^2 + y_{j+1}^2 - 2y_j y_{j+1}\cos\theta},\end{aligned} \tag{9.32}$$

which is actually the triangle inequality (with the Euclidean norm) for the vectors

$$\left((x_j - x_{j+1})\cos\frac{\theta}{2}, (x_j + x_{j+1})\sin\frac{\theta}{2}\right) \quad \text{and}$$

$$\left((y_j - y_{j+1})\cos\frac{\theta}{2}, (y_j + y_{j+1})\sin\frac{\theta}{2}\right).$$

The geometric meaning of (9.32) is that in Figure 9.3 $a_{12} \le a_1 + a_2$. (Actually, the geometric construction depicted in this figure provides an alternative simple proof.)

We can now prove the unimodality condition (7.7)

$$F_k(X+Y) \le \max(F_k(X), F_k(Y)).$$

Assume that $F_k(X) \le \gamma$ and that $F_k(Y) \le \gamma$. Then

$$\begin{aligned}&x_1 + \sum_{-k}^{0} \sqrt{x_j^2 + x_{j+1}^2 - 2x_j x_{j+1}\cos\theta} + y_1 + \sum_{-k}^{0} \sqrt{y_j^2 + y_{j+1}^2 - 2y_j y_{j+1}\cos\theta}\\ &\le \gamma(x_0 + y_0).\end{aligned} \tag{9.33}$$

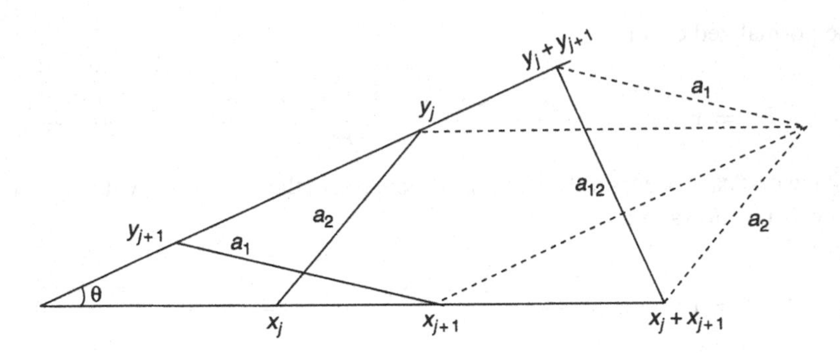


**Figure 9.3.**

By (9.32), the LHS of (9.33) is greater than or equal to

$$x_1 + y_1 + \sum_{-k}^{0} \sqrt{(x_j + y_j)^2 + (x_{j+1} + y_{j+1})^2 - 2(x_j + y_j)(x_{j+1} + y_{j+1}) \cos\theta},$$

which implies that $F_k(X + Y) \leq \gamma$.

Also, it is easily checked that both sides of (7.8) are $\infty$ as $a \to \infty$, since we are restricted to $a > 1$ we do not need the condition (7.9) as $a \to 0$.

Now, since all the terms in (9.31) are nonnegative, the monotonicity conditions (7.20) $(F_{k+1} \geq F_k)$ obviously holds. Thus, by (7.27) it follows that

$$\inf_S \sup_I \frac{x_{I+1} + \sum_{j=-\infty}^{0} \sqrt{x_{I+j}^2 + x_{I+j+1}^2 - 2x_{I+j}x_{I+j+1} \cos\theta}}{x_I}$$

$$= \inf_{a>1} \left( a + \sum_{j=-\infty}^{0} \sqrt{a^{2j} + a^{2(j+1)} - 2a^{2j+1} \cos\theta} \right)$$

$$= \inf_{a>1} \left( a + \sum_{j=-\infty}^{0} a^j \sqrt{1 + a^2 - 2a\cos\theta} \right)$$

$$= \inf_{a>1} \left( a + \frac{a}{a-1} \sqrt{1 + a^2 - 2a\cos\theta} \right). \tag{9.34}$$

The precise formula for the optimal $a$ is given in Baeza-Yates et al. (1993).

Note that if $\theta$ is small, then the optimal $a$ is $1 + \varepsilon$ with a small $\varepsilon > 0$. In the extreme case $\theta = 0$, the two rays coincide, and the problem is equivalent to searching on both sides of a street, assuming that the time to cross the street is negligible. In this case it is clear that the best plan is to use very small steps because there is no overhead when switching from one side of the street to the other side. The situation is entirely different when $\theta = \pi$. In this case the problem is equivalent to the search on the infinite line, discussed in Chapter 8, and the minimax $a$ is 2.

We have shown that a *geometric* bow–tie trajectory (with $a$ given by (9.34)) is optimal among all bow–tie trajectories. The dominance of the search pattern of bow–tie over other possible patterns follows from the fact that the worst cases occur when the target is at $x_i + \varepsilon$ so that it is always optimal for the searcher to go from point $x_i$ (on one ray) to $x_{i-1}$ (on the other ray) in the shortest possible path.

## 9.4 Search for a Point in the Plane

In this section, we use the continuous minimax result of Section 7.2.1 in order to find the minimax search trajectory for the following search problem in the plane. The (immobile) hider is located at a point in the plane, $H = (|H|, \beta)$, where $|H|$ is the distance from the origin $O$ and $\beta$ is the angle of $OH$ (with respect to the horizontal axis). The searcher chooses a trajectory starting from the origin. He discovers the hider at the moment when $H$ is covered by the area swept by the *radius vector* of his trajectory (see Figure 9.4).

We shall consider search trajectories $X(\gamma)$, which have the following "periodic and monotonic" property, i.e., the angle $\gamma$, $-\infty < \gamma < \infty$, of the radius vector $X(\gamma)$, is always increasing, and $X$ also satisfies $X(\gamma + 2\pi) \geq X(\gamma)$ for all $\gamma$.[1] In addition, we assume that $X'(\gamma)$ is piecewise continuous and bounded on any finite interval. (Note that here we present the search trajectory as a function of the angle not the time.)

The capture time $C(S, H)$ is the length of the trajectory traveled by the searcher until the hider is discovered. Thus,

$$C(S, H) = \int_{-\infty}^{\gamma(S,H)} \sqrt{X^2(\gamma) + X'^2(\gamma)}\, d\gamma$$

the upper limit $\gamma(S, H)$ of the integral being equal to the value of $\gamma$ at which the hider is to be discovered. The part of the trajectory used until the hider is found, is illustrated in Figure 9.4.

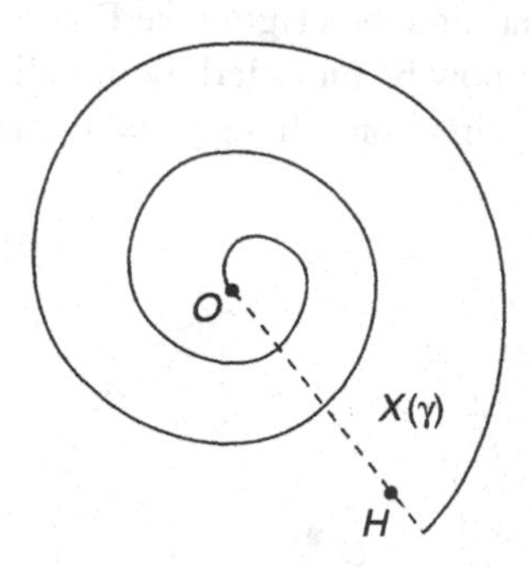


**Figure 9.4.**

[1] The result is probably valid even without this assumption. The required step is a continuous extension of the results in Section 9.2 that a minimax trajectory for the star search has to be periodic and montonic, a task that seems to be quite complicated.

Using the normalized cost function

$$\hat{C}(S, H) = \frac{C(S, H)}{|H|},$$

we wish to find a trajectory $S(\gamma)$ that minimizes $\sup_H \hat{C}(S, H)$. Thus we are interested in

$$\inf_X \sup_H \frac{\int_{-\infty}^{\gamma(S,H)} \sqrt{X^2(\gamma) + X'^2(\gamma)}\, d\gamma}{|H|}. \tag{9.35}$$

We note that $|H| > X(\gamma(S, H) - 2\pi)$, with the ($\varepsilon-$) worst case:

$$|H| = X(\gamma(S, H) - 2\pi) + \varepsilon$$

for some $\gamma(S, H)$.

Thus, (9.35) is equivalent to

$$\inf_X \sup_\tau \frac{\int_{-\infty}^{\tau} \sqrt{X^2(\gamma) + X'^2(\gamma)}\, d\gamma}{X(\tau - 2\pi)}. \tag{9.36}$$

Problem (9.36) is has the solution given by (7.31) (see Section. 7.2.1). Thus, the minimax solution is

$$\inf_{-\infty<b<\infty} \int_{-\infty}^{2\pi} \sqrt{e^{2b\gamma} + b^2 e^{2b\gamma}}\, d\gamma = \min_{b>0} e^{2\pi b} \sqrt{1 + \frac{1}{b^2}}. \tag{9.37}$$

Let $\bar{b}$ ($\bar{b} \doteq 0.16$) be the value of $b$ that minimizes (9.37). Then the minimax search trajectory is $X(\gamma) = \alpha e^{\bar{b}\gamma}$, where $\alpha$ is a positive constant. (Note that $\alpha = 1$ and $\alpha = e^{2\pi j \bar{b}}$, where $j$ is any integer, represent the same trajectory.) The solution is determined up to a multiplicative constant as in all the other functionals considered in Chapter 7.

It should be noted that, similarly to the previous problems, the spiral $X(\gamma)$ starts with small "oscillations." These oscillations can be avoided by assuming that the object is outside a circle of radius $\varepsilon$ around the origin (see Figure 9.5).

The optimal trajectory can now be modified by initially moving a distance $\varepsilon$ along a radius of this circle, in the direction where $\gamma = 0$ and then proceeding by using $X(\gamma) = \varepsilon e^{\bar{b}\gamma}$ for $\gamma \geq 0$.

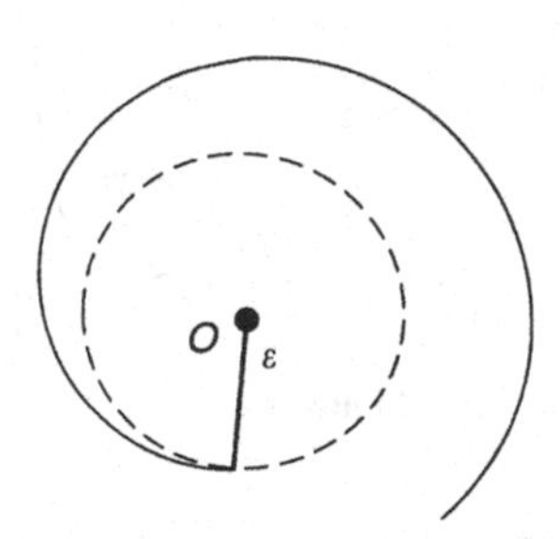


**Figure 9.5.**

**Remark 9.6** *The search game just described takes place in the plane. However, the object sought is not a point but an infinite ray. (This ray is represented, in polar coordinates, by* $(\alpha|H|, \beta)$, $\alpha \geq 0$.*) Thus, the preceding problem is essentially one dimensional. This explains the fact that the capture time is normalized by* $|H|$ *(see* (9.35)*) and not by* $|H|^2$.

## 9.5 "Swimming in a Fog" Problems

The type of problem to be described in this section was presented by Bellman (1956) as a research problem. A narrative-description of this problem is the following. A person has been shipwrecked at a point $O$ in a fog and wishes to minimize the maximum time required to reach the shore, given its shape and some information about its location around $O$. Similar problems can be considered in which the distribution of the location of the boundary (shore) line is given and one would like to minimize the expected time to reach it.

Gross (1955) considered the following formulation of that problem: Find a shortest plane trajectory with the property that if the origin of the trajectory is covered in any way by a given plane figure, some point of the trajectory lies on the boundary of the figure. He presented a discussion about the nature of the solution for the cases in which this figure is the circle, the equilateral triangle, a "keyhole"-shaped figure, and the infinite strip of unit width.

Isbell (1957) found the trajectory, which guarantees reaching an infinite line with unit distance from $O$ in minimum time, and briefly considered a two-line problem. The minimax search trajectory for a line at unit distance as described by Isbell (1957) is shown in Figure 9.6.

Being at $O$, imagine a clock face (Figure 9.6). Walk toward 1 o'clock for $\sqrt{4/3}$ units. Then turn on the tangent that strikes the unit circle at 2 o'clock. Follow the circle to 9 o'clock and continue on a tangent. Upon striking the line that is tangent to the unit circle at 12 o'clock, all the tangents to the unit circle have been swept. The maximum distance traveled in this minimax trajectory is $7\pi/6 + \sqrt{3} + 1 \doteq 6.4$ units.

Gluss (1961a, 1961b) presented a solution of the minimax search for a circle with a known radius and a known distance from $O$, and an approximate solution for minimizing the expected distance traveled in the case that the shore is a line of unit distance from

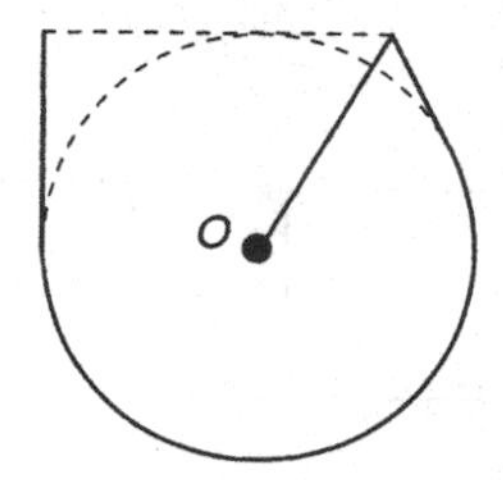


**Figure 9.6.**

$O$ uniformly distributed around $O$. An extensive survey on this problem is presented by Finch (1999).

The problem of swimming in the fog with a compass was considered in a general setting by Hassin and Tamir (1992). The problem can be formally described as follows: Given an open bounded set $Q$ in $R^2$ find a path $S$ of minimum "size" such that for every $z \in Q$ the set $\{z\} + S$ intersects the boundary of $Q$.

They consider two models with different size function. The first model is of a fisherman (say), in a small boat, lost at a big lake in a thick fog. The visibility is zero but the fisherman can navigate by dead reckoning. At each point in time he can choose a direction and a distance to travel. He has a map of the lake. His objective is to choose a path $S$ from his unknown starting point that minimizes the maximum distance that he must travel to shore. Here, the size of $S$ is its length.

In the second model a soldier is clearing a path outside of a mine-field under the same assumptions of the first model (zero visibility, unknown starting point, having a compass and a map) except that the objective now is to find a path $S$ with minimum (one-dimensional) measure.

The optimal solution of the two problems can be different, as the following simple example (Figure 9.7) shows. In this example, the minimal length (Figure 9.7*a*) is 2, while the minimal measure (Figure 9.7*b*) is only $1 + \sqrt{3}/2 \doteq 1.866$.

However, Hassin and Tamir show that if $Q$ is convex, then the optimal solution for both models is the same with the objective function equal to the width of $Q$, i.e., the minimum distance between a pair of distinct parallel lines that bound $Q$.

The minimax search for a line can be considered within the framework described in Section 6.1. We obtain the following research problem. Let $|H|$ be the distance of the line from the origin. Given the information that the expectation of $|H|$ satisfies $E|H| \leq 1$, what are the minimax search trajectory (pure strategy) and the optimal search strategy (mixed) of the searcher?

In contrast to the solution obtained by Isbell (1957), we expect a smooth minimax searching trajectory. If we could show that a result similar to the continuous theorem (7.31) of Section 7.2.1 holds for this problem, then the minimax trajectory would be an exponential spiral, and one could simply follow the technique presented in Section 9.4. However, the functional involved does not seem to satisfy the unimodality condition. Thus, an extension of those results is needed.

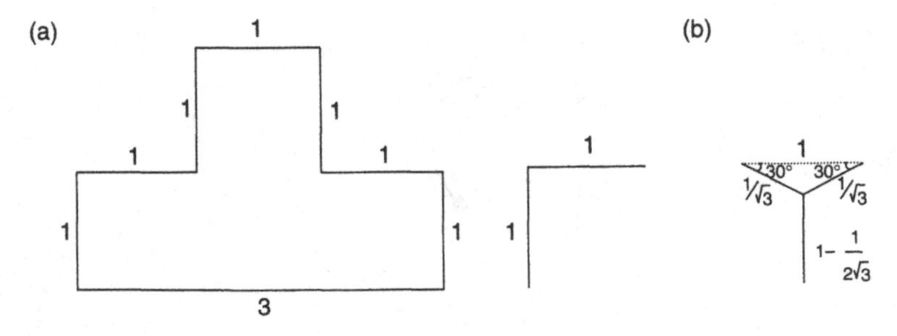


**Figure 9.7.**

## 9.6 Searching for a Submarine with a Known Initial Location

Assume that the search space is the entire plane. A submarine (the hider) starts moving, with speed not exceeding $w < 1$ at time $t = 0$ from the origin $O$. The searcher starts moving at $t = 0$ from a fixed point different from the origin (known to the hider). We make our usual assumptions that the maximal speed of the searcher is 1, that his discovery radius is $r$, and that he has no information about the path chosen by the submarine, except of its starting point. The searcher wins if and only if he captures the submarine at some time $t$. What are the optimal strategies of both players and what is the probability of ultimate capture?[2]

It is easy to verify that the maximin probability $P$ of ultimate capture satisfies

$$\begin{aligned} P &\geq 1 - \prod_{i=0}^{\infty}\left(1 - \frac{r}{\pi R}\frac{1-w}{w}\left(\frac{1-w}{1+w}\right)^i\right) \\ &\geq 1 - \exp\left(-\frac{r}{2\pi R}\frac{1-w^2}{w^2}\right), \end{aligned} \tag{9.38}$$

where $R$ is the distance of the searcher's starting point from the origin. The searcher can achieve the probability of capture appearing in (9.38) at each of the time instants

$$t_i = R\left(\frac{1+w}{1-w}\right)^i, \quad i = 0, 1, \ldots.$$

picking a random angle $\gamma_i$, $0 \leq \gamma_i < 2\pi$, where $\gamma_i$ has a uniform probability distribution and is independent of $\gamma_j$, $j < i$, moving in a straight line a distance $R_i = (w/1 - w)t_i$ from the point $O$, and moving back to the point $O$.

**Remark 9.7** *If*

$$\frac{Rw^2}{r} \ll 1,$$

*then the bound in* (9.38) *is close to* 1. *Thus, the above policy is nearly optimal for this case. It seems that this policy is nearly optimal in general, but this is still an open problem.*

It can be seen that it is not a good policy for the hider to move in a straight line using his maximal velocity. (In this case the searcher can capture him for sure by following an easily calculated trajectory.) A policy that does seem to be good for the hider is to move randomly for a certain period of time and only then to move in a straight line using his maximal velocity. This observation seems also true for the infiltration problem (with no safe zone) described in Section 5.3.

**Remark 9.8** *Thomas and Washburn (1991) considered 'dynamic search games' in which the hider starts moving at time* 0 *from a location (cell) known to the both players,*

[2]This problem was presented to S. Gal by Rufus Isaacs; problems with a similar flavor were considered by Koopman (1946) and Danskin (1968). See also Remark 9.7.

*while the searcher starts after a time delay known to both players. The distinguishing feature of this game is that for each $t$ the hider knows the trajectory of the searcher until time $t$. A classic example of this type is the still unsolved Flaming Datum problem in which a helicopter attempts to detect or destroy a submarine that has recently revealed its position by torpedoing a ship (see also Danskin, 1968; Koopman, 1946, and 1980). Search/Evasion games involving a submarine (the evader), which tries to penetrate through a line guarded by the searcher, were considered by Agin (1967), Arnold (1962), Beltrami (1963), Houdebine (1963), Langford (1973), Lindsey (1968), and Pavillon (1963).*

# Book II

# RENDEZVOUS SEARCH

# Chapter 10

# Introduction to Rendezvous Search

The *rendezvous search problem* asks how two unit speed searchers, randomly placed in a known search region $Q$, can minimize the time required to meet. Although originally posed informally by the first author (Alpern, 1976) about twenty five years ago, this problem has started to receive attention only in the last dozen years. The rendezvous search literature began with the pioneering article of Anderson and Weber (1990) on discrete location rendezvous and the general continuous formalization of Alpern (1995). Since then, the field has been expanding rapidly and has even received attention outside the academic literature (Mathews, 1995; Alpern, 1998). Book II (Rendezvous Theory) attempts to cover in detail most of the major developments in this area, expanding on the more concise treatment given in the survey article (Alpern, 2002a) on which it is based.

The rendezvous problem converts the min-max objectives of the search games considered in Book I into a min-min objective in which two players *both* wish to meet in least expected time. This problem may arise in practice when two people shopping in a supermarket (Alpern, 1998) realize that they have become separated, when parent penguins return with food for their offspring in a large colony, when two partners have to meet up after separately parachuting from a plane, or when rescuers search for a lost hiker who wants to be found. The rendezvous that is sought need not be purely spatial in nature, however. Two people with walkie-talkies containing a choice of several frequencies (channels) may be viewed as rendezvousing on channel space. In this sense the SETI (Search for Extraterrestrial Intelligence) program may be seen as a rendezvous problem in which strategies involve searching with radio telescopes and choosing both where to search and what frequency to search on.

In some ways rendezvous problems resemble their zero-sum search game predecessors: the dynamics of motion (unit bounds on the speeds), the objective function (the meeting, or capture time, $T$), the known domain of search $Q$. However, there are several new aspects that are present in rendezvous search theory that did not arise in the earlier search games. In zero-sum search games the possibility of communication or joint planning between the players could be safely ignored. However, in the case of rendezvous we must first distinguish two very different versions, depending on whether or not the players can meet in advance of play (before getting separated) to agree on the strategies each will adopt if rendezvous proves necessary. If they can meet in advance, then the optimization problem will allow them to choose *distinct* strategies (e.g., the *wait for mommy* strategy pair where one agrees to wait while the other carries out an optimal exhaustive search). This version is called the *asymmetric* (or player-asymmetric) rendezvous problem, and the associated least expected meeting time is called the *asymmetric rendezvous value*, denoted $R^a$. On the other hand, players may find themselves facing the rendezvous problem without having established a joint strategy. In this case the question we ask is what strategy should we advise the players to adopt, knowing that we have to give them both the *same* advice. For example, we may know that there are two hikers in a forest who need to meet (say, one has water, and the other food), and that they both have radios, but we don't know their names. So we cannot broadcast to them individual, distinct, instructions ("John do this, Jane do that"). Or we may have to write our advice in a handbook carried by all hikers. These cases are known as the *symmetric* (or player-symmetric) rendezvous problem, and the (generally larger) least expected meeting time is called the *symmetric rendezvous value*, denoted $R^s$. We note that in this version of the problem, the strategy that we advise both players may have to be a mixed strategy, involving independent randomization for the two players. (For example, if the search space is the line and the players are initially pointed in the same direction, then if they follow the same *pure* strategy, they will always remain the same distance apart and will never meet.)

Another essential ingredient in rendezvous problems is the amount of spatial symmetry possessed by the search space $Q$ and more important, the degree to which the players can "see through" this symmetry by common notions such as direction. For example, player-asymmetric rendezvousers on a circle might adopt the strategy of one going clockwise while the other goes counterclockwise. However, this is only possible only if they have a common notion of clockwise. Once this problem is appreciated, it can often be dealt with on an *ad hoc* basis by appropriately defining the set of feasible rendezvous strategies. However, the formalities required for a rigorous treatment of spatial symmetry problem involve the specification of a group $G$ of symmetries of $Q$, as spelled out in detail in Chapter 12. The reader may ask why the spatial symmetry problem did not arise in Book I. The answer is that in a zero-sum setting, if *either* player randomizes his motion with respect to available symmetry (e.g., equiprobably choosing between clockwise and counterclockwise in the circle), then it is the same as if *both* players do so. Consequently if the spatially symmetric form of the game (e.g., uniform initial location or equiprobable direction on the circle) is advantageous to either player, this is the form that will be played. In addition, it is usual to start the search games of Book I with a common knowledge initial location for the searcher. In this case the players might lack only a common sense of direction.

Much of the basic rendezvous theory involves the determination or estimation of the rendezvous values $R^a(Q, G)$ or $R^s(Q, G)$. These numbers represent the least expected time for two players to meet given that they are placed randomly in a known search space $Q$ and can choose distinct ($R^a$) or common ($R^s$) strategies among those that are feasible, given their common spatial notion summarized by the group $G$. Upper bounds on the rendezvous values $R^a$ and $R^s$ can be obtained by evaluating the expected meeting times for specific rendezvous strategies. Lower bounds require more subtle methods. In addition, rendezvous theory seeks to determine the (optimal) strategies that lead to these minimal meeting times.

To give the reader something a little bit more specific than the above generalities, we present the two motivating problems originally proposed in (Alpern, 1976). Both of these are player-symmetric problems, in that a common strategy for both players is required.

**Problem 10.1 (Astronaut Problem)** *Two astronauts land on a spherical body that is much larger than the detection radius (within which they can see each other). The body does not have a fixed orientation in space, nor does it have an axis of rotation, so that no common notion of position or direction is available to the astronauts for coordination. Given unit walking speeds for both astronauts, how should they move about so as to minimize the expected meeting time $T$ (before they come within the detection radius)?*

**Problem 10.2 (Telephone Problem)** *In each of two rooms, there are $n$ telephones randomly strewn about. They are connected in a pairwise fashion by $n$ wires. At discrete times $t = 0, 1, 2, \ldots$ players in each room pick up a phone and say "hello." They wish to minimize the time $T$ when they first pick up paired phones and can communicate. What common randomization procedure should they adopt for choosing the order in which they pick up the phones? (This problem is equivalent to spatial rendezvous on a complete graph.)*

It is interesting to note that while much progress has been made in rendezvous search over the last decade, these two problems that motivated the field are largely unsolved. Almost nothing has been done on the Astronaut Problem (although some work on planar rendezvous is discussed in Chapter 18). The Telephone Problem has received considerable attention, particularly in the pioneering paper of Anderson and Weber (1990), who obtained good upper bounds on the symmetric rendezvous value.

## 10.1 Relation to Coordination Games

Any work on rendezvous must mention the famous political science monograph, *The Strategy of Conflict* (Schelling, 1960), which initiated the discussion of coordination problems. Schelling considered the one-shot problem faced by two players who wish to meet at a common location. Each player makes a single choice of location, and if they fail to meet at the first try, then there are no more chances. He emphasized the importance of certain "focal points" that may stand out as prime candidates to players with similar cultural backgrounds.

As a practical matter, no one can deny the significance of common backgrounds and cultural conventions in aiding coordination, even of players who have never met before. Our treatment of rendezvous problems differs greatly, however, from that of Schelling. First, we normally (except for Chapter 13) will exclude the possibility of using focal points by assuming that the players have no common labeling of locations, usually due to symmetry properties of the search region $Q$. Second, our model is dynamic: early failure does not end the game because the players are allowed to search until they find each other. For these reasons we do not view Schelling's problems as true rendezvous *search* problems, although they certainly lie within the larger area of coordination games. In fact, we will briefly discuss certain dynamic extensions of Schelling's problem in Section 15.3.1.

## 10.2 Real-Life Rendezvous

The rendezvous problem has been faced by animals and people in the real world long before it was analyzed by mathematicians. In this section we review some examples that show the diversity of rendezvous situations. These actual problems may serve as a basis for models that try to incorporate various realistic features. We begin with animal players and go on to human players.

1. **Regrouping of animal kin groups** Animals that have fixed nests avoid the rendezvous problem. However, for migratory or undersea animals, such focal points may be lacking. Families or herds that need to periodically regroup under such circumstances may need advanced rendezvous strategies.

   Antarctic penguins live in very large colonies, and parents need to find their offspring when returning with fish (to regurgitate) for them. In these colonies there are no fixed family nesting sites, and after fishing parents apparently return to the last place they saw their offspring to begin their search. The spatial elements of this search have not received much attention, but the vocal recognition aspect has been studied by Aubin and Jouventin (1998) to determine the what we call the detection radius:

   > The king penguin, *Aptenodytes patagonicus*, breeds without a nest in colonies of several thousand birds. To be fed, the chick must recognize the parents . . . chicks can discriminate between the parental call and calls from other adults at a greater distance [than acoustic calculations would suggest]. This capacity termed . . . "cocktail party effect" . . . enhances the chick's ability to find its parents. . . . In the genus aptenodytes, the difficulty is enhanced because there is no nest . . . few if any landmarks to help in finding their partner.

2. **Search for a mate** For most animal populations, mating is more a problem of female choice and male competition than of spatial rendezvous. However, for some species with a low spatial density, simply *finding* a suitable (same species, opposite sex) partner is a nontrivial rendezvous problem. For example, R. Lutz and J. Voight are quoted in The Times (Issue 65084) that for the Octopus "the chances to mate are so rare . . . that they cannot risk missing any. . . . Living in the depths of the ocean, they seldom come across members of their own species and their lives are short."

   Griffiths and Tiwari (1995) describe a specific case of this type of problem played out by two birds, when a Spix's macaw (*Cyanopsitta spixii*) from a captive breeding program was released in the vicinity of the last observed wild bird of this species. The sex of the wild bird was determined to be male by DNA testing of a feather found by researchers, and a macaw of the opposite sex was released. The two birds faced a rendezvous problem on a line, "woodlands that are associated with the seasonal watercourses characteristic of the right bank of the Rio São Francisco" (Juniper and Yamashita, 1990). We do not know, however, whether males and females of this species have distinct strategies for searching for a mate, so the player-symmetric form is probably the appropriate model. By the way, the birds did rendezvous successfully. They were later seen flying together!

3. **Regrouping of hikers or ships at sea** When focal points are available but the group is moving, both the ordering and current location of a set of focal points must be known. Before radio, ships traveling in convoy along a land mass would signal when the next in an agreed sequence of identifiable shore points was operative. When traveling along a new land path, sometimes even the sequence of focal points cannot be agreed in advance. According to McNab (1994, p. 96), when going in a widely spaced single file patrol during the Gulf War, his SAS unit established and communicated the new rendezvous point (ERV) in the following manner.

   Every half hour or so we fixed a new ERV (emergency rendezvous), a point on the ground where we could regroup if we had a contact and had to withdraw swiftly. If we came to a prominent feature like a pile of old burial ruins, the lead man would indicate it as the new ERV by a circular motion of the hand and this would be passed down the patrol.

   Following a query on rendezvous at sea to a website specializing in nautical history, the following two useful email replies were obtained.

   Back in the late-16th, early-17th century, they [separated convoys] didn't always meet up again until they reached a pre-arranged rendezvous island, and sometimes not even then. From limited reading of early Dutch voyages to the Indies it seems that it was not very unusual for ships to become separated from their fleets in stormy weather at night, particularly if they suffered rigging damage and their progress became unpredictable to those who did not know of the damage. The jacht(s) of the fleet would search for the missing vessel while the rest of the fleet continued – the jachts were faster and could sail greater distance and catch up to the fleet (Burningham, 1999).

   My period of specialization is the American Civil War. Vessels that needed to rendezvous did so by pre-arrangement. Both sides in that war sometimes set up a pattern of rendezvous points which were to be followed in order, if not visited on a particular date. The practice was flawed because unforeseen delay or detention of one of the two vessels intended to meet might force a warship to wait for weeks at a particular location for a planned rendezvous. This time could have been more prosperously spent hunting prizes or running from pursuers. For instance the commerce raider CSS ALABAMA was to meet her collier AGRIPPINA in a series of ports for periodic refueling, but after the first several meetings AGRIPPINA's merchant captain took off for more interesting parts, leaving ALABAMA to her own devices for fuel. The accounts do not say what was done to LOOK for the other vessel at each place but rather seem to have consisted of steaming or sailing about the point of rendezvous for a while until the other vessel showed up (Foster, 1999).

Rendezvous search is also practiced by robots that are programed with strategies to help them meet. See the book of Dudek and Jenkin (2000) or the article of Roy and Dudek (2001) for more on robot rendezvous.

## 10.3 Rendezvous Strategies

Much of Book II is about rendezvous strategies. Often the strategies that are optimal in a given context are quite complicated and very specific to the context. However, certain types of generic strategies seem to occur very often and in a wide variety of contexts. We mention some of these here.

To help put the reader in the frame of mind of a rendezvouser, we quote the following exchange between Millie and Densher in the film version of Henry James' "Wings of the Dove," which recounts their mental states in their supposed earlier failed attempt to meet without prior planning (symmetric rendezvous) in Venice.

> "I thought I might see you wandering around."
> "I thought the same."
> "Where did you go?"
> "San Marco, the Rialto, all the places we went together."
> "Maybe we just kept missing each other, me turning the corner just as you went the other way."
> "I thought that, so I stayed in the same place and waited for hours."
> "And we still didn't see each other."

This passage illustrates a number of useful concepts. First, there is the real-life use of focal points (San Marco, the Rialto) based on cultural conventions and shared history. This is a strategic aspect of rendezvous, mentioned earlier in our discussion of Schelling, that we will normally exclude by making the search space spatially symmetric. However, in the case where the two players have a common labeling of points in the search space $Q$, a FOCAL strategy is feasible. This is simply an agreement to meet at an agreed point. That is, to go directly to that point and then wait there. While such strategies are in general not optimal, we may consider the problem of determining the FOCAL strategy with least expected meeting time (see Section 11.3).

The important idea of "staying in the same place" (we call this "waiting") arises in the above passage. In a player-asymmetric situation, a natural strategy pair to consider is what we call the *Wait For Mommy (WFM)* strategy, where one player (Child) stays still while the other (Mommy) carries out an exhaustive search, usually designed to find a stationary object in least expected time. (Apparently our name for this strategy is not so far-fetched, as it has been observed that the mother kangaroo teaches its baby that if they are separated it should find the nearest bush and stay under it.) Sometimes if the Child knows that Mommy is doing such a search and may reach him at certain known times, he can decrease the expected meeting time by moving so as to meet an approaching Mommy (if lucky), while returning to his start at all the known times. This type of generic strategy is called *Modified Wait For Mommy (MWFM).*

It turns out that WFM is rarely optimal. In fact it is never optimal when the search space $Q$ is the line (see Theorem 16.13 and Corollary 16.14). On the circle, it is indeed optimal when the players' initial placements are uniform and they have no common knowledge of direction (Corollary 14.11). But it is not optimal for any other initial placement (Theorem 14.12). However, it seems that the MWFM strategy is reasonably often optimal, as for the line when the initial distance $D$ between the players is known (see Definition 16.8 and Theorem 16.9). In this case the optimal strategy is for Mommy to go a distance $D$ in a random direction followed by $2D$ in the opposite direction, while Child first goes $D/2$ and back to start in a random direction; then $D$ and back in a random direction. The optimal strategy on a two dimensional grid found can also be viewed as a MWFM strategy (Theorem 18.1). Another variation on the WFM strategy is the *Alternating Wait For Mommy (AWFM)* strategy, in which the two players alternate between searching out possible initial locations of their partner and coming back to their own initial locations. This type of strategy is considered for rendezvous on an $n$-dimensional grid in Section 18.2. (The strategy described above, for two players on the line with known initial distance $D$ can also be viewed as an AWFM.)

Neither the WFM or MWFM strategies are feasible in the player-symmetric version of rendezvous search because they require an agreed allocation of the players to the two roles (Mommy, Child). However, if there is a constant time $\tau$ required for an exhaustive search of the search region $Q$, then the players may independently adopt the following *Randomized Wait For Mommy (RWFM)* strategy: In each period of length $\tau$, choose to exhaustively search with probability $p$ and to wait for the whole period with probability $1-p$. Note that if they make opposite choices in some period, then they have adopted the WFM and will definitely meet in that period. The probability of meeting when they both choose to search will depend on the search region $Q$. This type of symmetric strategy is used several times. See, for example, the strategy used in the proof of Theorem 12.4 (part 2), and the Anderson–Weber strategies $\pi_n$ for rendezvous on the complete graph $K_n$, as described in Section 15.3.

## 10.4 Outline of Book II

We now give a brief chapter by chapter outline of Book II, Rendezvous Search Theory.

Chapter 11 gives the reader an easy introduction to rendezvous by analyzing three specific easy versions of the rendezvous search problem. The problems that have been chosen require no general theory. They serve to indicate the main branches of rendezvous search and to motivate the subsequent material.

Chapters 12 through 15 (Part 3) are concerned with rendezvous on a compact search region. Chapter 12 is the most theoretical and formal. It gives a precise definition of the rendezvous values for a compact space and explains their dependence on the given group of symmetries of $Q$ that describes the common information available to the players. Some of this material can be skimmed in the first reading and returned to when required. Chapter 13 considers rendezvous on a labeled network. In most of Book II, it is assumed that the players have no common labeling system for the points of the search region $Q$. So for example FOCAL strategies are not feasible. However, in Chapter 13 only, we take the alternative assumption of a common labeling of points. The search region $Q$ is taken to be a continuous or discrete network. The special cases where $Q$ is the interval or the circle are considered. The analysis of rendezvous on the circle is continued in Chapter 14, without the assumption of a common labeling of points. The assumptions of a common notion of direction and a lack of such a notion are both considered. Chapter 15 considers some developments deriving from the Telephone Problem 10.2 mentioned earlier. This is the problem of discrete time rendezvous on a combinatorial graph. The players move to adjacent vertices in each time period, and are considered to meet when they first occupy the same vertex. It differs from the network analysis of the previous chapter in that here the players may transpose their positions on adjacent vertices without a meeting taking place.

The rendezvous problems mentioned above take place on a compact search region, and usually begin with an independent placement of the two players. The next three chapters (Part 4, Chapters 16 to 18) are concerned with rendezvous on an unbounded region. In these, the initial placement is usually arranged by specifying the distribution of the *vector difference* between the two players. Chapter 16 considers the asymmetric rendezvous problem on the line, denoted the ARPL. The play begins with Nature choosing the initial distance $d$ between the players from a known distribution $F$. The asymmetric rendezvous value $R^a(F)$ is determined (along with optimal strategies) for various classes of distributions $F$. The connection between rendezvous on the line and the related Double Linear Search Problem is exploited in this chapter. In Chapter 16, rendezvous search on the line is considered under various other assumptions such as unequal speeds for the two players, player symmetry (requiring a common mixed strategy for the players), bounded resources (each player has a bound on the total distance he can travel), an unknown distribution $F$ of the initial distance between the players, several players who must meet, and finally, the case where the initial location of Player I is known to Player II. Chapter 18 considers rendezvous in higher dimensions; most of the analysis is for the plane, but dimensions above 2 are also analyzed.

# Chapter 11

# Elementary Results and Examples

In this chapter we present some simple examples of rendezvous search that can be analyzed in a fairly self-contained manner. These examples will serve to introduce the reader to a sample of the variety of problems that will be considered in the rest of the book. In some cases we present here only the most preliminary results on a problem, indicating where further and stronger results can be found in subsequent chapters. The approach taken here will be more expository and less formal than in the later chapters.

## 11.1 Symmetric Rendezvous on the Line

Suppose that two players are placed on a road, a river, or more generally any search region that may reasonably be modeled as a line. Suppose further that they have not been able, prior to this placement, to discuss the roles that each will take in trying to find each other. This is what we call the player-symmetric, or simply symmetric, version of the rendezvous problem. We assume that each has a maximum speed of 1. What common advice should we give to the players to enable them to meet in least expected time? This advice might be in a book (maybe this one!), or maybe we would broadcast it over the radio (if we thought both would listen). In this problem neither player knows the direction to the other. In general (see Section 17.2), the initial distance between the players may be a random variable that depends on the method by which the players arrived on the line (or by which they became separated and lost). However, even the case we now consider, where the initial distance is a known constant $D$, presents significant difficulties. For simplicity, we will take $D = 2$. This distance might be known, for example, by the strength of some nondirectional signal that each might send to the other. However, in this setting we must assume that such a signal is not able to carry any information about direction, location, or planned strategy. So we may start the problem by placing one of the players (say, I) at the origin facing right, and the other (II) at either $+2$ or $-2$ and facing either left or right, with all four cases equiprobable.

The simplest symmetric rendezvous strategy for this problem is probably to go at unit speed 1 unit forward in a random direction and then back to your starting point, in each time period of length 2. With probability 1/4 (corresponding to both players going towards the other) the players meet 1 time unit into the period; otherwise they repeat the problem at the beginning of the next period. Hence the expected time $\hat{T}$ taken to meet satisfies the equation $\hat{T} = (1/4)1 + (3/4)(2 + \hat{T})$, so that $\hat{T} = 7$. One problem with this strategy is that when the players are back at their starting points at, say, time $t = 2$ (assuming they did not meet at time $t = 1$), they have information that they neglect to use – namely, that their partner is more likely to be in the direction not searched.

A better strategy for this problem is called the 1F2B strategy, which has period three: At the beginning of each period, choose a random direction to call forward (F). Go distance 1 in this direction and then distance 2 in the opposite backwards (B) direction. Repeat, with independent randomization, at the beginning of each period. All motion is carried out at speed 1. If both players adopt this strategy, how long will it take on average to meet?

To compute the average meeting time of the strategy 1F2B (adopted by both players), we first make the following observation: If they have not already met, the distance between the players at the beginning of period $k$ (time $t = 3(k - 1)$) is the same as their initial distance of 2. This is true by assumption for period $k = 1$. Assume it is true at the beginning of period $k$ and that they do not meet during this period. The assumption that they do not meet during period $k$ implies that they did not choose *opposite* directions as forward for this period because in that case they would either meet at time $3(k - 1) + 1$ (if their forward directions pointed toward each other) or time $3(k - 1) + 3$ (if their forward directions pointed away from each other). Consequently, they must have chosen the *same* direction as forward. This means they were moving in parallel throughout the period, at the same speed. So the distance between them remained constant, and is still 2 at the end of the period. This establishes the observation made at the beginning of the paragraph. Consequently, if they have not met at the end of the first period, their problem at time $t = 3$ is the same as their original problem. Thus if they both use this strategy, they meet at time $t = 1$ with probability 1/4, at $t = 3$ with probability 1/4, and they start the same problem again at time $t = 3$ with probability 1/2. Hence the expected meeting time $\hat{T}$ for the strategy 1F2B satisfies the equation

$$\hat{T} = \tfrac{1}{4}(1) + \tfrac{1}{4}(3) + \tfrac{1}{2}(3 + \hat{T}), \quad \text{with solution } \hat{T} = 5.$$

Since the symmetric rendezvous value $R^s$ for this problem is defined as the minimum of the expected meeting time, over all symmetric strategies, we have the following.

**Theorem 11.1** *The symmetric rendezvous value $R^s$ for the problem where the players are initially placed two units apart on a line and have no common notion of location or direction on the line, satisfies*

$$R^s \leq 5.$$

The upper bound of 5 can be improved and extended to known distributions (see Section 17.2) of initial distance or to unknown initial distance (see Section 17.4), and the problem can be analyzed for more than two players (see Section 17.5) and in higher dimensions (see Section 18.2.2). However, it is remarkable that even for the one-dimensional case where the initial distance is a known constant, determining the exact value of $R^s$ is still an open problem.

## 11.2 Multi-Rendezvous on the Circle

Some of the more interesting rendezvous problems consider the scenario where more than two players have to meet at a common location. Perhaps several players are needed to perform some task, such as putting up a tent or playing a baseball game. One such problem considers the optimal rendezvous strategy for $n$ players who are initially placed evenly around a circle and wish to minimize the time taken for all of them to meet. For multi-player problems of this type, information exchange is usually allowed between players who meet.

To make this problem more precise, we place the $n$ players $i = 1, \ldots, n$ consecutively at unit length arcs apart along a circle of circumference $n$. We specify that they do not have a common labeling of points on the circle or even a common notion of direction (e.g., clockwise). (The way in which these assumptions can be formalized using symmetry groups is described in Chapter 12.)

We consider the following strategy proposed in Alpern (2002a). It is to be used by each of the $n$ players, since we are considering the player-symmetric version. When the game begins, each player $i$ generates a random number $q_i$ uniformly from the unit interval and also sets his "first meeting time" $t_i$ to $\infty$. Subsequently $t_i$ will be set to the first time when Player $i$ meets another player. When play begins at time $t = 0$, each player follows a random walk with step length 1/2 until he meets another player. Since all motion is taken at speed 1, each step has time period 1/2 as well. When two players $i$ and $j$ meet, they follow these rules, depending on the relative size of their first meeting times $t_i$ and $t_j$ just prior to this meeting:

$t_i = t_j = \infty$: When it is a first meeting for both players, they both reverse their directions and reset both their $q$'s to $\min(q_i, q_j)$.

$t_i < t_j$: Player $i$ continues in the same direction he was going, and player $j$ sticks with $i$ and vanishes as a player, so that no future meetings with $j$ are considered.

$t_i = t_j < \infty$: The player with the larger value of $q$ follows the other one forever and vanishes as a player.

Let $T_0$ be the random variable describing the earliest time that a meeting occurs between two players. There will be at least two players who meet at time $T_0$ and maybe more. Of these, one will have the smallest value of $q$. Suppose this player $k$'s first meeting is with player $k+1$ at location $A$ at time $T_0 = t_k$. We will reset $q_{k+1}$ to the minimum $q_k$. So according to the rules given above these two players will proceed uninterrupted to the antipodal point $A'$ of $A$, and furthermore all other players will stick to one of these two. Thus all the players will be together at $A'$ at time $T_0 + n/2$, since $n/2$ is the time taken to travel from $A$ to $A'$. The expected value of $T_0$ is calculated as follows. We have $T_0 = 1/2$ except in the two cases that all the players initially go clockwise or they all go counterclockwise. Consequently, its expected value satisfies the equation

$$\hat{T}_0 = \frac{2^n - 2}{2^n}(1/2) + \frac{2}{2^n}\left(\frac{1}{2} + \hat{T}_0\right), \quad \text{or } \hat{T}_0 = \frac{2^{n-2}}{2^{n-1} - 1}.$$

So the expected meeting time is $\hat{T}_0 + n/2 = 2^{n-2}/(2^{n-1} - 1) + n/2$. Clearly, $(n - 1)/2$ is a lower bound on the $n$-player meeting time. If we denote by $R^s_{n,n}$, the least expected time for all $n$ of the original $n$ players to meet at a single point following the same (thus the $s$ for symmetric) mixed strategy, we obtain the following estimate showing that $R^s_{n,n}$ is asymptotic to $n/2$.

**Theorem 11.2** *The n-player symmetric rendezvous value $R^s_{n,n}$ for the circle is asymptotic to $n/2$. More precisely, it satisfies the inequality*

$$\frac{n-1}{2} \leq R^s_{n,n} \leq \frac{2^{n-2}}{2^{n-1} - 1} + \frac{n}{2}.$$

Most work on multiplayer rendezvous has been carried out in the context of the line. See Section 17.5.

## 11.3 FOCAL Strategies on the Line

When the players have a common labeling of all points in the search space, they can both go to an agreed point, waiting there until the other player arrives. Such a strategy is called a FOCAL strategy, as it resembles Schelling's notion of a focal point. In general, FOCAL strategies (see Section 10.3) are not optimal in the sense of yielding a meeting in least expected time. (An exception to this is when both players are uniformly distributed on a labeled circle, in which case every FOCAL strategy is optimal; see Theorem 13.11.)

When no FOCAL strategy is optimal, we may still seek the best focal strategy. Note that in general, simply picking a common meeting point may not fully specify a strategy, since there may be more than one way to reach it (multiple geodesics). However, for the case where the search space $Q$ is a line, a strategy that goes directly to a given point and waits there is well defined.

Suppose that $n$ players are placed on the labeled line independently according to a known joint distribution, not necessarily independently. As stated above, every FOCAL strategy for the line is completely specified by giving a point $\theta$ on the line. For such a strategy, the meeting time is the initial distance of the furthest of the players from $\theta$.

Let $X_{min}$ be the random variable denoting the location of the player with the minimum starting point and $X_{max}$ the location with the maximal starting point. Set

$$Y = \frac{X_{min} + X_{max}}{2}.$$

We now demonstrate an observation of S. Gal, that the optimal value for $\theta$ is the median of the random variable $Y$. The meeting time $T$ (for all players to be together at $\theta$) will be given by

$$T = \max(|X_{min} - \theta|, |X_{max} - \theta|).$$

Consequently,

$$T = \begin{cases} \theta - X_{min}, & \text{if } \theta \geq Y, \\ X_{max} - \theta, & \text{if } \theta \leq Y. \end{cases}$$

Suppose we move $\theta$ to $\theta + \Delta\theta$, where $\Delta\theta$ is a very small number. In the top alternative, we have $\Delta T = \Delta\theta$, and in the bottom alternative we have $\Delta T = -\Delta\theta$. Therefore the expected change in expected meeting time is given by

$$\Delta\hat{T} = (\Delta\theta)(Pr[\theta \geq Y] - Pr[\theta \leq Y]).$$

This formula neglects the small probability that a small change in $\theta$ changes the case. If the second factor (involving probabilities) is not zero, then we can make an improvement (reduction) in the expected meeting time by an appropriate sign for $\Delta\theta$. Consequently, a necessary condition for the optimality of $\theta$ is that

$$Pr[\theta \geq Y] = Pr[\theta \leq Y].$$

That is, $\theta$ must be a median of the distribution of the random variable $Y$.

Considering the analogous situation on the plane, we do not have any closed form solution for the optimal meeting point. This may be an interesting problem to investigate. Suppose, however, that the $n$ players are independently distributed according to a common distribution whose support is a compact set $K$. Then in the asymptotic case of a large number $n$ of players, we should choose a focal point $\theta$ that minimizes the maximum distance to a point in $K$.

A related problem arises in a worst case analysis of focal rendezvous strategies. Suppose a minimizer (rendezvous coordinator) picks a focal point $\theta \in K$, while a maximizing Nature picks the $n$ locations $x_i \in K, i = 1, \ldots, n$, as initial locations for $n$ rendezvousers. The payoff is given by

$$T = \max_{i=1,\ldots,n} |x_i - \theta|,$$

which represents the time required for $n$ player rendezvous assuming the focal strategy $\theta$ is used by all the players. This zero sum game has been analyzed for a general metric space $K$ for $n = 1$ by Gross (1964), who called the value the "rendezvous value of the metric space $K$." It is interesting that $n = 1$ is the only value of $n$ for which we would not use the word rendezvous.

## 11.4 Rendezvous in Two Cells

We now consider the only solved case ($n = 2$) of the "telephone problem" described in the previous chapter (Problem 10.2). This is equivalent to the problem of two players who are independently and equiprobably placed in two cells at time $t = 0$. At each time $t = 1, 2, \ldots,$ they may either stay (S) or move to the other cell (M). They are said to meet at the first time $T$, when they occupy the same cell. Their common aim is to minimize the expected value $\hat{T}$ of the meeting time $T$. For this case ($n = 2$) the 'dark' version (where the position of the other is not revealed after each move) is the same as the 'searchlight' version (where it is revealed), since each player in the dark version can infer the other's location from the fact that they have not met. In the same year two pairs of authors respectively analyzed the *dark* (Anderson and Weber, 1990) and *searchlight* (Crawford and Haller, 1990) versions of this problem, for a general number $n$ of locations. Their analyses of the case $n = 2$ coincided in establishing that the *random* strategy, of repeatedly choosing S or M equiprobably and independently of previous choices, is optimal.

**Theorem 11.3** *If two players are randomly placed into two cells, as described above, the optimal player-symmetric strategy for minimizing the expected meeting time is the random strategy.*

**Proof.** Associated with any strategy there will be a sequence of conditional meeting probabilities

$$p_i = Pr(T = i + 1 \backslash T > i),$$

from which the expected meeting time $\hat{T}$ may be calculated as

$$\hat{T} = \left[\tfrac{1}{2}\right](0) + \left[\tfrac{1}{2}\right]p_1(1) + \left[\tfrac{1}{2}(1 - p_1)p_2\right](2) + \tfrac{1}{2}(1 - p_1)(1 - p_2)p_3(3) + \cdots .$$

Note that this sum is decreasing in each $p_i$. Since the random strategy ensures a meeting with conditional probability $p_i = 1/2$ in each period, it suffices to show that for any mixed strategy simultaneously adopted by both players, we have

$$p_i \leq 1/2, \text{ for every } i = 1, 2, \ldots .$$

To see this, suppose that $T > i$, which means the players have not met by time $i$. In the case $n = 2$ (but *not* for higher $n$), this means that both players have followed exactly the same sequence of moving and staying (they have always made the same choice). Consequently, the mixed strategy that they are both adopting will give them the same probability $q$ of moving on step $i + 1$, and hence they will meet at time $i + 1$ with probability $2q(1 - q)$ (if one moves and one stays), which is no more than 1/2. ■

An analysis of the dark case for general $n$, stated in terms of rendezvous on a complete graph, is given in Section 15.3. The searchlight version is discussed in Section 15.3.1, under the heading of "revealed actions." In the searchlight version for $n = 3$, players who are initially at distinct locations should go on their next turn to the unique unoccupied location, where they will surely meet. For $n = 5$, they should each go to an unoccupied location. If they are still unlucky, they should go on their next move to the unique location that has never been occupied. This strategy turns out to be slightly better than using the random strategy on the two locations initially occupied.

# Part Three

# Rendezvous Search on Compact Spaces

# Chapter 12

# Rendezvous Values of a Compact Symmetric Region

This chapter gives a formal presentation of the rendezvous search problem on compact regions. The assumption that $Q$ is compact will restrict the application of these results to the spaces considered in Chapters 13, 14, and 15, so those and the present chapter are together called Part 3. In particular, we define the *rendezvous value* $R(Q)$ of a compact (closed and bounded) search region $Q$. The value gives the least expected time for two unit speed players to meet after a random (usually uniform) placement in the search region $Q$. They are said to meet when they come within a given detection radius $r$, which is usually zero for the one-dimensional search regions we will mainly study. In any case we always assume that $r$ is such that $Q$ can be searched exhaustively in finite time. We will show how the rendezvous value depends on a number of factors, some of which we will have to add to the notation $R(Q)$:

1. **Player symmetry** Are the players distinguishable in the sense that they can agree in advance on distinct strategies (the player-asymmetric version), or are they indistinguishable players (player-symmetric version) who must be told to adopt the same (possibly mixed) strategy?
2. **Spatial symmetry of search region** If the search region is symmetric (e.g., the circle), how much of this symmetry can the players 'see through' (e.g., common notion of direction)?

Player symmetry will be discussed in Section 12.1, where we show that there are in fact two rendezvous problems. The role of spatial symmetry will be explained in Sections 12.2 and 12.3, where we give both an intuitive discussion and a formalization based on symmetry groups. Sections 12.4 and 12.5 define the asymmetric and symmetric rendezvous values and establish the existence of these values. Further properties of these values are derived in Section 12.6. The material in this chapter is based mainly on the first author's original article (Alpern, 1995).

## 12.1 Player Symmetry

There are two distinct rendezvous values (and problems), depending on whether the two players can distinguish between themselves or equivalently whether they are allowed to use different strategies. If they can agree before the problem begins that, for example, Player I will stay still while Player II exhaustively searches $Q$, then we call this the *asymmetric rendezvous problem*. In this version any pair of pure strategies may be employed by the players. The least expected meeting time in this case is called the *asymmetric rendezvous value*, and denoted $R^a$. In the case that the players cannot distinguish between themselves (and must use the same mixed strategy), the problem they face (called the *symmetric rendezvous problem)* is more difficult, and has a generally larger least expected meeting time $R^s$, called the *symmetric rendezvous value*. The symmetry referred to here is player symmetry (which in other contexts is sometimes called anonymity). Clearly $R^a \le R^s$, since players in the asymmetric problem have a larger set of feasible strategies available, and the objective function (meeting time) is the same. In general we usually find that $R^a$ is strictly less than $R^s$. (An exception is noted in Section 13.3.1 for the case that $Q$ is an interval.)

For example, if there is a book carried by all hikers that says what to do if you become separated from your hiking partner, the writer of that part of the book would be dealing with a symmetric rendezvous problem: all hikers who read this book would follow the same strategy (though with independent randomizations if the advice given is mixed). On the other hand, if two paratroopers have a discussion in their plane before jumping regarding how they will meet up after landing, this is clearly an asymmetric problem. They can agree on distinct strategies. Sometimes there are symmetry breaking conventions that in practice convert apparently harder symmetric problems to the easier asymmetric one: The parent searches while the child waits; after a telephone disconnection, the original caller calls again. In problems of many-player rendezvous (not considered until Chapter 17) a pair of players may initially be indistinguishable (constrained to follow the same strategy), but when they meet, they can agree to adopt distinct strategies.

## 12.2 Spatial Symmetry

An intuitive example of the role played by spatial (geometric) symmetries in the search region was given in Alpern (1995). Although that example applies to the plane, which is not compact, we will nevertheless repeat it here to motivate the discussion. The problem concerns two players in the plane who wish to meet.

The first version of the problem has a flowing straight-line river (or canal), over which there is a single bridge. These features allow the two players to "see through" some of the symmetry of the plane (in fact, all of it). Each player can determine his position exactly with respect to a map common to both. For example, we may take the river as the $y$-axis, the bridge as the origin and assume that the river flows in the positive direction along the $y$-axis. This makes the search region "fully labeled" (as defined and described in Chapter 13), and the players can use strategies that depend on their initial positions.

The second version removes the bridge. This makes the problem slightly harder. Each player can now observe his initial distance from the river and the side of the river that he is on (the side from which the river flows to the right or to the left). A player's strategy could depend on this information.

The third version stops the river from flowing. Now each player knows only a single parameter – his initial distance from the river.

The fourth version removes the river. Players in this situation can still see through some symmetry of the plane in the sense that they presumably have a common sense of the clockwise direction (or the notion of "up"). In the asymmetric player version, for example, they could adopt distinct clockwise and counterclockwise spirals.

The fifth version puts the planar search region into Earth orbit, so that the players no longer have a common notion of "up." Assuming that they may land on this plane from either direction, they no longer have a common notion of clockwise.

How can we model these five versions of the same rendezvous problem (or 10 problems, if we consider both possible assumptions regarding player symmetry)? The answer, in this problem as well as more generally, lies in specifying a particular group of symmetries that the players cannot "see through." In the first problem there are no symmetries preventing the players from knowing their exact positions. The symmetry group (set of all symmetries) describing this situation is the trivial one consisting of just the identity transformation. In the second problem the group $G$ of symmetries that describes the players' uncertainty is just the real line $R$, which acts on the plane as translations in the direction of the positive direction of the river. If we view the canal as the $y$-axis with the bridge as the origin, each player knows his $x$-coordinate. His information space $\tilde{Q}$ is in general the quotient group $Q/G$, and in this case $R^2/R$, where the / indicates the equivalence classes of the set on the left of the / via the action of the group of symmetries on the right of the /. (Readers unfamiliar with these group theoretic notions need not worry: only very elementary notions of symmetry will be needed in this book, and geometric language will always be used in the first instance, with group notions added parenthetically.) In the third problem the players know their own distance $|x|$ to the river, the $y$-axis. (The relevant symmetry group additionally includes the 180° rotation, and the players' information set $Q/G$ is now the positive real axis $R^+ = [0, \infty)$.) In the fourth problem the symmetries they cannot see through consist of the group of all translations, rotations, and reflections. The information set is a singleton, meaning they have no initial information about their position. This will always be the case when the symmetry group, as in this case, is transitive. (The symmetry group $G$ is called transitive if given any two initial positions, there is a symmetry transformation in the group that takes the first position into the second.) In the rest of this chapter we will mainly be concerned with rendezvous problems where the symmetry group is transitive, and for this reason a strategy will not depend on the player's initial position. The fifth problem adds the reflection symmetries that interchange up and down, and removes the players' abilities to have a common sense of clockwise. This symmetry group is also transitive, but the set of strategies available to the players is smaller than in the fourth problem, which is also transitive.

In the rest of the book, we will always specify the relevant symmetry group for the rendezvous problem. In most cases we will not have to explicitly describe the given group of symmetries: often this will be implicit in our description of the search space (as in the terms *labeled* circle, *directed* circle, or *undirected* circle, in the next section).

## 12.3 An Example: The Circle

To illustrate the differing versions of the rendezvous problem that may be obtained by specifying whether there is player symmetry and by the given symmetry group, we consider the unit circumference circle $C$. We will consider three different versions of the problem: (i) the players have a common labeling of all points, (ii) the players have a common notion of clockwise direction but have no common labeling of points, and (iii) the players have no common notion of direction or labeling of points. These versions will be formalized by specifying three symmetry groups, which we call $G_1$, $G_2$, $G_3$. The first group $G_1$ consists of simply the identity transformation and consequently corresponds to the case where the circle is completely labeled: the players have a common notion of location. This is the case of no spatial symmetry. The symmetry group $G_2$ consists of all rotations of the circle and corresponds to the case of a "directed circle": the players have no common notion of location but do have a common notion of, say, clockwise. The symmetry group $G_3$ is the group of all symmetries of the circle: the rotations and reflections. This corresponds to an undirected circle where the players have no common notion of either location or direction. Note that both $G_2$ and $G_3$ act transitively on the circle, so that in these versions of rendezvous search the player strategies do not depend on their initial positions. In other words, the player acquires no new information upon "landing" on the search region, and so may choose his strategy earlier.

We now consider four strategies for rendezvous on the circle and consider for which situations (player symmetries and groups $G_i$) they are feasible.

**FOCAL** This is the strategy where the two players go directly at maximum speed 1 to a common (focal) point. This is feasible only for the case corresponding to group $G_1$, where the players have a common labeling of points. Since the two players do the same thing, it is feasible for either the player-symmetric or player-asymmetric problems.

**OP-DIR (opposite directions)** Player I goes clockwise at unit speed while player II goes counterclockwise at unit speed. This strategy pair is feasible only for the player-asymmetric version. Since it requires a common notion of direction, it is feasible only for the labeled or directed circle, groups $G_1$ or $G_2$.

**WFM (Wait For Mommy)** On the circle, this strategy pair has one player (the child) staying still, while the other (Mommy) goes around in a fixed direction at unit speed. This strategy pair is feasible for the asymmetric player versions corresponding to any of the symmetry groups. However, if there is a common notion of direction, the child could improve the meeting time by moving in the opposite direction to that of the mother.

**CO-HA-TO (coin half tour)** This is the strategy that is optimal for both players in the search game on the circle (see Section 4.3). Each player goes half way around the circle at speed 1 in a random direction in each time period of length 1/2. All the random choices are made independently. Of the four rendezvous strategies listed here, it is the only one that is a mixed strategy: it requires the use of (independent) randomization.

The following strategies were suggested in Alpern (1995) as possibly optimal for the six versions of rendezvous on the circle corresponding to player symmetry or asymmetry and the three spatial symmetry groups for the circle. The initial placement of the players is assumed to be uniform.

Rendezvous strategies for the labeled, directed, and undirected circle

| | $G_1$ (labeled) | $G_2$ (directed) | $G_3$ (undirected) |
|---|---|---|---|
| Symmetric, $R^s$ | FOCAL | CO-HA-TO | CO-HA-TO |
| Asymmetric, $R^a$ | OP-DIR | OP-DIR | WFM |

At this point, the above strategies are presented simply to aid the reader's intuition regarding the six versions of circle rendezvous and to illustrate the significance of the given symmetry group and the presence or absence of player symmetry. In four of the above six cases, the stated strategy has indeed found to be optimal: Theorem 13.10 (FOCAL), Theorem 14.7 (both cases of OP-DIR), and Corollary 14.11 (WFM). It is still an open question whether CO-HA-TO is indeed optimal in the two remaining cases.

## 12.4 The Asymmetric Rendezvous Value $R^a$

We are now ready to give a formal definition for the asymmetric and symmetric rendezvous values of a compact metric space $(Q, d)$. These values depend, of course, on the given detection radius $r$ and the given group of symmetries $G$. By a *symmetry* (or isomorphism) of $(Q, d)$, we mean a bijection of $Q$ that preserves distance ($d(g(x), g(y)) = d(x, y)$ for all $x, y$ in $Q$). Examples of symmetries are rotations and reflections of a circle.

We will only give a formal definition for the rendezvous values for the important case where the given symmetry group $G$ acts transitively on the search region $Q$, which essentially means that all points in $Q$ look alike, and the players receive no additional information when they are placed in $Q$ at the start of the problem. Consequently a (pure) strategy cannot depend on a player's initial position. With this in mind, we define the rendezvous strategy space $\mathcal{S}$ to consist of all continuous paths $s$ with maximum speed 1 that start at a given point $x_0$,

$$\mathcal{S} = \{s\colon [0, \infty) \to Q, s(0) = x_0, d(s(t), s(t')) \le |t - t'|\}. \tag{12.1}$$

Note that in Book II we represent pure strategies by lowercase letters, in keeping with the existing literature on rendezvous (rather than the upper case used to denote pure search or hiding strategies in Book I.) The *actual path* followed by a player will depend both on his chosen strategy $s$ and his initial placement in terms of both position and orientation in $Q$. The latter will be specified by the symmetry $g$ chosen from the group $G$, and the resulting path is $g(s(t)), t \ge 0$. For example, suppose that $Q$ is the circle

$C = [0, 1) \bmod 1$, $x_0 = 0$, and $s$ is the "forward" half-speed strategy $s(t) = t/2$. Then if the randomly chosen symmetry is $g(x) = 1/4 - x$, the actual position of the player at time $t$ is $1/4 - t/2$. It is important to keep in mind this distinction between strategies and paths.

The *meeting time* $T$ corresponding to two paths $s_1, s_2$ (with arbitrary starting points) is given by

$$T(s_1, s_2) = \min\{t\colon d(s_1(t), s_2(t)) \le r\},$$

where $r \ge 0$ is the given detection radius, usually taken to be zero in one-dimensional problems. If the players choose strategies $s_1, s_2 \in \mathcal{S}$, then their expected meeting time $\hat{T}$ is given by

$$\begin{aligned} \hat{T}(s_1, s_2) &= \int_{g_1 \in G} \int_{g_2 \in G} T(g_1(s_1), g_2(s_2))\, d\nu(g_1)\, d\nu(g_2) \\ &= \int_G T(g(s_1), s_2)\, d\nu(g), \end{aligned} \tag{12.2}$$

where $\nu$ denotes the invariant measure on the group $G$, which is simply the uniform distribution for all the spaces considered in this book. In some cases (see the general analysis for the circle in Chapter 13) we can let the initial placement of the players be an additional parameter of the problem. The reason that we need let $G$ act on only *one* of the paths is that for $g, h \in G$, $T(gs_1, hs_2) = T(h^{-1}gs_1, s_2)$. To make the abstract definition in equation (12.2) more concrete, observe that for $G = G_2$ (the rotation group on the circle) we have

$$\hat{T}(s_1, s_2) = \int_0^1 T(s_1 + a, s_2)\, da,$$

and when $G$ is the full group $G_3$ of all symmetries of the circle it becomes

$$\hat{T}(s_1, s_2) = \frac{1}{2} \int_0^1 T(s_1 + a, s_2)\, da + \frac{1}{2} \int_0^1 T(s_1 + a, -s_2)\, da,$$

corresponding to the random direction in that we need to face Player II.

In any case, once we have defined the expected meeting time $\hat{T}$, the asymmetric rendezvous value $R^a$ can be defined simply as

$$R^a = R^a(Q, G) = \inf_{s_1, s_2 \in \mathcal{S}} \hat{T}(s_1, s_2). \tag{12.3}$$

A strategy pair $(s_1, s_2)$ with $R^a = \hat{T}(s_1, s_2)$ will be called *optimal*. For compact search regions optimal asymmetric strategy pairs always exist. That is, the inf in definition (13.1) can always be replaced by a min.

**Theorem 12.1** *For compact search spaces, there exists an optimal asymmetric rendezvous strategy pair.*

**Proof.** Endow the strategy space $\mathcal{S}$ with the topology of uniform convergence on compact sets, under which it is compact. The expected meeting time function $\hat{T}(s_1, s_2)$ is lower semicontinuous in each variable. (This can be shown by a similar argument used for search games in Book I.) Hence a minimum is achieved. The minimum is finite because of our assumption that the search space can be fully searched in a finite time. Consequently, the strategy pair where one player searches exhaustively while the other remains stationary ensures that the meeting time is not more than the time required to search the whole space. ■

## 12.5 The Symmetric Rendezvous Value $R^s$

The definition of the symmetric rendezvous value is a bit more complicated, as mixed strategies may be required. As noted earlier, the symmetric problem models situations where a single plan (strategy) must be broadcast to two lost individuals whose names are not known. So one cannot say "Jim stay still and Jack search." In player-symmetric rendezvous problems, a pure strategy may not be very advisable. For example, if both players are lost on an undirected circle and facing in the same direction, then if they adopt the same *pure* strategy, they will never meet. So *mixed* strategies are in general needed. (An exception to this is the case of labeled networks as discussed in the next chapter. For example, the optimality of the FOCAL strategy on a labeled circle.) With this in mind, let $\mathcal{S}^*$ denote the set of all mixed search strategies, that is, the set of all Borel probability measures on $\mathcal{S}$. We then define the symmetric rendezvous value $R^s$ by

$$R^s = R^s(Q, G) = \inf_{s^* \in \mathcal{S}^*} \int_{\mathcal{S}} \int_{\mathcal{S}} \hat{T}(s_1, s_2)\, ds^*(s_1)\, ds^*(s_2), \quad \text{or} \tag{12.4}$$

$$= \inf_{s^* \in \mathcal{S}^*} \hat{T}(s^*, s^*), \quad \text{if we allow } \hat{T} \text{ to act on mixed strategies.} \tag{12.5}$$

If $R^s = \hat{T}(s^*, s^*)$, then $s^*$ is called an optimal symmetric strategy. As for asymmetric rendezvous, optimal strategies always exist. In general, a symmetric strategy is a pair of identical strategies.

**Theorem 12.2** *Optimal symmetric strategies exist for the symmetric rendezvous problem on any compact search region. That is, there is always an $s^*$ with $R^s(Q, G) = \hat{T}(s^*, s^*)$.*

**Proof.** Endow the mixed strategy space $\mathcal{S}^*$ with the so called *weak** topology that is defined by the sequential convergence $s_i^* \to s^*$ iff $s_i^*(f) \to s^*(f)$ for every pure strategy $f \in \mathcal{S}$. This makes the space $\mathcal{S}^*$ compact. The expected meeting time map $\hat{T}(s^*, s^*)$ is lower semicontinuous on $\mathcal{S}^*$ and consequently has a minimum. This minimum is finite because the expected meeting time for the "randomized wait for mommy" strategy (RWFM) is finite. This is the strategy where a player equiprobably searches the entire space in minimum time $a$, or waits for a time $a$ and repeats this randomization independently in each period of length $a$. ■

## 12.6 Properties of Optimal Strategies and Rendezvous Values

We consider the rendezvous search problem to be a team problem, because the two rendezvousers have identical utility functions. However, we can also look at the problem as a noncooperative game between the two rendezvousers. In this context it is obvious that an optimal asymmetric strategy pair is a Nash equilibrium: If either player could gain from a unilateral deviation from an optimal pair, then both would gain, and the new pair would contradict the assumed optimality of the original pair. It is less obvious but still true that the pair $(s^*, s^*)$ corresponding to an optimal symmetric strategy $s^*$ is also a Nash equilibrium.

**Theorem 12.3** *If a mixed strategy $s^*$ is an optimal strategy for the symmetric rendezvous problem, then the pair $(s^*, s^*)$ is a Nash equilibrium of the associated game.*

**Proof.** Suppose on the contrary that there is a mixed strategy $s'$ such that

$$\hat{T}(s', s^*) < \hat{T}(s^*, s^*). \tag{12.6}$$

By an arbitrarily small modification of $s'$ (for example, changing it to $s^*$ if the players have not met by some large time) we may assume that $\hat{T}(s', s')$ is finite. For any probability $p$, let $s(p)$ denote the mixed strategy that plays $s'$ with probability $p$ and $s^*$ with probability $1 - p$. Then the expected meeting time when both players adopt $s(p)$ is given by

$$\hat{T}(s(p), s(p)) = p^2\hat{T}(s', s') + 2p(1-p)\hat{T}(s', s^*) + (1-p)^2\hat{T}(s^*, s^*).$$

Let $\phi(p) = \hat{T}(s(p), s(p)) - \hat{T}(s^*, s^*)$. Observe that $\phi(0) = 0$ and

$$\phi'(p) = 2p\hat{T}(s', s') + [2 - 4p]\hat{T}(s', s^*) + [2p - 2]\hat{T}(s^*, s^*), \quad \text{and hence}$$

$$\phi'(0) = 2(\hat{T}(s', s^*) - \hat{T}(s^*, s^*)) < 0, \quad \text{by assumption.}$$

Consequently, for sufficiently small positive values of $p$, we have $\phi(p) < \phi(0) = 0$, or $\hat{T}(s(p), s(p)) < \hat{T}(s^*, s^*)$. This contradicts the assumed optimality of the strategy $s^*$. ■

It is often useful to have upper bounds on the expected time to rendezvous (that is, on the rendezvous values). If a search region $Q$ can be exhaustively searched in finite time, then regardless of player symmetry or spatial symmetry, the rendezvous value is finite. More precisely, we have the following upper bounds on the rendezvous values.

**Theorem 12.4** *Let $\bar{\mu}$ denote the length of a minimal Chinese Postman Tour (one that comes within the detection distance of any point) on the search space $Q$. Then for any transitive group $G$ of symmetries of $Q$ we have*

1. $R^a(Q, G) \le \bar{\mu}/2$.
2. $R^s(Q, G) \le 3\bar{\mu}/2$.

**Proof.** The first inequality is easily obtained by the WFM (Wait for Mommy) strategy in which one player remains stationary while the other follows a minimal Chinese Postman Tour equiprobably in either direction. If the stationary player is found at time $T$ going in one direction, then he will be found no later than $\bar{\mu} - T$ when going in the other direction. To obtain the second inequality, suppose both players adopt the randomized WFM strategy (RWFM) in which they choose their roles equiprobably (between the two in WFM above) in each time period of length $\bar{\mu}$. They make their choices independently of previous choices. Let $\hat{T}$ denote the expected meeting time for this symmetric strategy. In the first time period they will choose distinct roles with probability 1/2. In this case the same argument as for the first inequality shows that they will meet in expected time $\bar{\mu}/2$ from the beginning of the period. With probability 1/2, they will chose identical roles. Ignoring the possibility that they meet while both are searching, we find that at worst they begin again at time $\bar{\mu}$ and thus meet in expected time $\bar{\mu} + \hat{T}$. Consequently, we have that

$$\hat{T} \leq \tfrac{1}{2}(\bar{\mu}/2) + \tfrac{1}{2}(\bar{\mu} + \hat{T}), \qquad \text{or simply } \hat{T} \leq \tfrac{3}{2}\bar{\mu}.$$

■

Of course, a better estimate can be obtained in those cases where we can determine the probability that there is a meeting when both players choose to search rather than stay still.

# Chapter 13

# Rendezvous on Labeled Networks

As indicated in the previous chapter, most of the difficulties faced by the rendezvousers arise from their lack of a common labeling of location or a common notion of direction. These problems of spatial symmetry of the search region $Q$ were formalized through the notion of a given group $G$ of symmetries of $Q$. In this chapter, we will be concerned with the case where the two players have a *common labeling of all points of $Q$*, or equivalently, where the given group of symmetries of $Q$ consists simply of the identity transformation. This does not preclude the possibility that $Q$ may possess some symmetry, only that the players can "see through" any such symmetries via their labelings. For example, in the previous section we briefly considered the problem of rendezvous on a labeled circle (with the identity symmetry group called $G_1$). This is the only chapter where we will make this assumption. In most of Book II we will in fact assume that $G$ is transitive.

## 13.1 Networks and H-Networks

We will assume that our commonly labeled search region $(Q, \{identity\})$ is one-dimensional and refer to it as a *labeled network*. The problem of rendezvous on a labeled network may appear at first sight trivial, since (in either the asymmetric or symmetric contexts) the players can simply agree to meet at an agreed point. An example of such a strategy is the FOCAL strategy described in Section 12.3, where both players go directly to an agreed (focal) point. However, while such a simple strategy ensures a fairly quick meeting and may be a good "satisfising" solution, it is not in general optimal with respect to our least expected time criterion. This can be seen even in the case of the circle (for the player-asymmetric case only), where the FOCAL strategy has expected meeting time 1/3 while the OP-DIR strategy (clockwise–anticlockwise) reduces this time to 1/4. It turns out that for the labeled circle the FOCAL strategy is indeed optimal for the player-symmetric rendezvous problem (see Theorem 13.10). The optimization of FOCAL strategies on the line was discussed among the examples

Chapter 11. The reader should note that because the symmetry group implicit in labeled network rendezvous (that consists of just the trivial identity symmetry) is not transitive, the general analysis given in the previous chapter does not apply directly, although most of the ideas are still applicable. In particular, a player's strategy is now allowed to depend on his initial location.

In this chapter we will assume that the compact search domain $Q$ is a network. As in Chapter 3 of Book I, this means that $Q$ consists of a finite number of closed intervals (arcs) that meet only at their ends (nodes). We will obtain results on optimal rendezvous for general networks and sharper results for the particular cases of the interval and the circle. This chapter is based mainly on the first author's article (Alpern, 2002b), which is in turn a generalization of Howard's work on the interval and circle (Howard, 1999). Following those papers, we will consider in some cases a discrete approximation to the original network, which we call an *H-network*, in which the players move in discrete time among discrete locations (called nodes) placed along the original network.

For a labeled network, the players know their location at time zero and can base their subsequent motion on this knowledge. Consequently, a strategy $f$ is a collection of paths $f_x(t)$, $x \in Q$, such that $f_x(0) = x$ and $|f_x(t_2) - f_x(t_1)| \le |t_2 - t_1|$. The interpretation of such a strategy is that the player's location at time $t$ will be $f_x(t)$ if his initial random placement puts him at $x$. If Players I and II choose respective strategies $f$ and $g$ and their starting points are $x$ and $y$, their meeting time is given by

$$T_{x,y}(f, g) = \min\{t\colon f_x(t) = g_y(t)\}.$$

Their expected meeting time will be given by

$$\hat{T}(f, g) = \int_{Q\times Q} T_{x,y}(f, g)\, d\omega(x)\, d\nu(y), \tag{13.1}$$

assuming their initial distributions over $Q$ are $\omega$ and $\nu$, and the strategies $f$ and $g$ have sufficient regularity (say piecewise continuous in terms of the initial locations $x$ and $y$ in $Q$) so that $T_{x,y}$ is integrable.

The minimal expected meeting time for the asymmetric problem $\Gamma^a(Q, \omega\nu)$ is given by the asymmetric rendezvous value

$$R^a(Q, \omega, \nu) = \min_{f,g} \hat{T}(f, g).$$

The corresponding symmetric rendezvous value $R^s$ is defined by the optimization problem

$$R^s(Q, \omega) = \min_{f} \hat{T}(f, f).$$

Note that in the symmetric problem we assume a common initial distribution $\omega$ for the two players. Unlike the general player-symmetric problem discussed in the previous chapter, in the fully labeled case we do not need to consider mixed strategies. We will give an argument for this in the discrete context of the next section, where it is easier to understand.

In order to analyze rendezvous on labeled networks, we will use a discrete model developed in Alpern (2002b) based on Howard (1999). Given a network $Q$, add to each of the original arcs an odd number of additional nodes (of degree two). This ensures that the (integer) graph length of each of the original arcs is even and consequently that all circuits of the resulting discrete graph (based on the original nodes and additional nodes) have even length. By the length of a path in a graph, we mean the number of arcs it contains. We call a node "even" if it is an original node of $Q$ or a new node that is an even distance from an original node. Call the remaining nodes "odd." The reason for such a discretization is that if both players start (at the same time zero) at even nodes and move to a distinct adjacent node in each period (resting is excluded), then they cannot pass each other along an original arc without concurrently occupying a common node, and ending the game. Note that if the lengths of the original continuous network are all rational, then we can make the discrete length of each modified arc proportional to its original continuous length. This discrete model of rendezvous is very different from that discussed in Chapter 15 (Rendezvous on Graphs), and for this reason we will call the discretizations discussed here *networks*, rather than *graphs*, and use the terms *nodes* and *arcs* rather than *vertices* and *edges*.

## 13.2 Rendezvous on H-Networks

Let $Q$ denote a graph with nodes $\mathcal{N}$. Assume that all circuits have even length, so that by Theorem 2.4 of (Harary, 1972) we can partition the nodes into $\mathcal{N} = \mathcal{N}_e \cup \mathcal{N}_o$ such that every path in $Q$ alternates between $\mathcal{N}_e$ and $\mathcal{N}_o$. Such a (bipartite) graph, together with two given probability distributions $p$ and $q$ on the even node set $\mathcal{N}_e$, will be called an *H-network*. For example, the interval $H$-network $I[n]$ introduced in Howard (1999) has $\mathcal{N}_e = \{0, 2, 4, \ldots, 2(n-1)\}$, $\mathcal{N}_o = \{1, 3, \ldots, 2n-3\}$, with consecutive integers representing adjacent nodes in the network. The circle H-network $Circ[n]$ introduced in Alpern (2002b) is obtained from $I[n]$ by adding an additional node labeled $2n-1$ and making it adjacent to the nodes 0 and $2(n-1)$.

The asymmetric rendezvous problem $\Gamma^a(Q, p, q)$ for an H-network $(Q, p, q)$ begins at time $t = 0$ with Players I and II placed independently according to the positive distributions $p, q$ on the even nodes $\mathcal{N}_e$. In each period they must move to a *distinct* adjacent node. They are not allowed to remain at the same node in consecutive periods. This ensures that both players will always be on nodes of the same parity (even nodes at even integer times, odd nodes at odd integer times). Even if we allowed players to stay still, it would not be in their interest to do so. A strategy $s$ for a player on an H-network specifies for each even node $i \in \mathcal{N}_e$ a path $s_i(t)$ in the network $Q$ that starts at node $i$. This means that $s_i(t)$ and $s_i(t+1)$ are distinct adjacent nodes of the network, and $s_i(0) = i$. Let $\mathcal{S}$ denote the set of all strategies. Players adopting H-network strategies $f$ and $g$ and initially placed at even nodes $i, j$ will meet at time $T_{i,j}(f, g) = \min\{t\colon f_i(t) = g_j(t)\}$.

This discrete version has a positive probability that the game ends immediately, since for each even node $i$ we have $T_{i,i} = 0$, regardless of the chosen strategies. For strategy pairs $(f, g)$ with all the $T_{i,j}$ finite, the longest the game can continue is denoted by $t_{f,g} = \max T_{i,j}(f, g)$. The expected meeting time for two players adopting the strategy pair $(f, g)$ is the discrete analog of (13.1) given by

$$\hat{T}(f, g) = \sum_{i,j \in \mathcal{N}_e} T_{i,j}(f, g) p_i q_j. \tag{13.2}$$

As usual, any strategy pair minimizing (13.2) will be called *optimal.* The minimum expected time will be called the asymmetric rendezvous value of the H-network $(Q, p, q)$ and denoted $R^a(Q, p, q)$.

The symmetric rendezvous problem $\Gamma^s(X, p)$ is the same except that we assume that $q = p$ (players are placed with a common distribution) and that $g = f$ (the players must adopt a common strategy). There is no need for mixed strategies in the present context. A strategy $f$ minimizing $\hat{T}(f, f)$ will be called an *optimal symmetric strategy*, and the corresponding value of $\hat{T}$ will be called the *symmetric rendezvous value* of $(Q, p)$ and denoted $R^s(Q, p)$. Obviously, we have that $R^a(Q, p, p) \leq R^s(X, p)$. For the symmetric problem we shall adopt the same notations as defined above, except that a single variable is used; for example, $t_f$ denotes the last time $t_{f,f}$ that players using $f$ may meet.

With reference to a given strategy $s$ under consideration, it will be useful to think of the paths $s_i(t)$ as paths of a collection of *agents* of a player, with one agent starting at each even node. The formula (13.2) for $\hat{T}$ can be thought of as the expected time for an agent of the Player I to meet an agent of Player II. In the symmetric version of the problem, $\hat{T}$ is the expected time for the agents starting at various nodes to meet each other (simply one agent for each node). Since in this view agents at distinct nodes may use distinct strategies, the problem can be viewed as a player-asymmetric one, and consequently pure strategies will suffice.

For H-networks the set of strategies, or strategy pairs (for the asymmetric problem) is rather large. To simplify the search for optimal strategies, it is useful to restrict the search to a smaller set of strategies. With this in mind, we define *geodesic* and *sticky* strategies. The first notion applies to a pair of strategies, while the second applies to an individual strategy.

**Definition 13.1** *A pair $(f, g)$ is called a* **geodesic strategy** *if for any even node $i \in \mathcal{N}_e$, and any consecutive times $t_0$ and $t_1$ when agent $f_i$ first meets some distinct agents $g_j$ (at time $t_0$), and $g_{j'}$ (at time $t_1$) of the other player, we have*

$$d_{\mathcal{N}}(f_i(t_0), f_i(t_1)) = t_1 - t_0, \tag{13.3}$$

*where $d_{\mathcal{N}}$ denotes the graph distance (length of shortest path) in the network $Q$. The corresponding condition must also hold for $g$.*

Roughly speaking, a pair $(f, g)$ is geodesic if all agents of a player follow time-minimizing paths between times when they meet distinct agents of the other player. For example, on the interval H-network an agent in a geodesic strategy pair can change direction only when he meets an agent of the other player.

**Theorem 13.2** *Consider the asymmetric rendezvous problem on an H-network $(Q, p, q)$ with $p, q > 0$. A necessary condition for a strategy pair $(f, g)$ to be optimal is that it is geodesic. A necessary condition for a strategy $f$ to be optimal for the symmetric rendezvous problem is that $(f, f)$ is geodesic.*

**Proof.** Suppose that $(f, g)$ is an optimal strategy for the asymmetric problem that is not geodesic. Then, in the notation of the definition of geodesic, we have the following negation of (13.3):

$$d_{\mathcal{N}}(f_i(t_0), f_i(t_1)) < t_1 - t_0.$$

Observe that the integers on both sides of this inequality must have the same parity, since they are both lengths of paths from node $A = f_i(t_0)$ to node $B = f_i(t_1)$. Consequently, we have the stronger inequality

$$d_{\mathcal{N}}(A, B) \leq t_1 - t_0 - 2.$$

We now modify the path $f_i$ on the time interval $[t_0, t_1]$ as follows. First we ensure that the new path $f_i'$ is at $B$ at time $t_1 - 2$. Next observe that since $g_{j'}(t_1) = B = f(t_1)$, $g_{j'}$ must be at an adjacent node $B'$ to $B$ at time $t_1 - 1$. Now we also let $f_i'$ be at $B'$ at time $t_1 - 1$, sending it back to $B$ at time $t_1$. Notice that, in the modified strategy, $T_{i,j'}$ has been reduced by at least 1, while no meeting times have increased. Since $p_i$ and $q_j$ are assumed to be positive, it follows from (13.2) that the expected meeting time $\hat{T}$ has been reduced, which is impossible if $(f, g)$ was optimal. This establishes the result for the asymmetric problem. In the case that $g = f$ (the symmetric problem), we follow the modification of $f$ with an identical modification of $g$, which gives a new symmetric pair leading to the same contradiction. ■

Another important property, called *sticky*, applies to an individual strategy (rather than a pair).

**Definition 13.3** *A strategy $s \in \mathcal{S}$ is **sticky** if for any time $t$, and any $i, j \in \mathcal{N}_e$,*

$$s_i(t) = s_j(t) \text{ implies } s_i(t') = s_j(t') \text{ for } t' > t. \tag{13.4}$$

Roughly speaking, a strategy of a player is sticky if whenever two agents of that player meet, they stay together forever. Sticky strategies have the following optimality property with respect to the symmetric problem.

**Theorem 13.4** *For any symmetric rendezvous problem $\Gamma^s(Q, p)$ on an H-network $(Q, p)$, there exists an optimal symmetric strategy that is sticky.*

**Proof.** We view the symmetric problem for two rendezvousers as an asymmetric one for $n$ rendezvousers starting at the $n$ distinct even nodes. They have the common aim of minimizing the expected time taken for pairs of them to meet. Given any optimal strategy $f = (f_2, f_4, \ldots, f_{2n})$, we show how to successively modify it to make it sticky, while keeping the expected meeting times the same.

We begin by making sure that condition (13.4) holds for $t = 1$. To do this, suppose that some set of agents $\mathcal{A} \subset \mathcal{N}_e$ (denoting agents by their starting points) meet at time $t = 1$. Observe that since $f$ is optimal, the expected time for $f_a$ to meet the agents in $\mathcal{N}_e - \mathcal{A}$ must be the same for all $a \in \mathcal{A}$. So pick any $\tilde{a} \in \mathcal{A}$ and set $f_a = f_{\tilde{a}}$ for $a \in \mathcal{A}$. This does not change the expected meeting time. Doing the same modification for all groups of agents who meet at time $t = 1$, we can assume that the resulting strategy, that we again call $f$, satisfies the condition (13.4) for $t = 1$. Next suppose that two agents meet for the first time at $t = 2$, and denote by $\mathcal{B} \subset \mathcal{N}_e$ the set of all agents at that

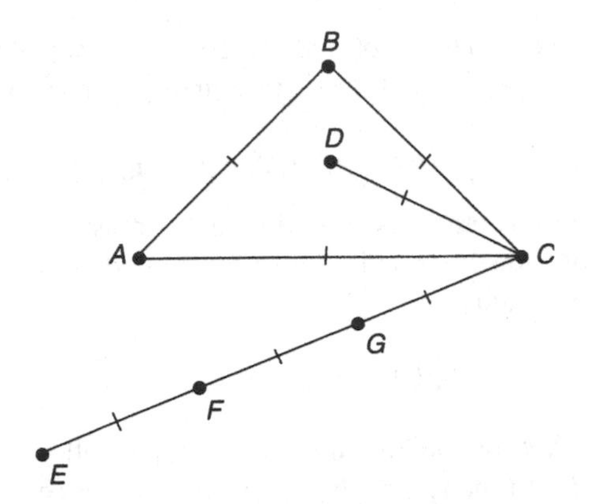


**Figure 13.1.** An H-network

meeting node. Now by the modification done for $t = 1$, we know that no agent in the set $\mathcal{B}$ has met any agent in the set $\mathcal{N}_e - \mathcal{B}$. Consequently, all agents in $\mathcal{B}$ have the same expected additional time (from $t = 2$) to meet those in $\mathcal{N}_e - \mathcal{B}$. So set all the paths (from time $t = 2$ onward) of agents equal to a particular one of these paths. This does not change the expected meeting time. Continuing in this fashion, we obtain a sticky strategy with the same expected meeting time as the original (optimal) one. (A more detailed proof can be found in Alpern, 2002b.) ■

Note that the optimality of sticky strategies holds only for the symmetric rendezvous problem. An H-network is given in (Alpern, 2002b) for which the asymmetric rendezvous problem has a unique solution in strategies that are not sticky. It is drawn in Figure 13.1. The even nodes $\mathcal{N}_e = \{A, B, C, D, E, F, G\}$ are named and indicated by small disks, while the odd nodes are indicated only by short lines. The initial distribution given there is not equiprobable, although it is the same for both players. Most of the probability is equally on $A$, $B$, and $C$, a small amount is on $D$, and a much smaller amount is on $E$.

The analysis given (Alpern, 2002b) shows that, up to labeling the two players, the unique optimal asymmetric strategy pair is as follows (the underscore represents the unlabeled intervening odd node, which it is simpler not to label):

$$\begin{array}{ll} f_A = (A, \_, B, \_C, \_, G) & g_A = (A, \_, C, \_G) \\ f_B = (B, \_, C, \_G) & g_B = (B, \_, A, \_, C, \_, G) \\ f_C = (C, \_, A, \_, B, \_C) & g_C = (C, \_, B, \_, A, \_, C) \\ f_D = (D, \_, C, \_, A, \_C) & g_D = (D, \_C, \_, B, \_C) \\ f_E = (E, \_, F, \_, G, \_, C) & g_E = (E, \_, F, \_, G, \_, C). \end{array} \tag{13.5}$$

In particular, note that $f_B(2) = f_D(2) = C$, but $f_B(4) = G \neq A = f_D(4)$. So the unique optimal symmetric strategy is not sticky. This shows that Theorem 13.4 cannot be extended to the asymmetric setting.

We give a brief heuristic argument to justify the claimed solution of (13.5). If all the probability is equal (1/3 each) on $A$, $B$, and $C$, then we know that the unique optimal asymmetric strategy is OP-DIR, in which $f$ goes clockwise and $g$ goes counterclockwise. Since nearly all the probability is on these nodes, the optimal strategy must still begin this way. By time $t = 2$, the agents starting at these three main nodes will have all met each other. The agent (of either player) starting at $D$ has no choice but to go to $C$. At time $t = 2$, agent $f_B$ and $f_D$ are both at node $C$. However, agent $f_B$ has met all the major agents of the other player, $g_A$, $g_B$, $g_C$, and $g_D$, and consequently should go along the line toward $E$ to meet the remaining agent $g_E$. However, agent $f_D$ cannot afford to do this, as he has not yet met the two major agents $g_B$ and $g_C$. Thus $f_B$ and $f_D$ must part after meeting at $t = 2$ at node $C$. A more rigorous presentation of this argument can be found in Alpern (2002b).

## 13.3 The Interval H-Network

The rendezvous problem on a finite interval was proposed by the first author. However no progress was made on this problem until the elegant work of Howard (1999), who solved the problem for certain initial distributions by modeling it by what we now call an *H-network*. We recall the definition of this H-network $I[n, p, q]$ from the previous section as $\mathcal{N}_e = \{0, 2, 4, \ldots, 2(n-1)\}$, $\mathcal{N}_o = \{1, 3, \ldots, 2n-3\}$, with consecutive integers adjacent in the network. Here $p$ and $q$ denote the given initial distribution of the players over the set of even nodes $\mathcal{N}_e$. While Howard considered the case of a common initial distribution, we will follow the more general presentation given by Alpern (2002b) that allows distinct distributions $p$ and $q$, as well as a given increasing (cost) function of the meeting time $T$. The very recent work of Chester and Tutuncu (2001), not discussed here, solves the version of this problem where the initial distribution is centrally symmetric and decreasing as one moves away from the center.

Howard called the agents starting at the terminal nodes 0 and $2(n-1)$ the *left and right sweepers*, respectively, and observed that each of these should go directly towards the opposite end, together with any other agents they meet. With this in mind, we may graph strategies for $I[n, p, q]$ in a triangle, using the horizontal coordinate for position and the vertical coordinate for time. The left and right sweeper will always meet at time $n-1$ and location $n-1$, which is the apex of the triangle. In Figure 13.2 we graph three strategies for the interval H-network: Howard's *right strategy* $\hat{s}$, the *central strategy*, and the *greedy strategy*. All these are sticky strategies and so may be unambiguously graphed as in the figure, with a single line style. In the right strategy $\hat{s}$, each agent moves to the right until he meets the right sweeper. In the central strategy, each agent moves to the center node $n-1$ and then (since remaining still is not allowed) oscillates between that node and $n$. In the greedy strategy, agents maximize the number of meetings at time $t = 1$, and then those at time $t = 3$, and so on. This family is well defined when $n$ is a power of 2 ($n = 2^3$ in the figure).

Most of the analysis of this section will concentrate on conditions for which the "right strategy" is optimal. We conjecture that the central strategy pair is optimal, for the asymmetric or symmetric problem, when the distributions $p$ and $q$ are single peaked at the center, and that the greedy strategy is optimal with respect to any concave cost function $c$ when the initial distributions are nearly uniform.

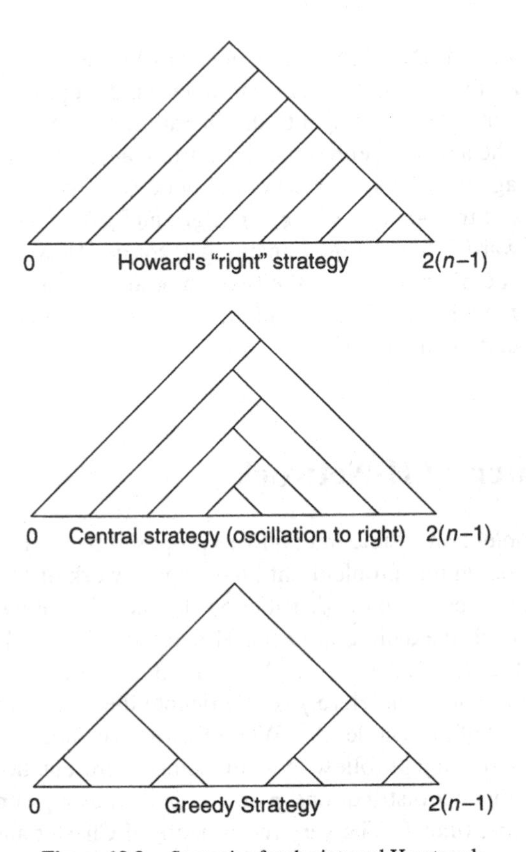


**Figure 13.2.** Strategies for the interval H-network

### 13.3.1 Nondecreasing initial distributions

Howard (1999) showed that the "right strategy" pair $(\hat{s}, \hat{s})$ is optimal for both the asymmetric and symmetric problems, when the initial distributions on the interval H-network are nondecreasing. More generally, the pair $(\hat{s}, \hat{s})$ has the following optimality property.

**Theorem 13.5** *For any n, the "right" strategy pair $(\hat{s}, \hat{s})$ is uniquely optimal for the asymmetric (and hence the symmetric) interval rendezvous problem $I[n, p, q]$, as long as both p and q are strictly increasing. This result holds more generally for any cost function c of the meeting time, as long as c is nondecreasing and convex.*

**Proof.** We shall outline the proof given by Alpern (2002b), which incorporates an earlier technique of Howard (1999). We will not prove the last part of the statement, which can be found in the original article. We will show that any strategy pair $(f, g)$ that is not equal to $(\hat{s}, \hat{s})$ cannot be optimal. We do this by slightly modifying one of the strategies (say $g$) so that $\hat{T}(f, g') < \hat{T}(f, g)$.

In fact, we only modify $g$ for a single coordinate $g_k$ (the agent starting at location $2k$), where $k$ is the smallest integer such that either $f_k$ or $g_k$ is not equal to $\hat{s}_k$. By relabeling Players I and II, if necessary, we may assume that it is $g_k$ that is not equal to $\hat{s}_k$. This means that all agents of both players who start at any of the nodes $0, 2, \ldots, 2(k-1)$

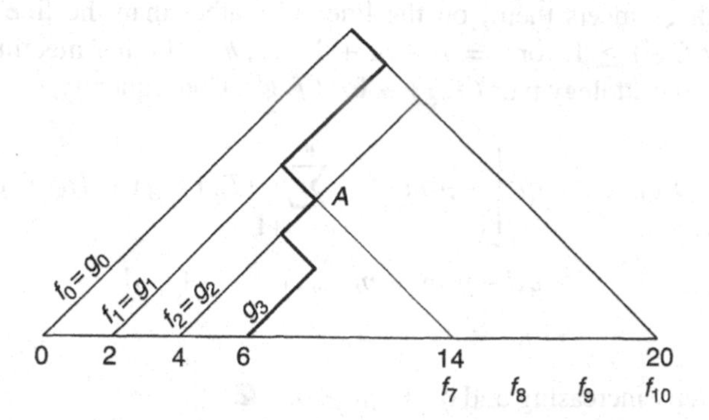


**Figure 13.3.** A nonoptimal interval strategy

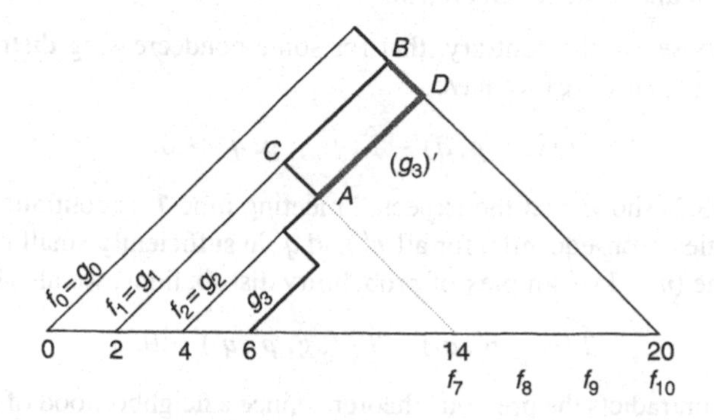


**Figure 13.4.** Modification of original strategy

follow the "right strategy" of going to the right until they meet the right sweeper, but that the agent ($g_k$) of Player II who starts at node $2k$ does not move in that way. In Figure 13.3, $k = 3$, and subscripts are raised for clarity. The six paths $f_0, g_0, f_1, g_1, f_2, g_2$ all agree with the paths given by the "right strategy" $\hat{s}$, but the path $g_3$ (the thick line starting at node $2 \cdot 3 = 6$) is not equal to $\hat{s}_3$. (It will not matter at all whether or not $f_3$ is equal to $\hat{s}_3$.) Since $g_k$ (in the figure, $g_3$) does not follow the "right strategy," it makes some left moves before reaching the right sweeper. Let $A$ denote the location where the last of these "left moves" is begun.

To obtain the modified strategy $g'$ we change $g_k = g_3$ after the point $A$ so that it moves to the right from $A$ until it meets the right sweeper at $D$ after $m$ such moves. The modification for Figure 13.3 is shown in Figure 13.4, where the modified path $g'_k = g'_3$ after $A$ is drawn in a thick grey line.

Note that the only path of $f$ that $g'_3$ meets later than $g_3$ does is the one we call $f_r$ ($f_1$ in the figure), that goes through $C$. In general, $g'_k$ meets $f_r$ at $B$, $m$ time units later than when $g_3$ meets it at $C$. On the other hand, $g'_3$ meets each of the $m$ rightmost paths (the three paths $f_8$, $f_9$, $f_{10}$ in Figure 13.4) *at least* one time unit earlier than the original path $g_k$ meets them, on the line $AD$ rather than the line $CB$. That is, $T_{ik}(f, g) - T_{ik}(f, g') \geq 1$, for $i = n - m + 1, \ldots, n$. All other meeting times are at least as large for the strategy pair $(f, g)$ as for $(f, g')$. Consequently,

$$\begin{aligned}\hat{T}(f, g) - \hat{T}(f, g') &\geq q_k \left[ -p_r m + \sum_{i=n-m+1}^{n} (T_{ik}(f, g) - T_{ik}(f, g')) p_i \right] \\ &\geq q_k[-p_r m + p_{n-m+1} + \cdots + p_n] \\ &> 0, \end{aligned} \tag{13.6}$$

because $p$ is strictly increasing and $q_k > q_0 \geq 0$. ■

From this result we can deduce the following earlier result of Howard (1999).

**Corollary 13.6** *For any $n$, the Howard strategy pair $(\hat{s}, \hat{s})$ is optimal for the asymmetric (and hence the symmetric) rendezvous problem $I[n, p, q]$ on the interval H-network, as long as both $p$ and $q$ are nondecreasing.*

**Proof.** Suppose, on the contrary, that for some nondecreasing distributions $p, q$, and some strategy pair $(f, g)$ we have

$$\hat{T}(\hat{s}, \hat{s}, p, q) - \hat{T}(f, g, p, q) > 0.$$

The formula (13.2) shows that the expected meeting time $\hat{T}$ is continuous in $p$ and $q$ for fixed strategies. Consequently, for all $p'$ and $q'$ in sufficiently small neighborhoods of $p$ and $q$ in the $(n - 1)$ – simplex of probability distributions, we also have that

$$\hat{T}(\hat{s}, \hat{s}, p', q') - \hat{T}(f, g, p', q') > 0.$$

However, this contradicts the previous theorem, since a neighborhood of a nondecreasing density contains strictly increasing densities. ■

Howard (1999) used the above result to obtain the following solution to the continuous version of the problem.

**Theorem 13.7** *Let $\Gamma^a([0, 1], p(x))$ denote the asymmetric continuous rendezvous problem on the unit interval $[0, 1]$, in which each player is independently initially placed according to a common nondecreasing density function $p$. An optimal strategy pair is the following continuous version of $\hat{s}$: each agent moves at unit speed to the right until meeting the right sweeper and then moves with him to the left at unit speed. The associated rendezvous value is*

$$R^a([0, 1], p, p) = \frac{1}{2}\left(1 - \int_0^1 (1 - F_p(x))^2 dx\right),$$

*where $F_p$ is the distribution function corresponding to the density $p$. Since the optimal strategy pair is symmetric, it is also the solution to the symmetric rendezvous problem, which consequently has the same minimal time. That is,*

$$R^s([0, 1], p, p) = R^a([0, 1], p, p).$$

**Proof.** We briefly outline the proof given by Howard. Suppose there were a strategy pair for $\Gamma^a([0,1], p(x))$ giving a lower expected rendezvous time than the continuous version of $\hat{s}$ described above. The interval [0,1] can be scaled to $[0, 2(n-1)]$ and the claimed better strategy could be approximated by a strategy that moves as in the discrete problem. This discrete strategy in $I[n, \tilde{p}, \tilde{p}]$ would do better than $(\hat{s}, \hat{s})$, where $\tilde{p}$ is the discrete version of the density $p$. However this is impossible by Corollary 13.6. ■

## 13.4 The Circle and the Circle H-Network

We now consider the rendezvous problem on a labeled circle, with the equiprobable initial distribution. Note that, under our assumption of a common labeling, the players can rendezvous at least as well on the circle as the interval because they could agree not to traverse some common location. This would effectively convert the circle rendezvous problem into one of rendezvous on the interval that results from "cutting" the circle. We will show that for the symmetric problem the players can indeed do this cutting without increasing their meeting time. On the other hand, for the asymmetric problem the players can exploit the circle topology by going in opposite directions.

As with the case of the interval discussed above, we attack the problem of rendezvous on a labeled circle by considering its discrete analog, the H-network $Circ[n, p]$. Recall that this is the H-network with node set $\mathcal{N}^C = \{0, 1, \ldots, 2n-1\}$ and even nodes $\mathcal{N}_e^C = \{0, 2, \ldots, 2(n-1)\}$. Nodes $i$ and $j$ are adjacent if $|i - j| = 1 \pmod{2n}$. We will call the direction from $i$ to $i+1$ "clockwise." We will show that, while in the asymmetric problem they can benefit from the additional transition from node $2n-1$ to node 0, in the symmetric problem they can make such a cut without increasing expected meeting times.

### 13.4.1 Symmetric rendezvous on the labeled circle

We now consider the player-symmetric rendezvous problem on the circle H-network $Circ[n, \bar{p}]$, with the equiprobable initial distribution $\bar{p}$ that gives probability $1/n$ to each of the $n$ even nodes of the network. Later in the section we will interpret our results in the context of the continuous unit circle by letting $n$ go to infinity.

First note that since the players could agree to avoid a common odd node, they can effectively choose to rendezvous on an interval network, and consequently we have

$$R^s(Circ[n, \bar{p}]) \leq R^s(I[n, \bar{p}]). \tag{13.7}$$

The main result of this section is the demonstration that in the symmetric problem the rendezvousers cannot do *better* on the circle than on the interval (so equality holds in (13.7). We do this by showing that *any* sticky symmetric strategy pair for the circle can be adapted (by relabeling of nodes) for use on the interval network. Consequently if we apply this process to an *optimal* sticky strategy pair for the circle network, we can enable rendezvousers on the interval network to meet in expected time $R^s(Circ[n, \bar{p}])$, so that $R^s(Circ[n, \bar{p}]) = R^s(I[n, \bar{p}])$.

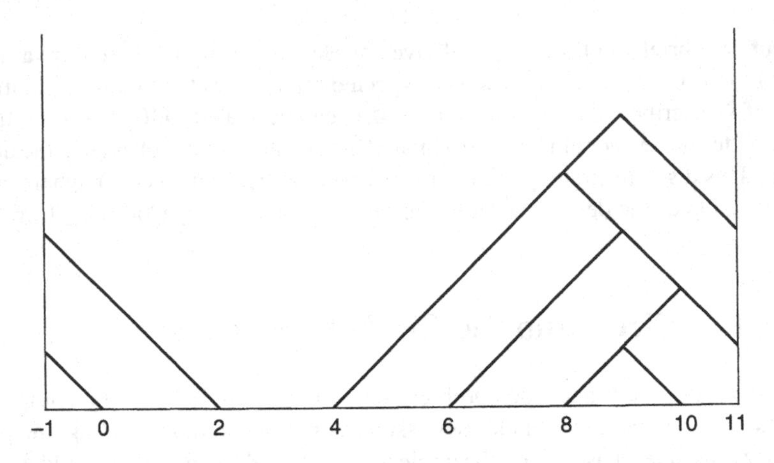


Figure 13.5. A symmetric strategy on Circ[6, $p$]

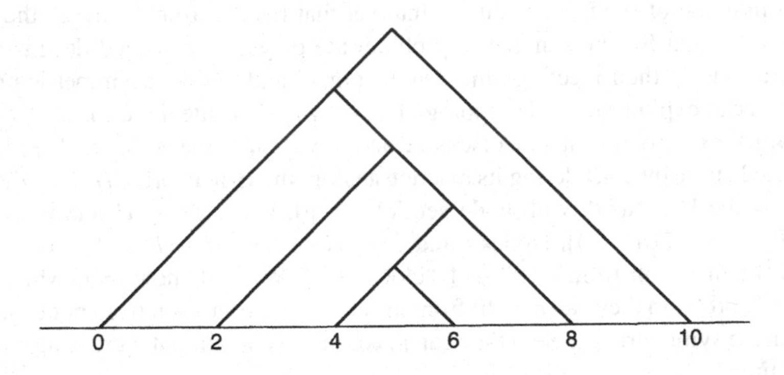


Figure 13.6. The same strategy with relabeling

To illustrate the relabeling algorithm, consider the symmetric sticky strategy for $Circ[6, p]$ drawn in Figure 13.5. Note that location $2n - 1 = 2(6) - 1 = 11$ is also drawn as location $-1$, with an identified line above each.

If we rotate this picture 4 units in the counterclockwise direction (leftwards, as drawn), then we get a strategy that is valid for the interval H-network $I$, as pictured in Figure 13.6. That is, the node originally numbered $j$ is now numbered $j - 4 \pmod{12}$.

The following result shows that the such a relabeling (rotation) is always possible.

**Theorem 13.8** *For every symmetric sticky strategy on the circle H-network* $Circ[n]$*, there is an even circular relabeling of the locations so that the resulting strategy is feasible for the interval network* $I[n]$.

**Proof.** We show how to relabel the nodes of $Circ[n]$ so that a given sticky strategy $f$ is a valid strategy for the resulting interval H-network. For each integer time $t = 1, \ldots, t_f$, let $P(t)$ denote the partition of $\mathcal{N}_e^C$ into sets (that will be intervals) whose agents have met by time $t$. That is, even nodes $i$ and $j$ belong to the same element of $P(t)$ if and only if $f_i(t) = f_j(t)$. For the strategy pictured in Figure 13.5, we have $t_f = 5$ and

$$P(1) = \{\{0\}, \{2\}, \{4\}, \{6\}, \{8, 10\}\},$$
$$P(2) = \{\{2\}, \{4\}, \{6\}, \{8, 10, 0\}\},$$
$$P(3) = \{\{2\}, \{4\}, \{6, 8, 10, 0\}\},$$
$$P(4) = \{\{2\}, \{4, 6, 8, 10, 0\}\},$$
$$P(5) = \{\{2, 4, 6, 8, 10, 0\}\}.$$

In the penultimate partition $P(t_f - 1)$ (i.e., $P(4)$ in the example) there are two intervals of agents (identified with their even starting points) that move together, one group going clockwise and one counterclockwise, to meet at time $t_f$ (equals 5 in the example). Let $b$ denote the most counterclockwise agent in the counterclockwise moving group, so that $b = 2$ in the example. Relabel the locations so that $b$ becomes the right end of the interval, $2(n-1)$ (in the example, 2 becomes 10). A more rigorous version of this proof is given in Alpern (2002b). Note that (unlike the illustrated example) the resulting strategy for the interval network may go past the right end node $2(n-1)$ or the left end node 0, but this is not a problem. If desired, the strategy could be further modified to keep it within the original interval. ■

If we combine the above result for the equiprobable density $\bar{p}$ with the existence (Theorem 13.4) of an optimal symmetric strategy for $Circ[n, \bar{p}]$ that is sticky, we see that $R^s(Circ[n, \bar{p}]) = R^s(I[n, \bar{p}])$.

**Theorem 13.9** *For the equiprobable distribution $\bar{p}$ giving probability $1/n$ to each even node, we have*

$$R^s(Circ[n, \bar{p}]) = R^s(I[n, \bar{p}]).$$

*An optimal symmetric strategy for $Circ[n, \bar{p}]$ is to agree to avoid a common odd node and then play any optimal strategy for the resulting interval network $I[n, \bar{p}]$. In particular, it is optimal for the players to go to a commonly agreed even node and then oscillate.*

The approach used here can also be used to reduce any symmetric rendezvous problem on $Circ[n, p]$, for an arbitrary $p$, to a family of problems on the interval.

By approximating the uniform distribution on a continuous circle by a large circle H-network with an equiprobable distribution we obtain the following continuous analog establishing the optimality of the FOCAL strategy for symmetric rendezvous on a labeled circle.

**Theorem 13.10** *Suppose two players are placed independently and uniformly on the unit circumference circle and allowed to move at unit speed. They cannot see each other but have a common labeling of all the points on the circle. An optimal symmetric rendezvous strategy is to move at unit speed, via the shorter of the two possible arcs, to a commonly agreed location. The symmetric rendezvous value equals 1/3.*

### 13.4.2 Asymmetric rendezvous on the labeled circle

We now consider the player-*asymmetric* rendezvous problem on a labeled circle $C$ (of unit circumference) with both initial distributions uniform. Howard showed that the OP-DIR strategy pair described in Section 12.3 is indeed optimal. Unlike the analysis given above for the symmetric case, we will not use discrete methods of approximation (H-networks) to obtain this result.

It will be useful to begin by considering a *half-sighted* variant of the rendezvous problem that gives the players additional, though asymmetric, information. For this version, we will assume that Player I can see Player II (this is equivalent to his simply knowing the starting position of Player II), but Player II cannot see Player I. If Players I and II start at respective positions $x$ and $y$ on the labeled circle, this means that I knows $x$ and $y$, while II knows only $y$. We are making this assumption here for the sake of argument, but in fact such information may be realistic. For example, in the model of Thomas and Hulme (1997) described later in Chapter 18, a helicopter is attempting rendezvous with a lost hiker in a forest. The helicopter can travel faster, but the hiker can spot the helicopter a long way off. In this sense the information of the hiker resembles that of Player I, at least when the helicopter is not out of sight.

We now formalize the half-sighted version of the player-asymmetric rendezvous problem on the circle. By the symmetry of the circle, this is equivalent to assuming that II starts at a fixed position 0 on the circle, while I starts uniformly. Given any Player II path $g_0(t)$ (a path starting at 0 and known to I), a simple calculation gives the optimal response $\bar{f}_x(t)$ for the agent of Player I starting at $x$. This will be a unit speed path with no change of direction, that intercepts $g$ at the earliest possible time

$$T_x(\bar{f}, g) = \min\{t\colon d(x, g(t)) = t\}.$$

Since it is not possible for two such paths (starting at distinct $x$'s) to meet, it follows that either all the paths $\bar{f}_x$, $x \in C$, go in the same direction, or there is a single starting point $x'$ with the following property: from all starting points $x$ on the clockwise arc from $x'$ to 0, $\bar{f}_x$ goes clockwise; on the similar arc clockwise from 0 to $x'$ it goes counterclockwise; from $x'$, either direction is optimal. Note that in particular for the stationary (constant) strategy $g(t) \equiv 0$, $x'$ is the antipodal point from 0; for the clockwise strategy $g(t) = t$ the optimal response is given by $\bar{f}_x(t) = x - t$, for all $x \in C$, so that all the paths go in the same direction (counterclockwise).

To calculate the expected rendezvous time observe that the intervals

$$J(t) = \{x \in C\colon T_x(\bar{f}, g) \le t\} = [g(t) - t, g(t) + t]$$

form a nondecreasing family of arcs of length (Lebesgue measure $\lambda$) $\lambda(J(t)) = 2t$ for $0 \le t < 1/2$, with $J(1/2) = C$. Consequently, the optimal expected meeting time for the strategy $g$ is given by

$$\int_{x \in C} t \, d\lambda(J(t)) = \int_0^{1/2} 2t \, dt = 1/4.$$

It follows that *every* strategy $g$ of the blind Player II is optimal for the half-sighted game, and that the rendezvous value for this version of the problem is 1/4. Summarizing these results for the half-sighted version, we have the following.

**Theorem 13.11** *For the half-sighted rendezvous problem on the circle, every strategy of the blind player is optimal, and the optimal responses of the sighted player go at unit speed without any changes of direction from every starting point. Furthermore, there is at most one point of the circle where the optimal response direction changes (and the response is nonunique). The asymmetric rendezvous value for the half-sighted version is 1/4.*

The asymmetric rendezvous value for the half-sighted version on the circle cannot be larger than the for the original version. Hence the asymmetric rendezvous value for the circle cannot be less than 1/4. On the other hand the OP-DIR strategy pair for the circle achieves an expected meeting time of 1/4, so OP-DIR must be optimal and 1/4 must be the asymmetric rendezvous value. So we obtain the following result originally due to Howard (1999).

**Corollary 13.12** *The optimal asymmetric strategy for the fully labeled circle C (symmetry group $G_1$ consisting only of the identity transformation), with initial uniform distributions for both players, is the OP-DIR strategy of Section 12.3: one player goes clockwise and the other counterclockwise, both at maximum (unit) speed. Assuming the circumference is 1, this gives an asymmetric rendezvous value of* $R^a(C, G_1) = 1/4$.

Since OP-DIR does not require that the players have a common labeling of points on the circle, it is also feasible (and hence optimal) for the common direction problem (with a smaller feasible set of strategies) discussed in the following chapter (see Corollary 14.11).

# Chapter 14

# Asymmetric Rendezvous on an Unlabeled Circle

The simplest compact search region that is one dimensional (a network) and possesses varied types of symmetry is the circle. The circle has also provided a very natural setting for studying the rendezvous problem, and we present the results of that study in this chapter.

In Section 12.3 six versions of circle rendezvous were presented, along with the original strategies suggested for them by the first author. These versions correspond to the player-asymmetric or player-symmetric settings on the one hand, and three types of common information of the other: common labeling of points, common notion of direction, and no common notion of direction. We refer to the second distinction by calling the problem rendezvous on the labeled circle, the directed circle, and the undirected circle. These can be formalized by the respective symmetry groups $G_1$, $G_2$, and $G_3$, as discussed earlier in Section 12.3.

Much of the analysis of the previous chapter concerned rendezvous on a labeled circle $C$ (presented as the unit interval [0, 1] mod 1), corresponding to the symmetry group $G_1$ consisting only of the identity transformation. That is, the players had a common labeling of all points. For the uniform initial distribution, we derived both the result of J. V. Howard (1999) that the optimal strategy for the asymmetric version is for the players to go at maximum speed in opposite directions (OP-DIR), and the result of the first author[12] that in the symmetric version an optimal strategy is for the players to go to an agreed location and then stop (FOCAL). These results agreed with the strategies suggested Section 12.3.

In this chapter we will be concerned with player-asymmetric rendezvous in the two cases in that no common labeling of points on the circle is available. The first (and easier) case we will analyze is where the players merely have a common notion of direction around the circle (e.g., clockwise). This corresponds to the rotation group $G_2$. This version can be reduced to a one-sided search problem in which a single player searches for a stationary object hidden on the circle (see Section 14.1). We will then study the more difficult case where even that information in not available, so the players do not even have a common notion of "up" (that would determine clockwise). This corresponds formally to the case of the full symmetry group $G_3$.

We now present the formal model for circle rendezvous, that applies both to the directed circle and the undirected circle. The initial positions of Players I and II are determined by choosing an arc of length $x, 0 \leq x \leq 1/2$, and placing them randomly at opposite ends. In the common-direction version, both players are initially placed with clockwise as forward, while in the no-common-direction version they are faced independently and equiprobably in either direction. The initial distance $x$ between the players is chosen according to a continuous positive density function $w(x)$ defined on the interval $[0, 1/2]$, with $\int_0^{1/2} w(x)\,dx = 1$. Note that in this notation, the uniform initial distribution corresponds to the constant density function $\bar{w}(x) = 2$. Note that the directed clockwise distance $y$ from Player I to Player II has a density function given by $w(y)/2$ and consequently a corresponding cumulative distribution function $W$ given by

$$W(y) = \frac{1}{2}\int_0^y w(x)\,dx, \quad \text{for } 0 \leq y \leq 1. \tag{14.1}$$

This formula assumes that the density function $w$ is defined for $x > 1/2$, and to do this we set $w(x) = w(1-x)$, because of our assumption that II is equally likely to be in either direction from I. Note that the cumulative distribution function of the directed distance $y$ satisfies the identity

$$W(y) = 1 - W(1-y). \tag{14.2}$$

We shall call a distribution with property (14.2) *symmetric*. As we shall see in this chapter, the common-direction rendezvous problem can be reduced to a one-sided search problem on a single circle, while the more difficult no-common-direction problem can be reduced to a problem in which two searchers with common aims coordinate so as to find a stationary object hidden on one of their respective circles.

The results of this chapter are taken from the first author's article [9], except for sections 14.2 and 14.1.1, which are new. The results of section 14.5 rely on Appendix B on Alternating Search, which is based on the first author's article with John Howard [23]. The first result in this area is due to Howard (1999), who considered the special case of the uniform initial distribution (our Corollary 14.11).

## 14.1 One-Sided Search on the Circle

In this section we consider a one-sided search problem in which a single unit speed searcher on the circle tries to find a stationary object that is hidden a clockwise distance $y$ from him, where $y$ is chosen according to a symmetric cumulative distribution function $W$. We do not need to assume that $W$ has a continuous density. While this problem is not in itself a rendezvous problem, its analysis will prove useful for both the common-direction and no-common-direction rendezvous problems on the circle.

A rendezvous search strategy for this problem is a continuous function $s$ with maximum speed (Lipshitz constant) 1, satisfying $s(0) = 0$. It possesses a derivative almost everywhere and the total distance it has traveled by time $r$ (its *total variation*) can be calculated simply as

$$[var\ s](r) = \int_0^r |s'(t)|\, dt. \tag{14.3}$$

Since $s$ has a maximum speed of 1, it follows that $[var\ s](r) \leq r$ for all $r$. The interpretation of a search strategy $s$ is that the searcher's position at time $t$ is $s(t)$ mod 1, with $s'$ positive corresponding to clockwise motion. Observe that every search strategy $s$ determines a distribution $F_s(t)$ giving the probability that the object will be found by time $t$ using search path $s$. We say that a strategy $\hat{s}$ *dominates* strategy $s$ if $F_{\hat{s}}(t) \geq F_s(t)$ for all $t \geq 0$. Note that in this case the expected capture time for $\hat{s}$ cannot be more than that for $s$. We will say a strategy is *dominating* if it dominates every other strategy.

It is intuitively obvious that the searcher should make full use of his maximum speed (here assumed to be 1) in that it is not optimal to move at a lower speed over a nontrivial time interval. It is easy to formulate this observation in a rigorous manner. The following Lemma will not be used until the next section.

**Lemma 14.1** *Consider the one-sided search problem on the circle for an object hidden according to a known distribution $W$. Any search strategy $s$ is dominated by a strategy $\hat{s}$ that moves at maximum speed. That is*

$$|\hat{s}'(t)| = 1 \quad a.e. \tag{14.4}$$

*Such a strategy will be called "fast." In particular, there is always a fast optimal search strategy.*

**Proof.** Assuming that the original strategy $s$ does not have this property, we simply define a new strategy that traces out the same path as $s$ but always moves at speed 1. It will thus reach any point that $s$ reaches and reach it no later than $s$ did. The new path $\hat{s}$ is formally defined by the formula

$$\hat{s}(r) = s([var\ s](r)).$$

This ensures that if $s$ has reached a location $y$ at a time $r$, by which time it has traveled a total distance $r_0 < r$ (so that $[var\ s](r) = r_0$), then $\hat{s}$ will reach location $y$ at the earlier time $r_0$. ■

We devote the rest of this section to determining a sufficient condition on $W$ such that the *Columbus* strategy $\tilde{s}(t) = t, 0 \leq t \leq 1$, (or its reverse) dominates every other strategy. (The strategy $\tilde{s}$ is the one that goes all around the circle in a clockwise direction at maximum speed 1, with corresponding distribution $F_{\tilde{s}}(t) = W(t), 0 \leq t \leq 1$.) Clearly, if $W$ is the uniform distribution corresponding to the *constant* density function $\bar{w} = 2$ on $[0, 1/2]$, the Columbus strategy $\tilde{s}$ is dominating. We will show moreover that the Columbus strategy is dominating for all initial distributions corresponding to densities on $[0, 1/2]$ that are *close* to constant. For a constant function the ratio of the maximum to the minimum is 1. With this in mind we define a notion that measures how close to constant a function is. (This term will always be applied to continuous functions on closed intervals, so the existence of the extrema below will not be a problem.)

**Definition 14.2** *A positive function is called* **slowly varying** *if the ratio of its maximum to its minimum does not exceed 2.*

We now derive conditions sufficient for the Columbus strategy to be a dominating search strategy. We would like to show that if by time $r$ a strategy goes a maximum distance $a$ clockwise and $b \leq a$ counterclockwise, it has less chance of finding the object than if it had gone clockwise the whole time, or $W(a) + W(b) \leq W(r)$. This motivates our sufficient condition (14.5) for the Columbus strategy to be dominating.

**Theorem 14.3** *Suppose that for all $x, r$ with $0 < x < r/3 \leq 1/3$, the distribution $W$ satisfies the condition*

$$W(x) + W(r - 2x) \leq W(r). \tag{14.5}$$

*Then the Columbus search strategy $\tilde{s}$, that goes all around the circle clockwise at speed 1, is dominating. The associated distribution of detection is*

$$F_{\tilde{s}}(r) = W(r). \tag{14.6}$$

**Proof.** Let $s$ be any other search strategy. Consider any time $r \leq 1$ when the searcher's distance from his starting point 0 is less than $r$. By changing the sign of $s$, if necessary, we can assume that $0 \leq s(r)$. (This means that his net motion is clockwise.) Let $a$ and $b$ denote, respectively, the maximum distances that the searcher has reached counterclockwise and clockwise of his starting point 0. It follows from our assumption on $r$ that $b < r$. Since the region searched by time $r$ is precisely an arc going from the point $a$ counterclockwise of 0 to the point $b$ clockwise of 0, the probability that the object has been found is given by

$$F_s(r) = W(a) + W(b). \tag{14.7}$$

Furthermore, the total distance (variation) $[var\ s](r)$ that the searcher has traveled by time $r$ satisfies

$$[var\ s](r) \geq 2a + b. \tag{14.8}$$

Since $s$ has maximum speed 1, it follows that $[var\ s](r) \leq r$, so that

$$2a + b \leq [var\ s](r) \leq r. \tag{14.9}$$

By (14.9) and the hypothesis (14.5) we have the required domination inequality with $x = \min(a, b)$:

$$F_s(r) \leq W(a) + W(b) \leq W(x) + W(r - 2x) \leq W(r) = F_{\tilde{s}}(r).$$

■

This result does not require that the initial distribution $W$ is given by a density function $w$. However, in that case we can say more.

**Corollary 14.4** *If the distribution $W$ is given by a continuous density function $w/2$ as in (14.1), and $w$ is either increasing on $[0, 1/2]$ or is slowly varying, then the Columbus search strategy $\tilde{s}$, that goes all around the circle clockwise at speed 1, is dominating.*

*In particular, this strategy is optimal in the sense that it minimizes the expected time to find the object.*

**Proof.** Let $a, r$ be given, with $0 < a < r/3 \leq 1/3$. Observe that

$$W(r) - [W(a) + W(r - 2a)] \tag{14.10}$$

$$= \int_0^r w(y)\,dy - \int_0^a w(y)\,dy - \int_0^{r-2a} w(y)\,dy \tag{14.11}$$

$$= \int_{r-2a}^r w(y)\,dy - \int_0^a w(y)\,dy. \tag{14.12}$$

So to establish the condition (14.5) of the previous theorem we need to show that (14.12) is positive. But since the ratio of the lengths of the intervals of integration in (14.12) satisfies

$$\frac{\text{length }([r-2a, r])}{\text{length }([0, a])} = 2,$$

the "slowly varying" hypothesis ensures that the expression (14.12) is non-negative. Now assume that $w$ is increasing on $[0, 1/2]$. If $r \leq 1/2$ the same observation (that would work even with ratio 1) shows that (14.12) is non-negative. If $r > 1/2$, the fact that $w(y) = w(1 - y)$ ensures that any interval (arc) on the circle, with length at least $a$ and not containing the point 0, has an $w$ integral exceeding that of $[0, a]$. Since this is true of the interval $[r, r - 2a]$, we are done. ■

It is worth noting that if $w(y)$ is decreasing on $0 \leq y \leq 1/2$, $w(1 - y) = w(y)$, and $w(1/2) > w(0)/2$, then $w$ is slowly varying.

Of course, the Columbus strategy $\bar{s}$ is not always optimal. For example, if the object is hidden at a distance 0.1 clockwise, 0.1 counterclockwise, and 0.2 clockwise with respective probabilities $1 - \varepsilon$, $\varepsilon - \varepsilon^2$, $\varepsilon^2$, then for sufficiently small $\varepsilon$ it is clear that an optimal strategy goes 0.1 clockwise, 0.2 counterclockwise, and finally 0.3 clockwise.

## 14.2 Asymmetric Rendezvous on a Directed Circle

We now consider the asymmetric rendezvous problem on the circle $C$ when the players have only a common notion of direction. This problem is variously referred to as *common-direction rendezvous on the circle*, or *rendezvous on a directed circle*. Formally, this corresponds to the case where the given symmetry group in the sense of Chapter 12 is the group $G_2$ of Section 12.3 consisting of all rotations. We assume that the distribution of the initial clockwise distance $y$ from I to II is a known function $W$. We may assume that both players are initially faced in the clockwise direction and that this is common knowledge. (Equivalently, they could both begin by *choosing* to face themselves in this direction.)

A strategy for a searcher is a unit speed function $f$ with $f(0) = 0$. The number $f(t)$ gives the net clockwise distance traveled by time $t$. If the players adopt strategies $f$ and $g$ and the initial placement is $y$, then they will meet at time

$$T(f, g; y) = \min\{t\colon f(t) - g(t) = y(\text{mod } 1)\}. \tag{14.13}$$

Consequently, their expected meeting time is given by

$$\hat{T}(f, g) = \int_0^1 T(f, g; y)\, dW(y). \tag{14.14}$$

An important general principle in the theory of rendezvous search is that when the two players have a common notion of direction, the asymmetric rendezvous problem is equivalent to a one-sided search problem (as in the previous section) where a single searcher seeks a stationary object hidden according to a known distribution. The maximum speed of the searcher is assumed to be twice the speed of the rendezvousers. The consequence of this observation is typically that the two rendezvousers should always move in opposite directions at their common maximum speed.

To illustrate this equivalence in the case of the directed circle, consider the search problem of the previous section, with the maximum speed of the searcher set at 2 (rather than 1, as assumed in that section). Suppose the searcher follows the strategy $z$ given by

$$z(t) = f(t) - g(t). \tag{14.15}$$

If the object is hidden a distance $y$ clockwise of the starting point 0 of the searcher, then the searcher following strategy $z$ will find it at time

$$T(z; y) = \min\{t: z(t) = y(\text{mod } 1)\},$$

that is, the same as the time $T(f, g; y)$ (14.13) taken for the rendezvousers to meet. Thus, we have shown the following equivalence on the circle of the common direction rendezvous problem and the one-sided speed 2 search problem.

**Theorem 14.5** *Consider the asymmetric rendezvous and search problems on a directed circle, with the parameter $y$ distributed in both cases according to a common distribution $W$. If the search strategy $z$ is related to the rendezvous strategy pair $(f, g)$ by the equation $z(t) = f(t) - g(t)$, then $z$ is optimal for the speed 2 search problem if and only if $(f, g)$ is an optimal pair for the common-direction rendezvous problem.*

Note that according to equation (14.15) the rendezvous pair $(f, g)$ determines the search strategy $z$ uniquely but that for a given $z$ there are in general many solutions $(f, g)$. However we know from Lemma 14.1 that there is always a "fast" optimal strategy $\hat{z}$ for the search problem, one that moves at maximum speed 2. For such a search strategy $\hat{z}$, however, $f$ and $g$ must have speed one in opposite directions. More precisely, we must have $f = \hat{z}/2$ and $g = -\hat{z}/2$. This analysis establishes the following.

**Corollary 14.6** *Consider the common-direction asymmetric rendezvous problem on the circle with the initial clockwise distance from Player I to Player II having distribution $W$. There is an optimal strategy pair of the form $(f, -f)$, where $2f$ is a "fast" optimal search strategy for the one-sided speed 2 search problem for a stationary object whose initial position is distributed according to $W$. Consequently, $f$ is optimal for the unit speed search problem.*

The reason that we established the connection between the two problems (rendezvous and search) is that in the previous section we obtained optimal solutions for the search problem under certain assumptions about $W$. We can now use the above Corollary to reinterpret those results in terms of common-direction rendezvous.

**Theorem 14.7** *Consider the asymmetric rendezvous problem on the directed circle C (assuming the two players have a common notion of direction). Suppose the initial distance between the players is chosen according to a positive continuous density function $w$. If $w$ is either increasing on $[0, 1/2]$ or is slowly varying, then an optimal strategy is for the players is the OP-DIR strategy of moving at maximum (unit) speed in opposite directions around the circle.*

**Proof.** Let $W$ be the symmetric cumulative distribution function corresponding to the density $w$. According to Corollary 14.4, the optimal search strategy under these assumptions is the Columbus strategy $\tilde{z}(t) = t$ that goes around the circle clockwise at unit speed. It follows from the last corollary that the pair $(\tilde{z}(t), -\tilde{z}(t)) = (t, -t)$ is an optimal strategy pair for the rendezvous problem. ■

In particular, since the uniform distribution is given by the slowly varying constant density $\bar{w} \equiv 2$, we have the following.

**Corollary 14.8** *For the asymmetric rendezvous problem on the circle, with a uniform initial distribution and a common notion of direction (symmetry group $G_2$), the OP-DIR strategy pair is optimal.*

This result was obtained by Howard (1999) as a consequence not of Theorem 14.7 but rather of Corollary 13.12, by the following argument: Since Corollary 13.12 shows that OP-DIR is optimal for the fully labeled circle (symmetry group $G_1$), and it is feasible for the problem with less (coarser) information considered here ($G_2$), it must be optimal for this problem too.

## 14.3 Asymmetric Rendezvous on an Undirected Circle: Formalization

In this section, taken (in part directly) from Alpern (2000), we consider the asymmetric rendezvous problem on a circle under the assumption that the players have no common labeling of points, nor any common notion of direction along the circle. We shall informally call this the *rendezvous problem on an undirected circle*. For those interested in a more formal description, we say that the given symmetry group is the full symmetry group $G_3$ (defined in Section 12.3) on the circle, the group generated by rotations and inversions. The initial placement of the players is the same as given in the previous section, except that after placement *they are randomly faced in either direction*.The distribution $W$ of the initial clockwise distance from Player I to Player II is known to both.

In the previous section, we showed how asymmetric rendezvous on a *directed* circle was equivalent to a problem where a *single searcher* seeks a single stationary object on a single circle. The case of rendezvous on an undirected circle is more complicated. We will show that it is equivalent to a problem in which *two searchers on two distinct circles* seek an object placed equiprobably on one of the circles. The two searchers can move at speed 2, but they move alternately (one at a time). This class of *alternating search* problems (based on the analysis of the first author and John Howard [23]) is discussed in Appendix B on Alternating Search, and some of the general results stated there will be used to solve the two-circle problem that arises in this section.

We now formalize the asymmetric rendezvous problem on an undirected circle. For this problem, a strategy $f$ for Player I is the same as in the previous section, except that $f(t)$ now measures the net distance that he has traveled by time $t$ *in the direction he was initially facing*. The meeting time corresponding to a strategy pair $(f, g)$ and an initial distance $y$ depends crucially on how the two players are initially faced. There are four cases to consider based on the four ways a pair of players may be faced. For example, if initially both players happen to be faced in the clockwise direction, then as in the previous section (14.13) they meet at the first time $t$ when $f(t) - g(t) = y (\text{mod } 1)$. The three other cases will be considered below. We now do a simple calculation to determine the probability $P_{f,g}(r)$ that players adopting strategies $f$ and $g$ will have met by time $r$, assuming the initial cumulative distribution $W$ (that is fixed in the analysis).

Define new variables $z_j$ by

$$\begin{aligned} z_1 &= f + g \\ z_2 &= f - g, \end{aligned} \tag{14.16}$$

and consider the four extreme values,

$$\begin{aligned} M_1(r) &= \max_{t \le r} z_1(t), \\ M_2(r) &= \max_{t \le r} -z_1(t), \\ M_3(r) &= \max_{t \le r} z_2(t), \\ M_4(r) &= \max_{t \le r} -z_2(t). \end{aligned} \tag{14.17}$$

If the two players start with I facing clockwise and II facing counterclockwise, then they will have met by time $r$ if either $M_1(r) \ge y$ or $M_2(r) \ge 1 - y$, where $y$ is the initial clockwise distance from I to II. In particular, they will definitely have met if $M_1(r) + M_2(r) \ge 1$, so we may restrict strategies $f, g$ so that $M_1(r) + M_2(r)$ never exceeds 1. In general, they will have met by time $r$ in this case with probability

$$p_1 = p_1(r) = W(M_1(r)) + [1 - W(1 - M_2(r))]. \tag{14.18}$$

There are four possible ways of initially facing the players, and the meeting probabilities in the other three are as follows:

$$\begin{aligned} p_2 &= W(M_2(r)) + [1 - W(1 - M_1(r))], \\ p_3 &= W(M_3(r)) + [1 - W(1 - M_4(r))], \quad \text{and} \\ p_4 &= W(M_4(r)) + [1 - W(1 - M_3(r))]. \end{aligned} \tag{14.19}$$

Since the four ways of initially facing the players are equally probable, it follows that the probability that players using rendezvous strategy $(f, g)$ will have met by time $r$

is given by $P_{f,g}(r)$, where

$$P_{f,g}(r) = (p_1 + p_2 + p_3 + p_4)/4, \quad \text{which by (14.18) and (14.19),}$$
$$= \frac{W(M_1(r)) + W(M_2(r))}{2} + \frac{W(M_3(r)) + W(M_4(r))}{2}. \tag{14.20}$$

The expected time required to meet for players using the strategy $(f, g)$ is then given by

$$E_{f,g} = \int_0^\infty (1 - P_{f,g}(r))\, dr. \tag{14.21}$$

## 14.4 Alternating Search on Two Circles

We now show that the problem of finding a rendezvous strategy pair $(f, g)$ that minimizes (14.21) is equivalent to the problem mentioned earlier of alternating search on two circles. To do this, we will adopt a change of variables and work instead with the equivalent variables $z_1$ and $z_2$ given in (14.16). We say these variables are equivalent in the sense that given $z = (z_1, z_2)$ in a suitable set $Z$, we may uniquely solve for $f$ and $g$ in (14.16) by the formulae

$$f = \frac{z_1 + z_2}{2},$$
$$g = \frac{z_1 - z_2}{2}. \tag{14.22}$$

As $f$ and $g$ range over all pairs of maximum speed one functions, $z$ ranges over the set $Z$ of pairs $(z_1, z_2)$ with *combined speeds* limited by 2:

$$Z = \{z \colon |z_1'(t)| + |z_2'(t)| \leq 2\}.$$

It is useful to think of $z_1(t)$ and $z_2(t)$ as the paths of two searchers for a single stationary hidden object who are located on distinct copies of the circle $C$, and move *one at a time* at speed 2. (As the periods of alternation become small, the limiting case has them both moving subject to a *combined* maximum speed of 2.) In this interpretation $z_i(t)$ measures the net clockwise distance that searcher $i$ has traveled (on circle $i$) by time $t$. The object is hidden equiprobably on either circle a clockwise distance $y$ from the relevant searcher (a location called 0), where $y$ is chosen from the symmetric distribution $W$. Nature's hiding the object on circle 1 corresponds to the case where Nature would place the two rendezvousers facing in opposite directions.

An equivalent linear version of this circular problem is obtained by having the two searchers start at the origins of two lines, and having Nature place *two* objects on one of the lines, one object at distance $y$ and the other in the opposite direction at distance $1 - y$, $0 \leq y \leq 1$. Note that after choosing the number $y$, there are in general two ways of placing the two objects, and we assume they are equally likely. The two searchers wish to minimize the expected time taken to find *one* of the objects.

Observe that if the objects are placed on line 1, and Searcher 1 has searched the interval $[-M_2, M_1]$ on line 1, then the probability that he has found an object is given by $W(M_1)+W(M_2)$. Similarly, if the objects have been placed on line 2 and Searcher 2 has searched the interval $[-M_4, M_3]$, then the probability he has found one object is $W(M_3)+W(M_4)$. Hence the probability that an object has been found by time $r$, using joint search strategies $z_1$ and $z_2$, is given by

$$\frac{W(M_1(r))+W(M_2(r))}{2}+\frac{W(M_3(r))+W(M_4(r))}{2},$$

which, of course, is the same as the second part of the formula (14.20) given in the previous section for $P_{f,g}(r)$. Thus we have shown that asymmetric rendezvous on the circle is equivalent to the alternating search problem on two circles, or its linearized version as the search for one of two objects placed on two lines. The importance of the new scenario is that we may use the solution concepts for alternating search problems established in Alpern and Howard (2000) and summarized in Appendix B on Alternating Search.

To illustrate the connection between the two problems, consider the asymmetric rendezvous problem on the circle with $y$ chosen uniformly on $[0, 1]$. In the two circle problem (this is easier to visualize than its linearized version) it seems intuitively clear (proofs will wait until the next section) that each searcher should pick a single direction, and that the alternation between the two searchers' motion is irrelevant. In any alternation scheme the object will be found by time $t$ with probability $t$, and hence in expected time 1/2. One such scheme is for both searchers to move at speed 1 for times $t$, $0 \leq t \leq 1$. That is, $z_1(t) = z_2(t) = t$. Solving for $f$ and $g$ from equation (14.16), we obtain $f(t) = t$ and $g(t) = 0$, or simply that Player II waits for Player I to find him. Another scheme in the alternating search problem is for Searcher 1 to search his circle first, and then for searcher 2 to searcher his circle. That is, $z_1(t) = 2t$ for $t \leq 1/2$ and $z_2(t) = 2(t - 1/2)$ for $t \geq 1/2$. This translates in the rendezvous problem to both rendezvousers first moving forward at speed $1 (f'(t) = g'(t) = 1$ for $t \leq 1/2)$, followed by rendezvous Player II reversing while rendezvous Player I continues in the same direction $(f'(t) = 1, g'(t) = -1$ for $t \geq 1/2)$. Actually neither of these solutions represents the general solution that we shall find later, as these are unstable solutions in that they occur only for the uniform distribution and vanish entirely if the uniform distribution is slightly perturbed (see Theorem 14.12).

In the alternating search problem it is quite possible that, for certain distributions $W$, neither searcher follows a path on his circle that would be optimal (in the sense of minimizing the expected search time) if he knew for sure the object was on his circle. However, as shown for more general problems of alternating search in Alpern and Howard (2000), we may assume that neither player follows a path that is strictly dominated. So assume from now on that $W$ is a Columbus distribution, one for which a searcher in the one-sided search problem on the circle (Section 14.1) should follow the Columbus strategy $\tilde{w}$ of going around the circle at maximum speed in a single direction. It follows that in the alternating search problem on two circles, all we need to find is the optimal *alternation function* $\alpha$. As described in Appendix B on Alternating Search, this function determines which of the two searchers should be moving at any given time or, more generally (if both are simultaneously moving), their individual speeds. In the case that the two searchers move one at a time, we simply take $\alpha$ to be a function whose derivative is step function taking the two values 2 and 0: when $\alpha'(t)$ is 2, searcher 1 is moving at speed 2, when $\alpha'(t)$ is 0, searcher 2 is moving at speed 2. More generally, $\alpha'(t)$ is Searcher 1's speed at time $t$, and $2 - \alpha'(t)$ is Searcher 2's speed at time $t$. Consequently, the formulae for the positions of the two searchers ($z_1$ and $z_2$) are given in terms of $\alpha$ by

$$z_1(t) = \alpha(t), \quad z_2(t) = 2t - \alpha(t). \tag{14.23}$$

This formula holds for a general alternation function $\alpha$ that is nondecreasing and satisfies

$$\begin{aligned} &\alpha(0) = 0, \quad \alpha(1) = 1, \quad \text{and} \\ &|\alpha(t_2) - \alpha(t_1)| \le 2|t_2 - t_1|. \end{aligned} \tag{14.24}$$

Such a function has a derivative almost everywhere, that equals the derivative of $z_1$ and determines the derivative of $z_2$. We have $\alpha(1) = 1$ because by time 1 both circles will be fully searched.

If an alternation rule $\alpha$ is adopted, then the object will be found by time $t$ with probability

$$\begin{aligned} &Pr(y \le z_1(t)) = Pr(y \le \alpha(t)) = W(\alpha(t)), \quad \text{or} \\ &Pr(y \le z_2(t)) = Pr(y \le 2t - \alpha(t)) = W(2t - \alpha(t)), \end{aligned}$$

depending on whether it was hidden on circle 1 or circle 2. The unconditional probability that it will be found by time $t$ is given by

$$F_\alpha(t) = \tfrac{1}{2}[W(\alpha(t)) + W(2t - \alpha(t))].$$

Consequently, we seek an (optimal) alternation rule that minimizes the expected time $E_\alpha$ required to find the object,

$$E_\alpha = \int_0^1 t\, dF_\alpha(t).$$

To briefly summarize this section, what we have done is the following. Given the problem of finding asymmetric rendezvous strategies $f, g$ to minimize $E_{f,g}$, we have reduced it to finding the optimal "alternating search" pair $z_1, z_2$. Then, for $W$ satisfying the Columbus conditions of Section 14.1, we have further reduced it to finding the optimal alternation function $\alpha$.

## 14.5 Optimal Rendezvous on the Undirected Circle

We now complete the determination of optimal asymmetric rendezvous strategies for the undirected circle for a class of initial distributions $W$ that includes the uniform distribution. This analysis combines three independent results:

1. The equivalence of asymmetric rendezvous on an undirected circle to the "alternating search" problem faced by two searchers seeking a single object hidden on one of their circles.

2. Sufficient conditions on a distribution $W$ for "Columbus search" to be dominating in finding an object hidden on a single circle.
3. Conditions stated in Appendix B on Alternating Search, based on Alpern and Howard (2000), giving optimal alternation rules $\alpha$ for certain kinds of distributions.

### 14.5.1 Monotone densities

Our first result, Theorem 14.9, determines an optimal rendezvous strategy for a large class of monotone densities $w$ that includes the uniform density $\bar{w}(x) = 2$. The strategy we determine is in fact *uniquely* optimal for all the densities we consider which are *strictly* monotone. The optimal strategies will have one player going all around the circle at maximum speed 1, while the other player will have to determine a single number $c$ that will determine the times when he changes direction. The time $c$ depends on whether $w$ is increasing or decreasing, as follows. If $w$ is increasing, define $c$ to be the smallest number in $[1/2, 1]$ that maximizes the quotient $W(y)/y$. If $w$ is decreasing, define $c$ to be the largest number in $[0, 1/2]$ that minimizes the quotient $(1 - W(y))/(1 - y)$. In both cases the number $c$ separates the two regions where $\bar{W} = W$ and where $\bar{W} > W$. Here $\bar{W}$ denotes the concavification of $W$, the smallest concave function that is pointwise at least as large as $W$.

**Theorem 14.9 (monotone densities)** *Consider the asymmetric rendezvous problem on the undirected circle $C$. (This corresponds to no common notion of location or direction, as given by the symmetry group $G_3$.) If the initial distance $x$ between the players is given by a monotone density $w(x)$, $0 \leq x \leq 1/2$, the following strategies are optimal, and ensure a meeting by time 1. In both cases one player (I) goes all around the circle at maximum speed one.*

*(i) If $w$ is increasing, then Player II moves at maximum speed in a random direction for time $c/2$, then reverses direction at maximum speed for time $c/2$, and finally stays still from time $c$ until time 1. If $w$ is strictly increasing, this strategy is uniquely optimal.*

*(ii) If $w$ is decreasing and $w(1/2) \geq w(0)/2$, Player II stays still for time $c$, then moves at speed one in a random direction for time $(1 - c)/2$, then reverses direction until time 1. If furthermore $w$ is strictly decreasing, this strategy is uniquely optimal.*

**Proof.** (i) Consider the alternating search problem on two circles with object distribution $W$ having density $w(x)/2$, where $w(y) = w(1 - y)$ for $y \geq 1/2$. The assumption that $w$ is increasing on $[0, 1/2]$ ensures (by the proof of Corollary 14.4) that condition (14.5) holds, and that the only undominated way of searching for a single object on one circle with distribution $W$ is the Columbus search, a simple search in a single direction at speed 1. So the analysis of Appendix B may be applied determine the optimal alternation rule $\alpha$. The assumption that $w$ is increasing, together with the definition of the cutoff time $c \geq 1/2$ for this case given above, implies that the concavification $\bar{W}$

of $W$ has the following form.

$$\bar{W}(y) = \begin{cases} yW(c)/c, & \text{if } 0 \le y \le c, \text{ and} \\ W(y), & \text{if } c \le y \le 1. \end{cases}$$

By Theorem B.5 in Appendix B on Alternating Search (part 1, with $0 = a$, $c = b$) we see that the intervals $[0, c]$ on the two circles must be searched alternately at speed 2. That is,

$$z_1' = 2,\ z_2' = 0 \quad \text{on } [0, c/2] \qquad \text{and} \qquad z_1' = 0,\ z_2' = 2 \quad \text{on } [c/2, c].$$

The interleaving problem that remains at time $c$ (normalized to time zero) has a concave function $W$ (that is strictly concave if $w$ is strictly increasing). Hence the second part of Theorem B.5 shows that an optimal strategy (uniquely if $w$ is strictly increasing) for the remaining time is for both searchers to move at speed one, so

$$z_1' = z_2' = 1 \quad \text{on } [c, 1].$$

If we interpret this solution in terms of the original rendezvous problem, by solving (14.15) for $f$, $g$ in terms of $z_1$, $z_2$, we get the claimed optimal strategy

$$f' = 1 \text{ on } [0, 1], \quad g' = 1 \text{ on } [0, c/2], \quad g' = -1 \text{ on } [c/2, c], \quad g' = 0 \text{ on } [c, 1].$$

(ii) If $w$ is decreasing and $w(1/2) \ge w(0)/2$, then it is slowly varying. Consequently, the Columbus strategy is undominated for searching an object on a single circle. Proceeding as in (i), we have in this case that $c \le 1/2$ and the concavification of $W$ is given by

$$\bar{W}(y) = \begin{cases} W(y), & \text{if } 0 \le y \le c, \text{ and} \\ W(c) + \dfrac{(y-c)(1-W(c))}{1-c}, & \text{if } c \le y \le 1. \end{cases}$$

$W$ is (strictly) concave on $[0, c]$ so Theorem B.5 (part 2) gives the (uniquely) optimal strategy $z_1' = z_2' = 1$ there. $W$ lies below $\bar{W}$ on the interior of $[c, 1]$ so Theorem B.5 (part 1) gives the (uniquely) optimal strategy $z_1' = 2$, $z_2' = 0$ on $[c, (c+1)/2]$ and $z_1' = 0$, $z_2' = 2$ on $[(c+1)/2, c]$. Solving (14.16) for the corresponding rendezvous strategy, we get the claimed solution

$$f' = 1 \text{ on } [0, 1], \quad g' = 0 \text{ on } [0, c],$$

$$g' = 1 \text{ on } \left[c, \frac{c+1}{2}\right], \quad g' = -1 \text{ on } \left[\frac{c+1}{2}, 1\right].$$

■

We complete this section with an example from Alpern (2000) of a monotone density $w$ on the initial undirected distance between the players, where we can determine unique and explicit optimal rendezvous strategies. We determine the number $c$ and the concavification $\bar{W}$.

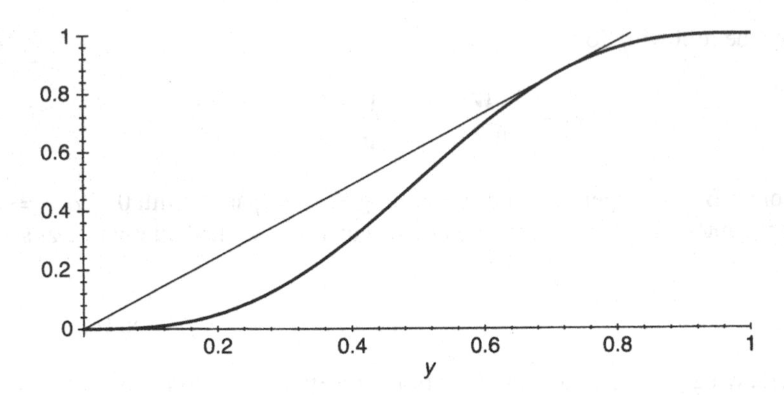


**Figure 14.1.** Analysis for density $w(x) = 2 + 2\sin(2\pi[x - 1/4])$

Consider first the density $w(x) = 2 + 2\sin(2\pi[x - 1/4])$ that is increasing on the interval $[0, 1/2]$. The corresponding symmetric density on $[0, 1]$ is given by $W(y) = y - \cos(2\pi[y - 1/4])/2\pi$, that is drawn in bold in Figure 14.1. We determine the cutoff time $c \approx 0.715$ by maximizing $W(y)/y$ on $[0.5, 1]$. For $y \leq c$ the concavification $\bar{W}$ of $W$ is given by the tangent line to $W$, $\bar{W}(y) = W(c)y/c \approx 1.2172y$ that is above $W(y)$, and for $y \geq c$ we have $\bar{W}(y) = W(y)$. The line $W(c)y/c$ that is tangent to $W$ at $x = c$ is drawn as a thin line.

Consequently, by Theorem 14.9 (i), Player I moves in a fixed direction at unit speed all the time, while II moves in a fixed direction at unit speed on the time interval $[0, 0.3075]$, moves in the opposite direction at unit speed during the time interval $[0.3075, 0.715]$, and remains stationary during the time interval $[0.715, 1]$.

## 14.5.2 Slowly varying densities

The next result applies to continuous densities $w$ on $[0, 1/2]$ that are *slowly varying* in the sense that *their maximum divided by their minimum is less than two*. Such densities are in a sense close to the uniform density, where this ratio has the minimum possible value of 1. An example of the occurrence of this type of distribution has been seen in the previous theorem, since the assumptions that $w$ is decreasing and $w(1/2) > w(0)/2$, together imply that $w$ is slowly varying.

**Theorem 14.10 (slowly varying densities)** *If the initial distance between the players is given by a slowly varying density, then any optimal strategy has one player going all around the circle at speed one, and the other player back at his starting point at time 1, by which time they have met.*

**Proof.** If $w$ is slowly varying, then Corollary 14.4 says that the Columbus strategy is optimal for searching for an object hidden according to $W$ on a single circle. Consequently, for any alternation rule $\alpha$ we have equation (14.23), and the alternating search problem has solution $z_1(t) = \alpha(t)$, $z_2(t) = 2t - \alpha(t)$. If we reinterpret this solution in terms of the asymmetric rendezvous problem using (14.16), we get $f(t) = 2t/2 = t$ and $g(1) = (\alpha(1) - (2t - \alpha(1)))/2 = (2\alpha(1) - 2)/2 = 0$, since $\alpha(1) = 1$. ■

We now consider how our last two theorems apply to the case of the uniform initial distribution, given by the constant density $\bar{w} = 2$. Since the constant $\bar{w}$ is both increasing and decreasing, both parts of Theorem 14.9 (on monotone densities) apply (with $c = 1/2$). Furthermore, a constant is certainly slowly varying, so Theorem 14.10 applies as well, and in particular, the WFM strategy suggested in Section 12.3 is optimal. Hence we obtain as a corollary the following result originally proved by Howard (1999) using a direct argument applying to the uniform distribution.

**Corollary 14.11 (uniform density)** *Consider the asymmetric rendezvous problem on the undirected circle $C$ (with its full symmetry group $G_3$). This means that the players have no common notion of location or direction on $C$. Assume that the players are initially uniformly distributed around the circle. Then any optimal strategy has one player going all around the circle at speed one, and the other player back at his starting point at time 1, by which time they have met. In particular, the WFM ("Wait for Mommy") pair is optimal.*

### 14.5.3 Arbitrary distributions

Corollary 14.11 says that the "Wait for Mommy" (WFM) strategy pair is optimal for the undirected circle when the initial distribution is uniform. This is in stark contrast to the general results of the authors and Anatole Beck (Theorem 16.13 and Corollary 16.14) that WFM is never optimal for asymmetric rendezvous on the line. The word "never" in this context means "for no initial distribution of the initial distance between the players." However the following result shows that the nonoptimality of WFM is still true on the circle in a generic sense.

**Theorem 14.12 (arbitrary distributions)** *Consider the rendezvous problem on the circle with no common locations or direction (symmetry group $G_3$). The only initial distribution of initial distance between the players for which the WFM strategy is optimal for asymmetric rendezvous is the uniform distribution.*

**Proof.** Suppose that $f' = 1, g' = 0$ (WFM) is the optimal strategy for some distribution $W$. Then in the alternating search problem on the circle, the solution is $z_1' = z_2' = 1$, that corresponds to $\alpha' = 1$ on all of $[0, 1]$. By Theorem B.5 (part 1) this implies that $W$ never lies below its concavification $W$, because in that case we would have $\alpha' = 0$ on some nontrivial interval. Since by definition $W$ is never above its concavification, it follows that $W$ must be concave. However, since $W(1 - y) = 1 - W(y)$, the concavity of $W$ on $[1/2, 1]$ implies the convexity of $W$ on $[0, 1/2]$. So we must have that $W$ is concave and convex on $[0, 1/2]$, which together with the fact that $W(1/2) = 1/2$ implies that $W(y) = y$ on $[0, 1/2]$. For $y \geq 1/2$ we then have $W(y) = 1 - W(1 - y) = 1 - (1 - y) = y$, since $1 - y \leq 1/2$. We have shown that if WFM is optimal, then the initial distribution must be the uniform distribution given by $W(y) = y, 0 \leq y \leq 1$. ■

It should be noted that even for the uniform distribution, WFM is not the unique optimal strategy.

# Chapter 15

# Rendezvous on a Graph

This chapter analyzes a discrete version of the rendezvous problem that includes and generalizes the "telephone problem" that we described in the Introduction to Book II (Chapter 10). At time 0 the two rendezvousers are randomly placed on the (finite) vertex set $\mathcal{V}$ of a graph $X = (\mathcal{V}, \mathcal{E})$. In this context, the search region $Q$ is identified with the vertex set $\mathcal{V}$. At each integer time $i$, each player stays still or moves to another vertex via an edge $e$ in the edge set $\mathcal{E}$. Both players wish to minimize the first time $i$ (denoted $T$) when they occupy the same vertex. Of course if the graph $X$ is not connected and the players start at vertices in different components of the graph then they can never meet. To avoid this problem we will assume that $X$ is connected. As in the continuous rendezvous problem formalized in Chapter 12, we need to consider how the symmetry of the graph affects the difficulty of rendezvous. So we assume that there is a given group $G$ of symmetries (isometries) of the graph $X = (\mathcal{V}, \mathcal{E})$. For a graph, an isometry is a bijection $g : X \to X$ that preserves the notion of adjacency in the sense that $(x, y) \in \mathcal{E}$ implies that $(gx, gy) \in \mathcal{E}$. For simplicity, we will restrict our attention to the case where $G$ is transitive. This implies in particular that the full group of all isometries of $X$ is transitive. Such a graph $X$, for which there is an isometry taking any vertex into any other vertex, is called *vertex-transitive*. Roughly speaking, this means that all vertices "look alike." In this chapter the letter $n = \#\mathcal{V}$ will always denote the number of vertices in the graph $X$. Unless otherwise stated, all the results of this chapter are taken from Alpern, Baston, and Essegaier (1999).

Let $\mathcal{P} = \mathcal{P}(X)$ denote the set of all paths in $X$, functions from the natural numbers into $\mathcal{V}$ that preserve adjacency. We interpret this to include paths that stay at the same vertex, by considering vertices to be self-adjacent. For pairs $(r, s)$ of paths we define discrete analogs of the meeting time and rendezvous value formulae given for continuous models in Chapter 12.

$$T(r,s) = \min\{i\colon r(i) = s(i)\},$$

$$\hat{T}(r,s) = \frac{1}{(\#(G))^2} \sum_{g_1,g_2 \in G} T(g_1 r, g_2 s) = \frac{1}{\#(G)} \sum_{g \in G} T(gr, s),$$

$$R^a(X,G) = \min_{r,s \in \mathcal{P}(X)} \hat{T}(r,s), \quad \text{and}$$

$$R^s(X,G) = \min_{\nu} \int_{\mathcal{P}(X)} \int_{\mathcal{P}(X)} \hat{T}(r,s)\, d\nu(r)\, d\nu(s),$$

where $\nu$ varies over probability measures on the path space $\mathcal{P}(X)$. Since the group $G$ is transitive, the definition of the expected meeting time $\hat{T}$ incorporates an equiprobable initial distribution. A more general formulation of rendezvous on graphs is developed in Alpern, Baston, and Essegaier (1999), including cases in which some vertices may be distinguishable (the symmetry group $G$ is not transitive). In that context the players may be allowed to choose their initial location as part of their strategy.

It is important to note that rendezvous on a graph is *not* a discrete version of network rendezvous. Recall that the notion of an H-network (see Chapter 13) was defined to ensure that players who pass each other along an arc of the original (continuous) network will be simultaneously at a common node of the discrete analog, so that they will meet. The situation in graph rendezvous context of this chapter is quite different, however. Players who are at adjacent vertices in one period and have transposed their positions at the next period are *not* considered to have met. This distinction was observed by Anderson and Weber (1990), who first proposed the graph formulation.

The "telephone problem" described in Chapter 10 can be viewed as a rendezvous problem on a graph. Consider the perspective of an observer who labels the set of paired (that is, wired together) telephones as $1, \ldots, n$, and notes the positions of the two players in terms of these labels. Since the players may move in any way they like from one telephone to another, their motions determine paths on the complete graph $K_n$ with vertices $1, \ldots, n$, and edges between *any* two vertices (the last condition defines what is meant by a *complete* graph). The possibility of generalizing this problem to arbitrary connected graphs, where not all transitions are allowed, was suggested in the pioneering article of Anderson and Weber (1990) (that mainly studied the problem on $K_n$), and their suggestion was carried out in a paper by Alpern, Baston, and Essegaier (1999), on which this chapter is mainly based.

We note that in the paper of Anderson and Weber (1990) on this subject, the problem begins at time $i = 1$ and it is assumed that the initial positions of the players are chosen randomly but distinctly. Consequently their formulae for meeting times differ from the ones given here (that are as in Alpern, Baston, and Essegaier, 1999), where the problem begins at time zero with the possibility of a common initial vertex (independent initial placement).

## 15.1 Asymmetric Rendezvous

The asymmetric rendezvous value of a graph can sometimes be determined exactly, as we show in this section. If a graph has a distinguishable location, it may be possible for the players to go directly there, and obtain an early meeting. However if such a possibility is precluded by an appropriate assumption of symmetry, then the following lower bound on the expected meeting time is obtained. This generalizes a similar result of Anderson and Weber (1990).

**Lemma 15.1** *Let $X = (\mathcal{V}, \mathcal{E})$ be a graph with a transitive isometry group G. Then*

$$R^a(X, G) \geq (n-1)/2, \quad \textit{where} \quad n = \#\mathcal{V}.$$

**Proof.** Let $A_j$ denote the event that the two players occupy the same vertex at time $j$. If Player I chooses the strategy (path) $r = (r_0, r_1, \ldots)$ and Player II chooses $s = (s_0, s_1, \ldots)$ then the event

$$A_j = \{g \in G\colon g(r_j) = s_j\},$$

has probability (normalized counting measure on $G$) $1/n$ regardless of $r_j$ and $s_j$. Hence $F(k)$, defined as the probability that they meet within the first $k$ steps $0, 1, \ldots, k-1$ (or that $T \leq k$) is less than or equal to $k/n$. Consequently, for any strategy pair $(r, s)$, the expected meeting time $\hat{T}$ satisfies

$$\hat{T} = \sum_{k=0}^{\infty}(1 - F(k)) \geq \sum_{k=1}^{n-1} \frac{n-k}{n} = \frac{n-1}{2}.$$

■

To present the remaining results of this section, we will need some definitions relating to paths on graphs. A path is called *Hamiltonian* if it includes every vertex exactly once. If a graph has a Hamiltonian path starting at any vertex, we will say that it is *efficiently searchable*. A graph is called *Hamiltonian* if it has a Hamiltonian circuit (a path whose final and initial vertices are adjacent is called a circuit). Certainly, any Hamiltonian graph is efficiently searchable, but the converse is not true. Consider, for example, the Petersen graph shown in Figure 15.1, that is efficiently searchable but not Hamiltonian.

Since this graph is vertex transitive, the existence of a single Hamiltonian path implies that there is one starting at any vertex. The following result is true more generally without the transitivity assumption (see [15]) but we will need it only in that case, where the proof is simple. Given the transitivity assumption, the hypothesis of efficiently searchable is equivalent to the existence of a single Hamiltonian path.

**Lemma 15.2** *Let G be a transitive symmetry group on an efficiently searchable graph X. Then the asymmetric rendezvous value satisfies $R^a(X, G) \leq (n-1)/2$.*

**Proof.** The proof uses a "Wait for Mommy" strategy, where one of the players is stationary. Suppose that the moving player traverses a Hamiltonian path starting at his

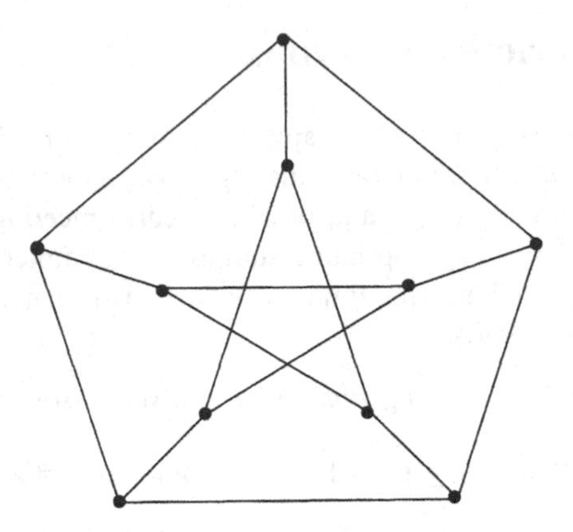

**Figure 15.1.** The 10-vertex Petersen graph

initial vertex. The transitivity assumption implies that the stationary player's position is random (equiprobable), so he will be met by the moving player with probability $1/n$ at each of the times $0, 1, \ldots, n-1$, with an expected meeting time of $(n-1)/2$. ∎

Combining our two lemmas, we have the following result (Alpern, Baston, and Essegaier, 1999).

**Theorem 15.3** *If X is a graph with a Hamiltonian path, and G is a transitive group of isometries of X, then the asymmetric rendezvous value is given by*

$$R^a(X, G) = (n-1)/2.$$

This result was proved earlier by Anderson and Weber (1990) for the special case where $X$ is a complete graph and $G$ is the group of all permutations of the vertex set. The results of this section may be seen as extension of their work. Note that this result applies to the Peterson graph, with its full group of isometries. (A Hamiltonian path for the Petersen graph can be obtained by first going around the five outer vertices, going to the adjacent inner vertex, and then following the nonrepeating path along the inner vertices.)

At this point it might seem appropriate to give an example of a graph $X$ that has no Hamiltonian path, together with a transitive group $G$ of isometries, where $R^a(X, G) > (n-1)/2$. However, if $G$ is transitive, then the graph $X$ must be vertex transitive. And as far as we know, Lovasz's conjecture (Lovasz, 1970), that all vertex transitive graphs have Hamiltonian paths, is still open. If this conjecture is true, then the above theorem holds without the assumption that $X$ has a Hamiltonian path. The conjecture has in fact been established for many classes of graphs, for example if $n$ is prime. See Alspach (1981) for more in this direction. In any case, the following is a weaker conjecture then that of Lovasz.

**Conjecture 15.4** *If $G$ is a transitive group of isometries of the graph $X$, then the asymmetric rendezvous value is given by*

$$R^a(X, G) = (n-1)/2.$$

## 15.2 Symmetric Rendezvous

In the symmetric rendezvous context, the "Wait For Mommy" (WFM) strategy is not available. A natural strategy to consider is a simultaneous random walk, where in each period the players pick an adjacent vertex randomly. Tetali and Winkler (1993) have shown that this always gives an expected meeting time no more than $16n^3/27$ and give an example where the expected meeting time is about $2n^3/27$. Of course, if the agents cooperate, they can do better. To see this, observe that every connected graph with $n$ nodes can be searched from any starting point within $2n$ steps (double every arc of a minimal spanning tree and follow an Eulerian circuit). Suppose that in each time interval of length $2n-1$ the agents search exhaustively with probability 1/2 and wait with probability 1/2. Then in each time interval of length $2n-1$, the probability that one player searches and one player waits (ensuring a meeting) is 1/2. Consequently, the players will meet in expected time no more than $(1 \cdot 1/2 + 2 \cdot 1/4 + 3 \cdot 1/8 \cdots)(2n-1) = 2(2n-1)$. Note that this weak estimate neglects the possibility that two moving players may meet. When this possibility is incorporated for a particular graph, the optimal moving probability will exceed 1/2. Consider a circuit of minimal length $\bar{\mu}$ ($\leq 2n$) that includes all of the vertices. Let $\tilde{\mu}$ denote the maximum, over all vertices, of the minimum length of a path starting at that vertex that visits all other vertices. If $X$ is efficiently searchable, $\tilde{\mu} = n$, and in any case $\tilde{\mu} \leq \bar{\mu} \leq 2n$. Substituting $\tilde{\mu}$ for $2n$ in the previous estimate, we obtain the following estimate for symmetric rendezvous on a graph.

**Theorem 15.5** *If $G$ is any given group of isometries on a connected graph $X$, then the symmetric rendezvous value satisfies $R^s(X, G) \leq 2(\tilde{\mu} - 1)$.*

In the next two sections we will concentrate on two classes of graphs: the complete graphs $K_n$ (the telephone problem) and the cycle graphs $C_n$.

## 15.3 Symmetric Rendezvous on a Complete Graph

As mentioned in the introduction to this chapter, the "telephone problem" (see Chapter 10) can be viewed as one of symmetric rendezvous on the complete graph $K_n$, with the full group $G$ of all permutations of the vertices. The first author mentioned this important problem to E. J. Anderson, who subsequently with his colleague R. Weber produced an elegant analysis that constitutes the first published work on rendezvous search theory. Anderson and Weber (1990) proposed the following family of strategies: if the search has not been successful by time $(n-1)k$, the next $n-1$ steps consist either of remaining still or going randomly to the $n-1$ remaining locations, with these two probabilities optimized to minimize the expected meeting time in the case that both searchers use it. We call this the *Anderson–Weber strategy* and denote it by $\pi_n$. Note that for $n = 2$, it stays still or searches with probabilities $(1/2, 1/2)$ (we call this the *random strategy*) and in the case $n = 3$ these optimal probabilities are $(1/3, 2/3)$.

Note that on $K_2$ each player knows the location of the other player (namely, the other node) after each period before the meeting time. Consequently, the symmetric rendezvous problem on $K_2$ studied by Anderson and Weber (1990) is equivalent to the repeated coordination game with revealed actions studied by Crawford and Haller (1990) (see the following subsection for more on the revealed action version). For this reason the following 1990 result, discussed and proved earlier as Theorem 11.3, may be attributed to either pair of authors.

**Theorem 15.6** *$R^s(K_2) = 1$, and the random strategy (stay or move equiprobably) is optimal.*

For general $n$ the family $\pi_n$ gives the following upper bound (Anderson and Weber, 1990) for $R^s(K_n)$.

**Theorem 15.7** *$R^s(K_n) \leq \hat{T}(\pi_n, \pi_n)$, which is asymptotic to $0.82888497n$.*

Anderson and Weber (1990) claimed (but did not rigorously prove) that $\pi_3$ is optimal for $K_3$. It has been shown subsequently (Alpern, Baston and Essegaier, 1999; Alpern and Pikounis, 2000) that $\pi_3$ is indeed optimal for $K_3$ if the strategy space available to the players is limited in certain natural ways, that we outline below.

Their strategy $\pi_3$ is what we call a 2-Markov strategy, in which the same probability vector $v = (v_1, \ldots, v_5)$, over the five possible 2-step strategies $\sigma_i$ listed below, is used for each of the time intervals 1, 2 and 3, 4, and more generally $2k + 1, 2k + 2$.

$$\sigma_1 = [a, a], \sigma_2 = [b, c], \sigma_3 = [a, b], \sigma_4 = [b, a], \sigma_5 = [b, b].$$

This notation, due to Anderson and Weber, is explained as follows. The location $a$ is the location at time $2k$. The location $b$ is a randomly chosen location from the other two possibilities. The location $c$ is the unique remaining location after $b$ has been determined in the first step. Thus the Anderson–Weber strategy is given by $\pi_3 \equiv (1/3, 2/3, 0, 0, 0)$. Let $\mathcal{S}_{n,m}$ denote the set of all strategies for $K_n$ that are $m$-step Markovian in the above sense. The following result Alpern, Baston and Essegaier (1999) established the optimality of the strategy $\pi_3$ within such a class, and we follow the original proof.

**Theorem 15.8** *The Anderson–Weber strategy $\pi = \pi_3$ is optimal with respect to the class $\mathcal{S}_{3,2}$.*

**Proof.** Given any probability vector in the simplex

$$\Delta_4 = \{v = (v_1, \ldots, v_5) : v_i \geq 0, \ \Sigma_{i=1,5}\, v_i = 1\},$$

it is easy to determine the expected meeting time $\hat{T}(v)$ in terms of the probabilities $p = p(v)$ and $q = q(v)$ that the agents will meet after one or two steps, respectively. So $r = 1 - p - q$ is the probability that the original problem reoccurs at the end of period 2. Then the expected meeting time satisfies the equation

$$\hat{T} = 1p + 2q + (\hat{T} + 2)r, \quad \text{that has solution } \hat{T} = (2 - p)/(p + q).$$

It is easy to check that $p(v) = v(P/4)v'$ and $q(v) = v(Q/4)v'$, where $'$ denotes transpose and

$$P = \begin{pmatrix} 0 & 2 & 0 & 2 & 2 \\ 2 & 1 & 2 & 1 & 1 \\ 0 & 2 & 0 & 2 & 2 \\ 2 & 1 & 2 & 1 & 1 \\ 2 & 1 & 2 & 1 & 1 \end{pmatrix}, \quad Q = \begin{pmatrix} 0 & 2 & 2 & 0 & 0 \\ 2 & 1 & 0 & 1 & 1 \\ 2 & 0 & 1 & 1 & 1 \\ 0 & 1 & 1 & 0 & 2 \\ 0 & 1 & 1 & 2 & 0 \end{pmatrix}.$$

(Note that in this chapter $Q$ does not denote the search region, so we may use it for the above matrix.) We claim that $p(v)$ and $q(v)$ both have maxima of 1/3 at $v = \pi$, so that $T$ is minimized there, with $T(\pi) = (2 - 1/3)/(1/3 + 1/3) = 5/2$. Note that

$$\pi P \pi' = \pi Q \pi' = (1/3 \quad 2/3) \begin{pmatrix} 0 & 2 \\ 2 & 1 \end{pmatrix} \begin{pmatrix} 1/3 \\ 2/3 \end{pmatrix} = \frac{4}{3}.$$

To establish the first claim, set $t = v_2 + v_4 + v_5$, so that $vPv' = 4t - 3t^2$, which has a maximum of 4/3, that equals $\pi P \pi'$. For the second claim, set $\alpha = v_2 + v_3$, $\beta = v_4 + v_5$, $w = \alpha + \beta$, and observe,

$$\begin{aligned} vQv' &= 2(v_2 + v_3)(2v_1 + v_4 + v_5) + v_2^2 + v_3^2 + 4v_4v_5 \\ &\leq 2\alpha(2 - 2\alpha - \beta) + \alpha^2 + \beta^2 \quad \text{(using standard inequalities)} \\ &= 2\alpha(2 - \alpha - w) + \alpha^2 + (w - \alpha)^2 = 4\alpha(1 - w) + w^2 \\ &\leq 4w(1 - w) + w^2 \leq 4/3 = \pi Q \pi' \quad (\text{since } \alpha \leq w). \end{aligned}$$

■

Recall that we have discussed in Chapter 12 the fact that a mixed strategy solution $(\nu, \nu)$ to the symmetric rendezvous problem must be a symmetric Nash equilibrium in the game where each player has utility given by the expected meeting time function $\hat{T}$. So to check that a strategy is a solution to the symmetric problem we can first ask if it is Nash. If it fails this easier test, it is not the symmetric solution. M. Pikounis and the first author have applied an easier version of this test to the symmetric Anderson–Weber strategy $(\pi_3, \pi_3)$ – that it passed. For the weaker test, we see if any of the primary strategies $\sigma_k$ do better against a proposed mixed strategy $s$ than does $s$ itself. If so, $s$ cannot be Nash and hence cannot be optimal. If none of the primary strategies does better against $s$ than $s$ itself, we will say $s$ is *locally Nash.* The justification of this term is given in Alpern and Pikounis (2000), where the following result is proven as Theorem 3.

**Theorem 15.9** *The Anderson–Weber strategy $\pi_4$ is locally Nash with respect to the class $\mathcal{S}_{4,3}$.*

We note that Anderson and Weber (1990) do not make any optimality claims at all for $\pi_4$. In fact they conjecture that it is not optimal.

We return now to a consideration of the strategy $\pi_3$. It was shown above that $\pi_3$ is optimal among 2 step Markov strategies. Since a two-step Markov strategy is also a four step Markov strategy, it is natural to ask whether it is still optimal in the larger class of four step Markov strategies. This question has not been answered, but the following weaker form of optimality within this class has been established in Alpern and Pikounis (2000).

**Theorem 15.10** *The Anderson–Weber strategy $\pi_3$ is locally Nash with respect to the class $\mathcal{S}_{3,4}$.*

We conclude this section with what should probably be called the Anderson–Weber conjecture, since they were the first to suggest the optimality of $\pi_3$.

**Conjecture 15.11** *The Anderson–Weber strategy $\pi_3$ is optimal on $K_3$.*

### 15.3.1 Revealed actions

Rendezvous search may be seen as a dynamic version of Schelling's version of the rendezvous problem described in Chapter 10. Unlike Schelling's version, where failure to meet in the first try ends the game, our version of rendezvous *search* has continued efforts by the players, until they eventually meet. The word search is italicized to emphasize that we have adapted the Schelling problem to the context of dynamic search theory. A different dynamic version of Schelling's problem was proposed by Crawford and Haller (1990), who adapted it to the context of repeated coordination games with revealed actions. In our terminology, Crawford and Haller considered the symmetric rendezvous problem on the complete graph $K_n$ under the assumption that each player's position is revealed at the end of each period. We may think of this as a search problem in which a searchlight illuminates the search domain at repeated intervals (or after each period of the discrete problem). Such searchlight versions have been studied in search theory, for example by Baston and Bostock (1991).

In some cases (e.g., the line), the illumination provided by a searchlight would make the subsequent optimal play quite trivial. However, this is not true for graph problems such as $K_n$, where the *telephone problem* with revealed actions might correspond, for example, to having the called phones ring (but too late for them to be answered). However, each player would know the phone that was called in the previous period.

The work that we describe below, taken from Ponssard (1994) and covering results of Crawford and Haller (1990) and Kramarz (1996), belongs to the significant economic literature on coordination games, that is a different offshoot of Schelling's version of rendezvous. It concerns a repeated game model of $n \times n$ coordination, where players wish to use the same strategy but have no common labeling of strategies (rows, columns). After each period, the strategies that were used are revealed. While not so appropriate for the spatial search models considered in rendezvous search, this revelation (searchlight, in our terms) accurately represents information available to players when the strategic choice made in each period are publicly visible signals such as prices set by firms or the choice of products to supply.

If $n = 2$, the random strategy is optimal, as in Anderson and Weber's result of $R^s(K_2) = 1$ given in the previous section. (For $n = 2$ the searchlight version is identical to the original, as players who have not met can obviously deduce the location of their partner, even though it is not explicitly revealed.) For $n = 3$ the optimal symmetric move from the position $(1, 0, 1)$ (from the point of view of an observer who puts a 1 in occupied positions) is clearly to go to the (middle) unoccupied position, that results in a definite meeting in the next time step. For $n = 5$ consider the position $(1, 1, 0, 0, 0)$ (there is really only one position, up to symmetry). Two strategies immediately present themselves. The first is to coordinate on the two occupied positions, giving an expected additional time $\hat{T}$ to meet of

$$\hat{T} = 1(1/2) + (1 + \hat{T})(1/2), \quad \text{with } \hat{T} = 2.$$

The second is to coordinate on the three unoccupied positions. In this case the players will choose a common location with probability 1/3 and otherwise definitely meet in the following period (by coordinating on the unique location that has always been unoccupied). This strategy gives the lower (better) expected additional time of

$$\hat{T} = 1(1/3) + 2(2/3) = 5/3.$$

Crawford and Haller (1990) show that the analyses given for the cases up to $n = 5$ are sufficient for the case of general $n$. Thus the searchlight version (*ringing* telephone problem) of symmetric search on $K_n$ does not present all the difficulties of the unsolved original (nonringing) telephone problem, that is still open for $n = 3$. However, the multi-agent version of this problem is only partially solved, as discussed in Chapter 17.

## 15.4 Symmetric Search on $C_n$

The cycle graph $C_n$ is given by the vertex set $\{0, 1, 2, \ldots, n-1\}$ with $i$ adjacent to $j$ if $|i - j| = 1 \pmod n$, and is best pictured by arranging these numbers (vertices) around a circle. Search games on this class of graphs were first studied by Ruckle (1983a, 1983b). In this section we consider the problem faced by two rendezvousers who are initially placed randomly on the vertices. We will first follow Ruckle, in assuming the players are restricted to directionally symmetric Markovian strategies. Since such strategies go equiprobably in either direction, their use implicitly assumes that the players have no common notion of direction around the circle. In this context we can derive optimal rendezvous strategies. Then we relax Ruckle's assumption and see how much this reduces the expected meeting time.

### 15.4.1 Symmetric Markovian strategies

Ruckle considered the search game on $C_n$ where the players are initially randomly placed and are constrained to use (directionally) symmetric Markovian strategies that in each period move clockwise with probability $p$, anticlockwise with probability $p$, and remain still with probability $1 - 2p$ . In Ruckle's analysis the searcher and hider chose their respective probabilities $p$ and $q$ to respectively minimize or maximize their expected meeting time. Here, we follow the analysis of Alpern (1995), in assuming both players must choose a common probability $p$, $0 \leq p \leq 1/2$, to minimize the expected rendezvous time $T_n(p)$ corresponding to the Markov chain in which both players move with probability $p$.

In order to evaluate the function $T_n(p)$ we consider a reduced Markov chain with $n$ states, where the state is the clockwise distance from Player I to Player II (mod $n$). The state 0 is absorbing; otherwise successive states are obtained by independent randomization of the players, with a common "moving" probability $p$. For $n > 4$, the transition probabilities $A = \{a_{i,j}\}_0^{n-1}$ for the reduced $n$-state Markov chain are given by the formulae

$$a_{ij} = \begin{cases} 1 & \text{if } i = j = 0, \\ 0 & \text{if } i = 0,\ j \neq 0, \\ p^2 & \text{if } j = i \pm 2 \ (\text{mod}\, n), \\ 2p(1-2p) & \text{if } j = i \pm 1 \ (\text{mod}\, n), \\ (1-2p)^2 + 2p^2 & \text{if } i = j \neq 0, \\ 0 & \text{otherwise.} \end{cases} \tag{15.1}$$

If $n = 4$, there is a slight modification to the above probabilities. Namely,

$$a_{i,j} = 2p^2 \quad \text{if } j = i \pm 2 \ (\text{mod } 4).$$

The expected meeting time $T_n(p)$ is the same as the expected number of periods that the reduced Markov chain is in any of the nonabsorbing states $\{1, 2, \ldots, n-1\}$. If we denote by $B = B(p)$ the $n-1 \times n-1$ submatrix of $A$ corresponding to these states, and let $e$ and $e'$ denote the column and row vectors consisting of all 1's, then the expected meeting time can be calculated as in the following lemma, taken from Alpern (1995). The full justification of a more complicated version of the formula in the lemma is given in Ruckle (1983b), where it is also shown that $T_n(p)$ is convex.

**Lemma 15.12** *The expected rendezvous time for the Markovian strategy p on the cycle graph $C_n$ is given by the formula*

$$T_n(p) = \frac{1}{n} \sum_{k=0}^{\infty} e' B^k e = \frac{1}{n} e'(I - B)^{-1} e.$$

Using a direct analysis for $n = 3$, and the above lemma together with either form of (15.1) for the entries of $B$, we obtain the following.

**Theorem 15.13** *Let $\hat{p}_n$ and $\hat{R}_n^s$ denote the optimal moving probabilities and symmetric rendezvous values for the cycle graph $C_n$, when the players are restricted to a common, directionally symmetric, Markovian strategy. Then*

$$\hat{p}_3 = \frac{1}{3}, \qquad \hat{R}_3^s = 2,$$

$$\hat{p}_4 = \frac{1}{3}, \qquad \hat{R}_4^s = \frac{27}{8},$$

$$\hat{p}_5 = \frac{2}{3}\left(1 - \frac{1}{\sqrt[3]{10}}\right), \qquad \hat{R}_5^s = T_5(\hat{p}_5) \doteq 4.88.$$

**Proof.** First consider the case $n = 3$, that we discuss without any of the above notation. Suppose the players are at distinct vertices of $C_3$ and they each move to another vertex with probability $2p$. With probability $\frac{1}{4}(2p)^2$ they will both move to the unoccupied vertex and meet there. With probability $(1 - 2p)(2p)$, one will stay still and the other will find him. Consequently the probability that they meet in the next period is $\frac{1}{4}(2p)^2 + (1 - 2p)(2p) = -3p^2 + 2p$, that has a maximum of $1/3$ when $p = \hat{p}_3 = 1/3$. This is the random strategy, in which a player goes equiprobably to any of the three locations, independent of his previous location. If they don't meet, they will be in the same situation in the following period. The minimum expected additional meeting time $x$, given that they start at distinct locations, consequently satisfies the following equation:

$$x = \tfrac{1}{3}(1) + \tfrac{2}{3}(1 + x).$$

Therefore, $x = 3$ and $\hat{R}_3 = \frac{1}{3}(0) + \frac{2}{3}(x) = 2$.

For $n = 4$ or 5 we use the Markov chain analysis and the previous lemma. First consider the case $n = 4$. Using the $n = 4$ version of (15.1), we get

$$T_4(p) = \frac{5 - 9p}{4(2p - 6p^2 + 4p^3)},$$

that has a minimum on $[0, 1/2]$ at $\hat{p}_4 = \frac{1}{3}$, and a symmetric rendezvous value of $\hat{R}_4 = 27/8$. When $n = 5$, we use the main transition probability formula (15.1), giving

$$T_5(p) = \frac{4 - 6p}{4p - 10p^2 + 5p^3}.$$

This has a minimum on $[0, 1/2]$ at $\hat{p}_5 = \frac{2}{3}\left(1 - 1/\sqrt[3]{10}\right)$, with a corresponding rendezvous value of about 4.88. ■

Numerical approximations of $\hat{R}_n$ and $\hat{p}_n$ for $n \leq 15$ are given in Alpern (1995).

If the above strategies are modified so that with probability $p$ the searcher moves not just one step but $k$ steps in a fixed direction, the expected meeting times can be reduced. The following least expected meeting times for symmetric rendezvous strategies on $C_n$ were calculated by S. Essegaier (1993) in his M.Sc. dissertation, supervised by Alpern:

| Step Length | $C_4$ | $C_5$ | $C_6$ | $C_7$ | $C_8$ |
|---|---|---|---|---|---|
| 1 | 3.37500 | 4.87655 | 6.64000 | 8.57987 | 10.7400 |
| 2 | 4.24141 | 4.12982 | 7.68495 | 7.61800 | 11.6400 |
| 3 | 3.37955 | 4.03499 | 6.26342 | 6.85714 | 8.8175 |
| 4 | 4.02473 | 3.60000 | 8.80062 | 6.55032 | 9.8037 |
| 5 | | 4.00000 | 5.84508 | 6.13841 | 9.1100 |
| 6 | | | 6.53709 | 5.57143 | 9.5330 |
| 7 | | | | 6.00000 | 8.3319 |
| 8 | | | | | 9.0472 |

A few observations can be made from this small table. Aside from $n = 4$, the smallest number comes from a step length of $n - 1$. For odd $n$, the expected meeting time decreases in step length up to $n-1$. For even $n$, odd step lengths perform better than even lengths.

Using the above analysis to concentrate on step lengths of $n-1$, and fairly complicated calculations, the following bounds for the symmetric rendezvous value of $C_n$ were obtained in Alpern, Baston, and Essegair (1999).

**Theorem 15.14** $R^s(C_n) \leq \begin{cases} \dfrac{(n-1)(2n-1)}{2n}, & \textit{if } n \textit{ is odd,} \\ 1.254 + O(1), & \textit{if } n \textit{ is even.} \end{cases}$

A sharper bound for even $n$ of $1.15n + O(1)$ is also mentioned in Alpern, Baston and Essegaier (1999).

# Part Four

# Rendezvous Search on Unbounded Domains

# Chapter 16

# Asymmetric Rendezvous on the Line (ARPL)

Up to this point we have been assuming that the search region $Q$ on which the rendezvousers have to meet is compact (closed and bounded). We now consider an important case where the search region is unbounded. Namely, the real line **R**. We assume here that the players have no common labeling of the points and (except in Section 16.2) no common notion of a positive direction on the line. The undirected unlabeled line is the search region that has attracted the most attention in the literature, both for the asymmetric and symmetric player cases. This search region models the examples where two people become separated on a road or a beachline. Another non-spatial example concerns the situation where two people carrying walkie-talkies with rotary frequency dials attempt to find a common frequency to talk on, since the set of frequencies form a line.

This chapter will be devoted to the (player-)*asymmetric* rendezvous problem on the line (ARPL). The symmetric version is still unsolved even for the simplest cases – the best results so far obtained will be presented in the next chapter.

Rendezvous on the line is similar to two previous search regions: the interval and the circle. The main change from rendezvous on the interval (as studied in Chapter 13) is that *absolute* location is no longer relevant because of the translational symmetry of the full line. It is replaced by *relative* location. This is because the players do not have a common labeling of points. In particular, the game will begin with a known distribution of the *distance between* the players, rather than a joint distribution over two locations. This is similar to the initial setup for rendezvous on the unlabeled circle (as studied in Chapter 14). In particular, it is clear that if two rendezvousers on the circle know they are very close together (relative to the circumference), then they can play as if they are on a line, using the analysis of this chapter rather than that of Chapter 14.

We consider the following scenario, originally proposed by Alpern (1995) for rendezvous on the line. At time $t = 0$, Nature places the two players a distance $d$ apart on the line and faces them independently and equiprobably in either direction. Each player regards the direction he is initially facing as "forward." The initial distance $d$ is drawn from a known cumulative distribution function that we call $F$. The random initial facing of the players is just a way of specifying that they have no common notion of direction along the line. The players each have unit speed and wish to meet in the least expected time. The set $\mathcal{S}$ of rendezvous strategies for the case of the line may be written as

$$\mathcal{S} = \{f : [0, \infty) \to \mathbf{R} : f(0) = 0,\ |f(t) - f(t')| \le |t - t'|\}.$$

The number $f(t)$ gives the player's net forward motion at time $t$ relative to his starting point. For example, if his initial position is $b$ and he is initially facing left, his actual path is $b - f(t)$.

## 16.1 Asymmetric Rendezvous Value $R^a(F)$

The *asymmetric* rendezvous problem on the line (ARPL) was first analyzed by Alpern and Gal (1995) (after the earlier introduction of the symmetric version). In the asymmetric player version, where the players may adopt distinct strategies, we denote the least expected meeting time by $R_F^a$; in the symmetric version by $R_F^s$.

The subsequent analysis is similar to that given earlier for the circle. We adopt the same change of variables (see Section 14.3) from a rendezvous strategy pair $(f, g)$ for the line to a new pair of variables $(z_1, z_2)$:

$$\begin{aligned} z_1 &= f + g, \\ z_2 &= f - g. \end{aligned} \tag{16.1}$$

The variable $z_1$ will be relevant when the rendezvousers are initially facing in opposite directions; $z_2$ when they are initially facing in the same direction. To determine whether the rendezvousers have met by time $u$, we consider the four extreme values,

$$\begin{aligned} M_1(u) &= \max_{t \le u} z_1(t), \\ M_2(u) &= \max_{t \le u} -z_1(t), \\ M_3(u) &= \max_{t \le u} z_2(t), \\ M_4(u) &= \max_{t \le u} -z_2(t). \end{aligned} \tag{16.2}$$

If the two players using strategies $f$ and $g$ were initially facing each other, then they will have met by time $u$ if and only if for some $t \le u$ their net relative motion towards each other, $z_1(t)$, is at least as large as their initial distance or, equivalently, if $d \le M_1(u)$. Similarly, if they were initially facing away from each other, they will have met by time $u$ if $d \le M_2(u)$. If Player I was initially facing Player II, but II was facing away from I, they will have met by time $u$ if $d \le M_3(u)$, and in the opposite case if $d \le M_4(u)$. Since each of these four "facing" cases is equally likely, and the initial distance $d$ is drawn from the cumulative distribution function $F$, it follows that the probability that players adopting the pair $(f, g)$ will have met by time $u$ is given by

$$P_{f,g}(u) = \frac{1}{4} \sum_{i=1}^{4} F(M_i(u)). \tag{16.3}$$

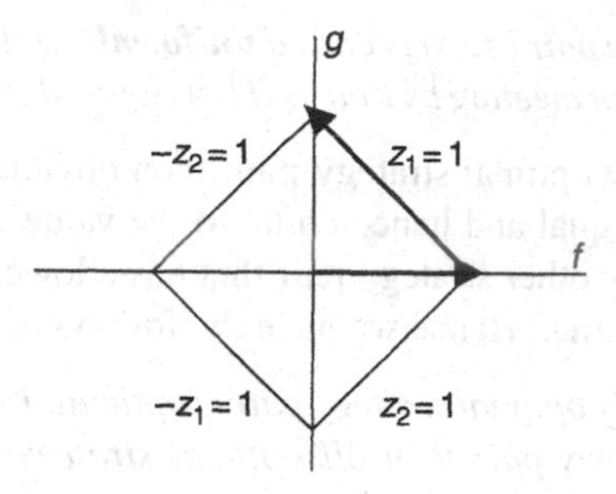


**Figure 16.1.** Strategy pair in *f,g* plane

For a given strategy pair $(f, g)$ the expected meeting time is given by $\int_0^\infty u\, dP_{f,g}(u)$. Consequently the asymmetric rendezvous value is given by

$$R^a(F) = \min_{f,g} \int_0^\infty u\, dP_{f,g}(u). \tag{16.4}$$

The existence of the minimum (rather than infimum) is justified by the same general arguments used in Chapter 12. Any pair $(f, g)$ for which the minimum is attained will as usual be called *optimal.*

To illustrate some of these ideas, suppose that $F$ represents the atomic distribution with all the mass at 1. That is, the initial distance between the players is known to be 1. In this case $F(M_i(u))$ is 1 if $M_i(u) \geq 1$; otherwise it is 0. Take $u = 2$. That is, we consider the probability that agent adopting some strategy pair will have met by time 2.

First consider the strategy pair $(\tilde{f}, \tilde{g})$, where $\tilde{f}$ moves forward at speed 1 for one time unit and then reverses at speed 1, while $\tilde{g}$ stays still up to time 1 and then moves forward at unit speed. In Figure 16.1 we draw the parametric path $(\tilde{f}(t), \tilde{g}(t))$ for $t \leq 2$ in $f, g$ space, together with the square determined by the four lines $\pm z_i = 1, i = 1, 2$.

Since the path has intersected three of the four sides of the square, the probability that the players have met by time 2 is $3/4$, $P_{\tilde{f},\tilde{g}}(2) = 3/4$. In general, for the atomic distribution, the meeting probability for any time $t$ can take only the values $i/4$, where $i \in \{0, 1, 2, 3, 4\}$. The integer $i$ is the number of the sides of the square determined by the lines $z_1 = \pm 1$, $z_2 = \pm 1$, that the path has met by time $t$.

It is sometimes useful to determine the maximum probability $\bar{P}(u)$ that the players can meet by a certain time $u$.

$$\bar{P}(u) = \bar{P}(u; F) = \max_{f,g} P_{f,g}(u). \tag{16.5}$$

It follows from the fact that for any pair $(f, g)$ we have $\bar{P}(u) \leq P_{f,g}(u)$ and the formula (16.4) that

$$\int_0^\infty t\, d\bar{P}(t) \leq R^a \leq \int_0^\infty t\, dP_{f,g}(t). \tag{16.6}$$

Note that even if $(f, g)$ is optimal we will in general have $\bar{P}(t) > P_{f,g}(t)$ for some values of $t$. However, it is useful to consider strategy pairs where this never happens.

**Definition 16.1** *A strategy pair $(f, g)$ is called* **uniformly optimal** *if for every time $u$, it maximizes the probability of meeting by time $u$. That is, for all $t$ we have $\bar{P}(t) = P_{f,g}(t)$.*

If $(f, g)$ is a uniformly optimal strategy pair, then obviously both extremes of the inequality (16.6) will be equal and hence equal to the value $R^a$. Consequently, $(f, g)$ is optimal. In this case any other strategy pair that has a lower meeting probability for some time $t$ cannot be optimal. Hence we have the following.

**Theorem 16.2** *A uniformly optimal strategy pair is optimal. Furthermore, if there exists a uniformly optimal strategy pair then all optimal strategy pairs must be uniformly optimal.*

## 16.2 Finiteness of $R^a(F)$

Of course, since the search region is unbounded, there may be no strategy pair that gives a finite expected meeting time. However, we can obtain information on the finiteness of $R^a(F)$ by comparing rendezvous on the line with the special case where Player II is stationary. That is, where $g$ is identically zero. This problem, originally posed by Bellman (1963) and Beck (1964) is known as the Linear Search Problem (LSP). It is shown Chapter 8 (Book I) that the stationary object (Player II in our application) can be found in finite expected time if and only if $F$ has finite mean. The following result of Alpern and Gal (1995) establishes a connection between the Linear Search Problem and the asymmetric rendezvous problem on the line (ARPL).

**Theorem 16.3** *Let $F$ be the distribution of initial distance in the asymmetric rendezvous problem on the line. If the two players have a common notion of direction along the line, then the rendezvous problem is equivalent to the Linear Search Problem. For the rendezvous problem without common direction, the rendezvous value satisfies*

$$L(F)/2 \leq R^a(F) \leq L(F), \tag{16.7}$$

*where $L(F)$ denotes the least expected time to find Player II in the Linear Search Problem (where he is required to stay still). Consequently, $R^a(F)$ is finite if and only if the distribution $F$ has finite mean.*

**Proof.** We begin by noting that the inequality $R^a(F) \leq L(F)$ is trivial, as the minimum problem defining $R^a(F)$ has a larger feasible set ($g$ need not be identically zero). To obtain the lower bound for $R^a(F)$, consider the related rendezvous problem on the line where the two players have a common notion of direction along the line (or equivalently are known to be faced in a common direction at time zero). Since this "common direction" problem gives the players more information, its associated rendezvous value cannot be larger than $R^a(F)$. However, for this common direction problem we may assume (as in the case of the directed circle in Chapter 14) that the two players always go in opposite directions, that $g = -f$, or $f - g = 2f$. Consequently, the times when the rendezvousers meet, when $2f = f - g = \pm d$, are the same as when a linear searcher following path $2f$ (with speed 2) finds the hidden stationary object. The least time for the latter problem is $L(F)/2$ and hence also for the equivalent common direction rendezvous problem. Hence we have established the required inequalities. Since the extreme values $L(F)/2$ and $L(F)$ are finite if and only if $F$ has finite mean, the same must be true for the intermediate rendezvous value $R^a(F)$. ■

We note that in fact the right-hand inequality of (16.7) is strict because the Wait For Mommy (WFM) strategy is never optimal (see Corollary 16.14 below).

## 16.3 The Double Linear Search Problem (DLSP)

In the previous section we showed that the asymmetric rendezvous problem on a *directed* line is equivalent to the Linear Search Problem, where a single searcher seeks a stationary object that is hidden according to a known distribution. We now demonstrate that the asymmetric rendezvous problem on the *undirected* line (where agents have no common notion of direction) is equivalent to the following problem (the DLSP) posed by Alpern and Beck (1999b). These equivalences use the same ideas as those given for the circle in Chapter 14, where it was shown that the asymmetric rendezvous problem was equivalent to a search problem on two circles.

**Definition 16.4** *In the Double Linear Search Problem (DLSP) a single object is randomly placed a distance d from the origin of one of two given lines. That is, with probability* 1/4 *in each of these four locations. The distance d is drawn from a known cumulative distribution function F. Two searchers, starting at the origins of the two lines, move until one of them has found the object. They may move alternately at speed* 2 *or more generally with combined speed* 2. *Their common aim is to find the object in least expected time.*

The DLSP can be defined more generally for an arbitrary distribution of the object over the two lines that is not symmetric between the lines or on each line. But the above doubly symmetric version will be adequate for our needs. The reader should also note that in the original version of the DLSP given in Alpern and Beck (1999b), the searchers had a combined speed of 1 rather than 2.

Figure 16.2 shows the four possible equiprobable locations of the object at $O_i \pm d$, once an initial distance $d$ is selected. Of course $d$ is generally not known to the searchers (except when the distribution $F$ has a single atom).

As in the case of the search problem on two circles discussed in Chapter 14, we denote the actual paths of the two searchers on their respective lines by $z_1(t)$ and $z_2(t)$,

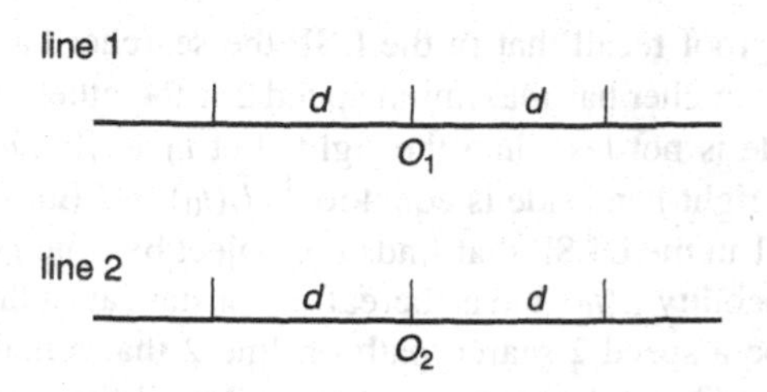


**Figure 16.2.** Double linear search problem

with speed restriction $|z_1'(t)| + |z_2'(t)| \le 2$, and initial positions $z_1(0) = z_2(0) = 0$. There is a one to one correspondence between asymmetric rendezvous pairs $(f, g)$ and DLSP strategies $(z_1, z_2)$ that is given by equation (16.1). The object in the DLSP will have been found by time $u$ by one of the two searchers if and only if it lies in the interval $[-M_2(u), M_1(u)]$ on line 1 or the interval $[-M_4(u), M_3(u)]$ on line 2. The probability it has been found by time $u$ is therefore given by $P^{DLSP}_{z_1,z_2}(u) = \frac{1}{4}\sum_{i=1}^{4} F(M_i(u))$ which by equation (16.3) is the same as the probability $P_{f,g}(u)$ that rendezvousers using $(f, g)$ will have met. Consequently, we have

$$P^{DLSP}_{z_1,z_2}(u) = P_{f,g}(u). \tag{16.8}$$

Since these probabilities respectively determine the expected finding (in the DLSP) or meeting (in the ARPL) times, we have the following.

**Theorem 16.5** *For any cumulative distribution function $F$, the rendezvous value $R^a(F)$ for asymmetric rendezvous on the (undirected) line is equal to the least expected time in the Double Linear Search Problem. Optimal strategy pairs in either problem lead to optimal pairs in the other according to equation* (16.1). *In this sense the two problems are equivalent.*

Note that the DLSP can be solved approximately, for any cumulative distribution function $F$, by the DP algorithm described in Book I.

## 16.4 Meeting-Probability Maximization

In this section we consider the problem of evaluating $\bar{P}(u)$, the maximum probability of rendezvousing by a given time $u$, as defined in (16.5). We do this by comparing it with the maximum probability of finding the object in the Linear Search Problem (with searcher speed 1) by time $u$, which we denote by $\bar{L}(u)$. We assume, of course, a common cumulative distribution function $F$ in the two problems. The study of the maximum probability $\bar{L}(u)$ was initiated by Foley, Spruill, and Hill (1991) and continued in Alpern and Beck (1997). The following is taken directly from Alpern and Beck (1999b).

**Lemma 16.6** *For any distribution $F$ and any $u \ge 0$, we have*

$$\bar{P}(u) = \tfrac{1}{2} \max_{0 \le t \le 2u} (\bar{L}(t) + \bar{L}(2u - t)). \tag{16.9}$$

**Proof.** (For this proof recall that in the LSP the searcher has maximum speed 1, while in the DLSP a searcher has maximum speed 2 if the other is still.) We first show that the left-hand side is not less than the right. Let $t_0 \in [0, 2u]$ be some time that the maximum on the right-hand side is equal to $\frac{1}{2}(\bar{L}(t_0) + \bar{L}(2u - t_0))$. Let $z_1(t)$ be a speed 2 path on line 1 in the DLSP that finds the object by time $t_0/2$ (if it is on line 1) with the optimal probability $\bar{L}(t_0)$ and is thereafter constant at its final location $z_1(t_0/2)$. Similarly, let $z_2(t)$ be a speed 2 search path on line 2 that remains still at the initial location 0 until time $t_0/2$ and then moves at speed 2 until time $u$ (a total moving time of $(2u - t_0)/2$) to find the object (if it is on line 2) with probability $\bar{L}(2u - t_0)$. Then

$P^{DLSP}_{z_1,z_2}(u) = \frac{1}{2}(\bar{L}(t_0) + \bar{L}(2u - t_0))$. But by (16.8) this is the same as $P_{f,g}(u)$ when we solve for $f, g$ in terms of $z_1, z_2$. Since $\bar{P}(u) \geq P_{f,g}(u)$, we are done.

To show that the right-hand side is at least as large as the left, assume that $\bar{P}(u) = P_{f,g}(u)$ for some pair $f, g$. Let $z_1, z_2$ denote the corresponding DLSP strategy pair with $P^{DLSP}_{z_1,z_2}(u) = \bar{P}(u)$. For $i = 1, 2$, let $\alpha_i(s) = \int_0^s |z_i'(t)|\, dt$ be the total variation of $z_i$ up to time $s$. Let $l_i(s)$ be the unit speed linear search path defined up to times $\alpha_i(u)$ by $l_i(s) = z_i(\alpha_i(s)/2)$. The path $l_i$ covers the same interval during times $[0, \alpha_i(u)/2]$ as the path $z_i$ covers during times $[0, u]$. Hence

$$\bar{L}(\alpha_1(u)/2) + \bar{L}(\alpha_2(u)/2) \geq \bar{P}(u).$$

But since the pair $z_1, z_2$ has combined speed bounded by 2, it follows that $\alpha_1(u) + \alpha_2(u) \leq 2\,u$. Consequently, the right-hand side of the equation to be established is at least as big as the left-hand side $\bar{P}(u)$. ■

We can use this result to solve the "minimax rendezvous problem" on the line, for bounded distributions $F$, that was done by alternative means in Lim and Alpern (1996). This problem tries to find the asymmetric strategy pair of rendezvous strategies that minimizes the maximum time that it might take to meet. (Of course, for unbounded distributions, where the initial distance $d$ can be arbitrarily large, there is no such maximum time.) So we restrict our attention to the case where $D = \inf\{x\colon F(x) = 1\}$ is finite.

**Theorem 16.7** *If $F$ is bounded then the "Wait For Mommy" (WFM) strategy is a minimax strategy for the asymmetric rendezvous problem on the line and the minimax rendezvous time is $3D$. WFM is the strategy pair where one player stays still while the other (Mommy) goes at speed 1, first a total distance $D$ to the right and then reverses and goes a distance $2D$ to the left.*

**Proof.** Let $\bar{t}$ denote the minimax rendezvous time, so that $\bar{P}(\bar{t}) = 1$. It follows from (16.9), with $u = \bar{t}$, that for some $t \leq 2\bar{t}$ we have $\bar{L}(t) + \bar{L}(2\bar{t} - t) = 2$. Since $\bar{L}$ is a probability, and hence cannot exceed 1, it follows that both $\bar{L}(t)$ and $\bar{L}(2\bar{t} - t)$ must both equal 1. Since either $t$ or $2\bar{t} - t$ must be less than or equal to $\bar{t}$, it follows from the nondecreasing nature of $\bar{L}$ that $\bar{L}(\bar{t})$ is also equal to 1. This means that if one player remains stationary, there is always a (linear search) strategy for the other that will still find him in the minimax time $\bar{t}$. But this is just another way of saying that Wait For Mommy is a minimax strategy. ■

It is worth noting that any strategy for the "child" that puts him back at his starting point at time $D$ and $3D$ is also minimax, as long as "Mommy" searches as indicated above.

## 16.5 Atomic Distribution

We now assume that the initial distance between the players is a known number $D$. It turns out that in this case there is a strategy pair $f, g$ that is uniformly optimal, and consequently by Theorem 16.2 also optimal. In such a situation it is easiest to establish the strong property of uniform optimality.

For the atomic distribution with $d = D$ the optimal and uniformly optimal strategy for the Linear Search Problem is obvious: go right to $D$, then left to $-D$. (We will denote this by $FBB$ meaning forward $D$, back $D$, back $D$.) The resulting probability maximizing function $\bar{L}$ is given by

$$\bar{L}(t) = \begin{cases} 0, & \text{if } s < D, \\ 1/2, & \text{if } D \le s < 3D, \\ 1, & \text{if } 3D \le s. \end{cases}$$

We can now calculate the probability maximizing function $\bar{P}$ for the asymmetric rendezvous problem using Lemma 16.6 (a maximizing value of $t$ is also given).

$$\bar{P}(u) = \begin{cases} 0 = \frac{1}{2}(0+0) & & \text{if } u < D/2, \\ \frac{1}{4} = \frac{1}{2}\left(\frac{1}{2}+0\right) & \text{with } t = D & \text{if } D/2 \le u < D, \\ \frac{1}{2} = \frac{1}{2}\left(\frac{1}{2}+\frac{1}{2}\right) & \text{with } t = D & \text{if } D \le u < 2D, \\ \frac{3}{4} = \frac{1}{2}\left(\frac{1}{2}+1\right) & \text{with } t = D & \text{if } 2D \le u < 3D, \\ 1 = \frac{1}{2}(1+1) & \text{with } t = 3D & \text{if } 3D \le u. \end{cases}$$

It follows from (16.6) that the rendezvous value $R^a$ for the point distribution $d = D$ satisfies

$$R^a \ge \frac{1}{4}(D/2 + D + 2D + 3D) = \frac{13D}{4},$$

with equality if there is a uniformly optimal strategy pair. We can now derive one such a pair. An optimal interleaving of the undominated strategies on each line. Namely, $(F_1, B_1, B_1)$ and $(F_2, B_2, B_2)$ is given (determined by the above values for $t$) by the double linear search strategy $(F_1, F_2, B_2, B_2, B_1, B_1)$. Here, for example, the second $B_2$ means that in the fourth period of length $D/2$ (since speeds at 2 here) the searcher on line 2 moves back (leftward). This strategy can be described by giving the searchers' motion (derivative) in each of four (unequal) time periods. The bottom row gives the corresponding motions $f, g$ of the rendezvousers.

| Times | $(0, D/2)$ | $(D/2, D)$ | $(D, 2D)$ | $(2D, 3D)$ |
|---|---|---|---|---|
| $(z_1', z_2')$ | $(2, 0)$ | $(0, 2)$ | $(0, -2)$ | $(-2, 0)$ |
| $(f', g')$ | $(1, 1)$ | $(1, -1)$ | $(-1, 1)$ | $(-1, -1)$ |

The Double Linear Search strategy $(z_1, z_2)$ is described as follows: Searcher 1 goes to $+D$; Searcher 2 goes to $+D$, Searcher 2 goes to $-D$; Searcher 1 goes to $-D$. (The order of the first two stages and of the second two stages may be reversed without affecting optimality.) The optimal rendezvous strategies that correspond to these optimal DLSP strategies via (16.1) are the "Modified Wait For Mommy" strategies for the line, as defined below.

**Definition 16.8** *The "Modified Wait For Mommy" (MWFM) strategy pairs for rendezvous on the line with known initial distance $D$ are as follows, with all moves taken at maximum speed 1. One of the players (Mommy) searches as if looking for a stationary child. That is, by going $D$ is one direction and then $2D$ in the other. This is the same as in WFM. However, the other player (Child) is no longer stationary. Instead, he moves a distance $D/2$, hoping to meet his Mommy earlier, and then returns to his start to try to meet her there. If he has not met her by time $D$, then she is distance $2D$ away at that time. So he moves a distance $D$ in either direction, and then returns again to his starting point at time $3D$. The equiprobable meeting times for these strategy pairs are $D/2$, $D$, $2D$, $3D$, with expected time $13D/8$. If the distribution has a maximum $D$ but the mean is $\lambda$, then the expected meeting time is $(9D + 4\lambda)/8$.*

An optimal strategy for the atomic distribution with $D = 1$ is drawn in Figure 16.3.

We may also view the working out of the optimal MWFM strategy pair $(f, g)$ for the atomic distribution $D = 1$ by placing Player I at the origin. Figure 16.4 shows

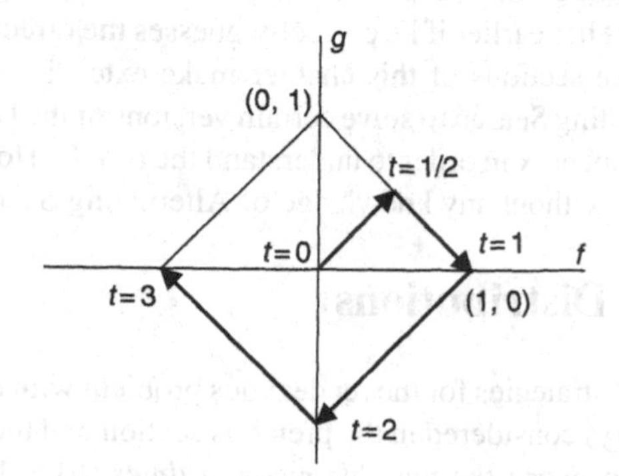


**Figure 16.3.** Optimal strategy for $D = 1$

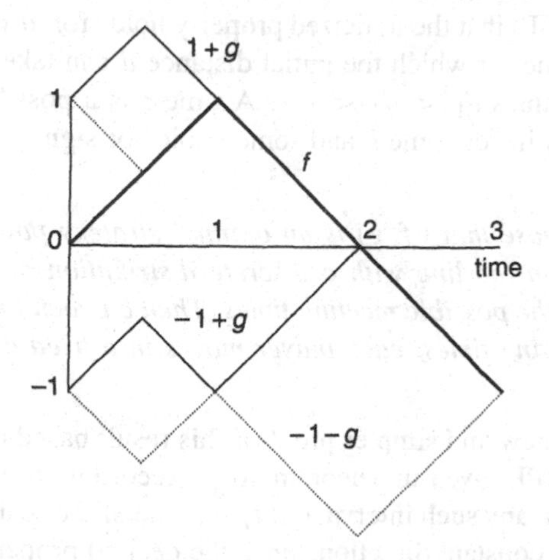


**Figure 16.4.** Player 1 and four paths of Player II

the actual path $f$ of Player I in a thick line, and the four possible paths $\pm 1 \pm g(t)$ of Player II in dotted and thin lines, drawn only until they intersect with Player I.

We see that $T(f, 1+g) = 1, T(f, 1-g) = 0.5, T(f, -1+g) = 2$, and $T(-1-g) = 3$, with $\hat{T}(f, g) = (0.5+1+2+3)/4 = 13/8 = 13D/8$. We see that $(f, g)$ is uniformly optimal. Summarizing the results for the atomic distribution, we have the following.

**Theorem 16.9** *If the initial distribution between the players is a known number D, the asymmetric rendezvous value is given by* $R^a = 13D/8$. *The unique optimal rendezvous strategies are the MWFM strategies. Furthermore, these strategies are uniformly optimal. If F is a bounded distribution with mean* $\lambda$ *and maximum D, then* $R^a \leq (9D + 4\lambda)/8$.

This result was originally obtained by the authors in Alpern and Gal (1995), but the presentation given here via the DLSP comes from Alpern and Beck (1999a). We call the optimal strategy pair the *Modified Wait For Mommy* (MWFM) strategy. One player (I, the Mommy) follows an optimal search path for an immobile hide, while the other follows a path that returns to his start point at any time Mommy might find him there but moves to possibly meet her earlier if he correctly guesses the direction of her approach.

The following three sections of this chapter make extensive use of the results of Appendix B on Alternating Search to solve certain versions of the DLSP. So the reader is advised to read that Appendix in order to understand the proofs. However, the statement of results will be clear without any knowledge of Alternating Search.

## 16.6 Discrete Distributions

Recall that the optimal strategies for the rendezvous problem with the atomic initial distribution ($d = D$ surely) considered in the previous section had the following property: *In the time intervals between the possible meeting times* ($\{0.5, 1, 2, 3\}$ *in that case*), *each player moves in a single direction at maximum speed.* More generally, Alpern and Gal (1995) proved directly (without reference to the later established equivalence of the ARPL and the DLSP) that the italicized property holds for *any* discrete distribution function $F$. That is, one for which the initial distance $d$ can take on only a countable number of possible values $d_1 < d_2 < \cdots$. A time $t$ is a possible meeting time for the strategy pair $(f, g)$ if, for some $k$ and some choice of signs, it is the first time that $\pm f(t) \pm g(t) = d_k$.

**Theorem 16.10** *Suppose that* $(f, g)$ *is an optimal strategy pair for the asymmetric rendezvous problem on the line with a discrete distribution of initial distance. Let* $t_1 < t_2 < \cdots$ *denote the possible meeting times. Then on each time interval* $(t_j, t_{j+1})$ *between possible meeting times, each player moves in a fixed direction at maximum speed.*

**Proof.** We give a new and simpler proof of this result based on the equivalence of the DLSP and the ARPL given in Theorem 16.5. According to this equivalence, it is enough to show that on any such interval $(t_j, t_{j+1})$, one of the searchers $z_i$ in the DLSP moves at speed 2 in a constant direction, since the desired property of the rendezvous strategy pair $(f, g)$ then follows from (16.1).

Suppose that $(z_1, z_2)$ is the optimal DLSP strategy corresponding to $(f, g)$ according to (16.1), and that $z_i$ moves during this time period. Since $z_i$ corresponds to an undominated way of searching line $i$ for an object hidden at some point $\cdots - d_2 < -d_1 < d_1 < d_2 \cdots$, it follows that it will be moving in a single direction during time $(t_j, t_{j+1})$. By the definition of the $t_j$, it follows that it will not find the hidden object during this time. Consequently, Theorem B.3 of Appendix B on Alternating Search shows that $z_i$ is moving at speed 2 during this interval. ■

The above result is similar in spirit to the result on the geodesity of optimal strategies on labeled networks and can also be applied when the speeds of the players are unequal.

## 16.7 Arbitrary Distributions

We now obtain some weaker optimality results, valid for any distribution $F$ of the initial distance that has a finite mean (and hence finite value $R^a(F)$). It is easier to first analyze the Double Linear Search Problem and then reinterpret the results for the Rendezvous Search Problem.

Before considering the DLSP we first consider the simpler LSP for a single line, in which a single searcher seeks an object hidden according to a known distribution. For this problem we define an important class of strategies that we call *oscillation strategies*. (See also Section 8.1 of Book I.) Every strategy is dominated by one of these, which is why they are important. Roughly speaking, an oscillation strategy simple goes right and left from the origin, always going further in each direction than previously. Obviously, there is no way to avoid traversing intervals that have already been searched (corresponding to what will be called the "old part").

**Definition 16.11** *Consider a sequence $\{x_i\}_{i=-\infty}^{\infty}$ with $x_i \geq 0$ and $F(x_{i+2}) > F(x_i)$. These numbers determine the **transits** $Tr_i$ in which the player moves at unit speed from $(-1)^{i-1}x_{i-1}$ to $(-1)^i x_i$ during the time interval from $t_{i-1}$ to $t_i$, where $t_j = \sum_{i=-\infty}^{j} x_i$. In some cases there is an **initial transit** from* 0 *(that we label $x_0$) to $x_1$. Every non-initial transit $Tr_i$ begins with an **old part** (from $(-1)^{i-1}x_{i-1}$ to $(-1)^{i-2}x_{i-2}$) that retraces ground already covered, and ends with a **new part** (from $(-1)^{i-2}x_{i-2}$ to $t(-1)^i x_i$) that searches new territory. The path consisting of the successive transits is called an **oscillation strategy**.*

An optimal strategy for the DLSP will be made up of an oscillation strategy $w_i$ for each line $i$, together with an alternation rule $\alpha$ (see Appendix B on Alternating Search) that determines which of the two searchers moves at any given time, or more generally what their respective speeds are. That is, $w_1$, $w_2$, and $\alpha$ determine DLSP strategies

$$z_1(t) = w_1(\alpha(t)), \quad \text{and} \quad z_2(t) = w_2(2 - \alpha(t)).$$

In general it is possible that the optimal way of interleaving two oscillation strategies may involve the interruption of a player's transit. However, the following result shows that in this case only the "new part" of the strategy (searching new territory) may be so interrupted. Once the DLSP strategy $z$ begins to retrace an interval already searched, it continues at full speed 2 until it reaches the end of that interval.

**Theorem 16.12** *Consider the DLSP problem for any distribution $F$. Given any oscillation strategies $w_1$ and $w_2$ for the respective lines, any optimal alternation rule will leave the "old part" of any non-initial transit uninterrupted. That is, for some time $t^*$, we will have $\alpha(t^*) = s_i$ and $\alpha(t^* + t_i + t_{i-1}) = s_i + 2(t_i + t_{i-1})$. In particular, for each player there are nontrivial time intervals (namely, the "old parts" of all non-initial transits) when he moves in a fixed direction at his maximum speed of 2.*

**Proof.** Every non-initial transit begins with a "first part" that covers an interval that has already been searched. The corresponding cumulative distribution $F_w(t)$ (that gives the probability that $w$ has found the object if it is on the same line) consequently has zero density during this period. That is,

$$F_w(t^* + t_i + t_{i-1}) = F_w(t^*).$$

The fact that an optimal alternation rule will not interrupt an interval of zero density is stated in Theorem B.4 of Appendix B on Alternating Search. ■

During any time interval when one of the DLSP searchers $z_i$ moves in a single direction at speed 2, the corresponding rendezvous strategies $f$ and $g$ each move in a single direction at speed 1. Consequently the previous result for the DLSP has the following implication for the ARPL.

**Theorem 16.13** *Consider the asymmetric rendezvous problem on the line with an arbitrary initial distribution $F$. Then for any optimal strategy pair there are nontrivial time intervals on which each player is moving in a single direction (not necessarily the same) at unit speed.*

In the "Wait For Mommy" (WFM) strategy pair the "child" never moves, so this strategy cannot be optimal. This was first proved directly by Gal (1999), but we obtain it here as a corollary of the previous result (Theorem 16.13) from Alpern and Beck (1999b).

**Corollary 16.14 (Gal)** *For the asymmetric rendezvous problem on the line, there is no distribution $F$ for which Wait For Mommy is an optimal strategy pair.*

## 16.8 Convex Distributions

Suppose that the initial distribution $F$ has support $[0, D]$ and is convex on this interval. We first consider the DLSP problem with this $F$. In this case we can show that for any oscillation strategies $w_1$ and $w_2$ on the two lines, there is always an alternation rule $\alpha$ such that the full transits of each searcher are uninterrupted by motions of the other searcher. Furthermore, the individual strategies $w_1$ and $w_2$ can be assumed to each go directly to $+D$ and then to $-D$. Combining these results of Alpern and Beck (1999b), it follows that the strategy pair that is optimal for the atomic distribution is also optimal for convex distributions.

**Theorem 16.15** *Consider the DLSP with a convex initial distribution $F$ on the interval $[0, D]$. The optimal search strategy pairs $z_1, z_2$ all have the following form: First one* *player goes to an end. Then the other player goes to an end. Then either player goes to his opposite end. Then the other player goes to his opposite end.*

If we reinterpret this DLSP in terms of the equivalent ARPL, we obtain the following result originally conjectured by Baston and Gal (1998) and proved by Alpern and Beck (1999a).

**Theorem 16.16** *Consider the asymmetric rendezvous problem on the line with a convex initial distribution $F$ on $[0, D]$. Then the MWFM strategy is optimal. In particular, it follows from the expected meeting time formula of $(9D + 4\lambda)/8$ in Definition 16.8 (of MWFM) that the asymmetric rendezvous value for the uniform distribution (with mean $\lambda = D/2$) is $11D/8$.*

It is worth noting that, unlike the situation for the atomic distribution summarized in Theorem 16.9, the MWFM strategy is no longer *uniformly* optimal for general convex distributions.

Note that in Section 8.7 we have given an algorithm for finding approximate solutions to the DLSP (and hence the ARPL) for a general distribution of initial distance. Other algorithmic approaches are given by C. DiCairano (1999).

# Chapter 17

# Other Rendezvous Problems on the Line

In the previous chapter we analyzed the basic form of the asymmetric rendezvous problem on the line (ARPL). That problem may be summarized as follows, italicizing those aspects of the problem which we will alter in this chapter.

> *Two* players are placed a distance $d$ apart on the line and faced in random directions. The distance $d$ is drawn from a *known* distribution $F$. A strategy for each player is a function measuring his net motion in the forward direction. Both players have the *same* maximum speed (normalized to 1). The total distance that such a strategy may travel is *unbounded*. The players choose two strategies $f$ and $g$, *not necessarily the same*, so as to minimize the *expected* meeting time.

Many variations on the basic problem have been studied, including very recent work not discussed here, on rendezvous when marks are left at the player's starting points (Baston and Gal, 2001). In this chapter each section will consider how the problem changes when the italicized qualifiers are changed in one of the following ways.

1. **Unequal speeds** Player I has maximum speed 1, while Player II has a maximum speed $M \leq 1$.
2. **Player symmetry** We consider the player-symmetric version of the problem, where both players are constrained to adopt the *same* mixed strategy. Player symmetry is also considered in some of the following sections.
3. **Bounded resources** Here we assume that Players I and II have respective resources $a$ and $b$ that bound the total distance they may travel. There are two cases for this version: If the distribution $F$ (of the players' initial distance) is bounded, and the resource pair $(a, b)$ is sufficiently large to ensure a meeting, we consider how the expected meeting time can be minimized. On the other hand, if the resources are too small to guarantee a meeting, we consider how the players may maximize the probability of meeting.

4. **Unknown Initial Distribution** We suppose that the initial distribution $F$ is not known to the players. In this case they choose strategies (of various types) so as to minimize the maximum (relative to $F$) of the ratio of $\hat{T}/E(F)$, where $\hat{T}$ is the expected rendezvous time and $E(F)$ is the mean of the distribution $F$.
5. **Multi-player Rendezvous** We consider the problem $\Gamma(n, m)$, in which $n$ players are randomly placed at consecutive integer positions on the line, and faced in random directions. They move at unit speed until the first time $T_m$ that $m$ of them occupy a common position.
6. **Asymmetric Information** We suppose that one of the players (say Player II) knows the initial position of the other. This problem may arise when Player I starts from a known position such as a base camp and is called on a cell phone by Player II who is mobile but needs medical attention. Under what conditions is it optimal for Player I to wait for II to reach him?

## 17.1 Unequal Speeds

We will now analyze the version of the ARPL problem where Player I (called "Fast") has unit speed while Player II ("Slow") has speed $M \leq 1$. We will assume the simplest distribution $F$ of initial distance. Namely, the atomic distribution where the initial distance is known to be 1. Recall from the previous chapter that the solution to this problem for the original (common) speed bound $M = 1$ was the MWFM (Modified Wait For Mommy) strategy pair. We choose a coordinate system with Fast starting at the origin, and the four agents of Slow starting at $\pm 1$ and facing in either direction. Note that two of the agents of Slow start at each of the points $\pm 1$. If Slow chooses the strategy $g(t)$, the paths of his four agents will be $\pm 1 \pm g(t)$.

Depending on the four ways that the two players may initially be faced, there will be four possible meeting times, the times when Fast meets each of the four agents of Slow. We denote these by $t_1 \leq t_2 \leq t_3 \leq t_4$. (For the MWFM strategy, Definition 16.8, with $M = 1$ and $D = 1$, these are $0.5 \leq 1 \leq 2 \leq 3$.) Without loss of generality, we may assume that the first meeting of Fast with Slow will take place at time and location $a$, for some value $1/(1 + M) \leq a \leq 1$. Note that for the case $M = 1$ discussed in the previous chapter, we have $a = 1/2$. Theorem 16.10 shows that Fast and Slow should go to location $a$ at maximum speed. So Fast meets one of the agents of Slow who started at $+1$ at time

$$t_1 = a. \tag{17.1}$$

At this time $t_1$, the other agent of Slow who started at $+1$ will be at location $1 + (1 - a)$. So if he reverses at time $t_1$ and moves at maximum speed $M$, he can meet Fast in additional time

$$t_2 - t_1 = \frac{2(1 - a)}{1 + M}. \tag{17.2}$$

At this time Fast knows the direction of Slow and turns to find him. Slow should move so that the closer of his two remaining agents to Fast will meet him first. That is, he should change directions. This will lead to a meeting at time $t_3$ satisfying

$$-1 + (1 - a) + M(-(t_2 - t_1) + (t_3 - t_2)) = 2t_2 - t_3 \tag{17.3}$$

$$\text{at } x_3 = -(t_3 - 2t_2) = 2t_2 - t_3. \tag{17.4}$$

At time $t_3$ the distance between Fast and the remaining agent of Slow is $2(x_3-(-1))$. Hence Fast meets the remaining agent of Slow (who turns at time $t_3$) at time $t_4$, in additional time

$$t_4 - t_3 = \frac{2(2t_2 - t_3 + 1)}{1 + M}, \quad \text{at } x_4 = -(t_4 - 2t_2). \tag{17.5}$$

Starting with $t_1 = a$, we may successively solve the numbered equations for the times $t_2$ through $t_4$, obtaining,

$$\begin{aligned}
t_1 &= a \\
t_2 &= \frac{-a + aM + 2}{1 + M} \\
t_3 &= \frac{aM - a + 4}{1 + M} \\
t_4 &= \frac{-3a + aM^2 + 6M + 6 + 2aM}{(1 + M)^2}
\end{aligned}$$

The expected meeting time $\hat{T}$ is given by

$$\begin{aligned}
\hat{T} &= (t_1 + t_2 + t_3 + t_4)/4 \\
&= \frac{(M^2 + M - 1)a + 3(M + 1)}{(M + 1)^2}.
\end{aligned} \tag{17.6}$$

Since the parameter $a$ appears linearly, we minimize $\hat{T}$ by setting it equal to its two extremes $1/(1 + M)$ or 1, according as its coefficient $M^2 + M - 1$ is, respectively, negative or positive. Since this coefficient is zero when $M$ is equal to the "golden mean" or "golden section" $\gamma = (\sqrt{5} - 1)/2 \doteq 0.61803$, we have the following.

**Theorem 17.1** *Consider the asymmetric rendezvous problem on the line, with the players starting a distance 1 apart, and with maximum speeds 1 (for the Fast player) and $M \leq 1$ (for the Slow player). For $M \geq \gamma$, both players should always move at their maximum speeds, with Fast reversing at time $(1 + 3M)/(M + 1)^2$ and Slow reversing at times $1/(M + 1)$, $(1 + 3M)/(M + 1)^2$, and $(5M + 3)/(M + 1)^2$. For $M \leq \gamma$, Fast always moves at his maximum speed 1, reversing at time 1. Slow stays still until time $t = 1$; thereafter he moves at his maximum speed $M$, reversing at time $1 + 2/(1 + M)$. The asymmetric rendezvous value for this problem is given by*

$$R^a = \begin{cases} \dfrac{4M^2 + 7M + 2}{(M + 1)^3}, & \text{if } M \geq \gamma, \\ \dfrac{1}{2}(1) + \dfrac{1}{4}\left(1 + \dfrac{2}{1 + M}\right) + \dfrac{1}{4}\left(1 + \dfrac{2}{1 + M} + \dfrac{4M}{(1 + M)^2}\right) = \dfrac{2 + 4M + M^2}{(1 + M)^2}, & \text{if } M \leq \gamma. \end{cases}$$

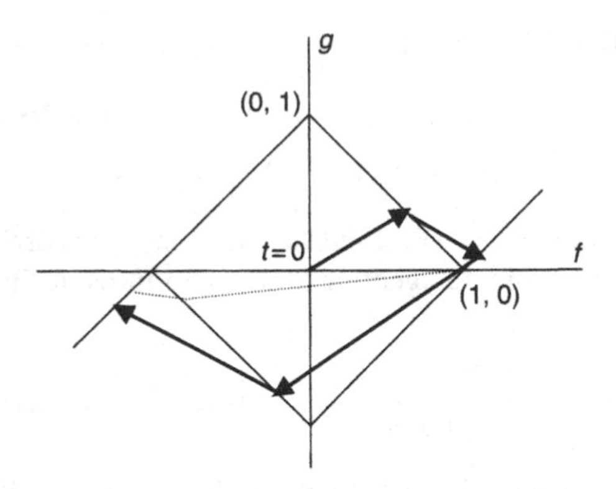


**Figure 17.1.** Optimal strategies for $M > \gamma$ and $M < \gamma$

The two classes of strategies are shown in Figure 17.1 using a parametric drawing in $f$, $g$ space. The bold line shows the optimal strategy pair for $M > \gamma$ and the dotted line shows the optimal path for $M < \gamma$.

## 17.2 Player Symmetry

We now consider the player-symmetric version of rendezvous on the line. We will assume that the initial distribution has a maximum distance $D$ and that the players move in "steps" of length $D/2$. That is, players move at unit speed at all times and may change direction only at times that are integer multiples of the step length $D/2$. A pure strategy may be described in the following notation, introduced in Anderson and Essegaier (1995):

$$n_1 A n_2 B n_3 A n_4 B \ldots$$

This corresponds to picking a random direction to call forward (*A*head), going in this direction for $n_1$ steps, then *B*ackward for $n_2$ steps, and so on. In this notation the first author's simple strategy discussed in Chapter 11 as 1F 2B would be written as $1A2B$, with infinitely many independent repetitions until rendezvous is achieved. Note that this is a mixed strategy because of the independent repeated choices of the forward direction at time $3nD/2$. A better strategy was obtained by Anderson and Essegaier (1995) using an optimal mixture of the following four 6-step motions

$$2A4B,$$
$$1A3B2A,$$
$$1A2B1A2B,$$
$$1A1B1A3B.$$

This strategy, like $1A2B$, has the property that if the players have not met after the period of the strategy (six steps), then their distance at that time is the same as their initial distance. The "distance preserving" property ensures that the expected *additional* meeting time after failure to meet at the end of a period is the same as the original expected meeting time. This makes calculations of expected meeting times relatively easy.

However if the knowledge of failure to meet at the end of a period (6 steps) gives the players additional information, then it may be that they can do better than simply repeating with independent randomization. This was observed by Baston (1999), who began his analysis by considering the following four 6-step strategies:

$$\alpha_1 = 1A2B2A1B,$$
$$\alpha_2 = 1A3B2A,$$
$$\beta_1 = 1A1B1A3B,$$
$$\beta_2 = 2A4B.$$

Observe that if the players have not met by time $3D$ (the period) then they must have chosen the same direction as forward and the same letter strategy (both $\alpha$ or both $\beta$). Furthermore, if they chose $\alpha_1$ and $\alpha_2$, then the former started in front of the latter; if they chose $\beta_1$ and $\beta_2$, then the latter started in the common forward direction. Consequently if the probabilities of the two subscripts in each case are nearly the same, and so there is a good chance that they chose distinct subscripts, then they may wish to have the forward player move back (add an additional $B$) while the backward player moves forward (add an $A$). This gives a new improved set of four 7-step strategies,

$$1A2B2A2B,$$
$$1A3B3A,$$
$$1A1B1A3B1A,$$
$$2A5B.$$

If the players have not met by time $7D/2$, then they must have chosen the same direction as forward and adopted the same strategy. So the problem they face at time $7D/2$ is the same as the original problem. Optimizing the probability distribution over these four strategies leads to the following result of Baston (1999), that gives the best known bound for the symmetric problem.

**Theorem 17.2** *For any bounded distribution $F$ of initial distance, with maximum $D$ and mean $\lambda$, the symmetric rendezvous value for the line satisfies*

$$R^s(F) \leq 1.701D + \lambda/2.$$

This represents a considerable improvement over the first author's initial estimate of $2D + \lambda/2$ when posing the problem in Alpern (1995), and a smaller improvement over the estimate of $1.78388D + \lambda/2$ obtained subsequently by Anderson and Essegaier (1995). This symmetric problem appears difficult and it is not at all clear what properties the optimal mixed strategy will possess. For example, we can ask how far the players may move from each other or from their initial positions in an optimal strategy. In Baston's strategy, they move at most 3 units apart (if both choose $2A5B$ with different forwards) but may move arbitrarily far from their initial positions ($3D/2$ further each time $2A5B$ is picked).

It is worth noting again that this seemingly easy rendezvous problem, of two indistinguishable players who know they are a unit distance apart on the line, is still unsolved. The problem looks harder now than when it was initially posed.

## 17.3 Bounded Resources

The assumptions of the rendezvous search problem usually allow the players to search for each other indefinitely until they succeed in meeting. But what if they are traveling in cars with finite fuel supplies or are hikers with limited stamina. This problem was modeled by Alpern and Beck (1997, 1999b) by assuming individual limits for the total distance (more precisely, the total variation) each player can travel. We call these distance bounds $a$ and $b$, for Players I and II, respectively. Without loss of generality we may assume that $a \geq b$. Let $Var(s) = \lim_{r \to \infty} (var\ s)(r)$ denote the total variation of a function $s: R \to R$ over its full domain, and define (recalling that $\mathcal{S} = \mathcal{S}_\infty$ is the set of unrestricted strategies)

$$\mathcal{S}_a = \{s \in \mathcal{S}: Var(s) \leq a\} \quad \text{and} \quad \mathcal{S}_{a,b} = \mathcal{S}_a \times \mathcal{S}_b.$$

We shall be considering two types of optimization that players with bounded resources may adopt: expected time minimization and rendezvous probability maximization. The first of these is reasonable only when the players can ensure that they will meet before running out of resources. Obviously, this presumes that the distribution $F$ of their initial distance is bounded, with some maximum value $D$. So what are the conditions on the resource bounds $a$ and $b$ for which players can ensure a meeting even if their initial distance is $D$ (or less)? This question was answered in Alpern and Beck (1997) as follows.

**Theorem 17.3** *Suppose that two players are placed a distance $D$ apart on the line, faced in random directions, and can travel respective total distances of $a$ and $b$, where $a \geq b$. They can guarantee a meeting if and only if*

$$3a + 5b \geq 15D.$$

The sufficiency of this condition will be established in the following subsection, where moreover expected time minimizing strategies with these resource bounds will be given. The proof that these conditions are necessary can be found in Alpern and Beck (1997).

This section now divides into two subsections, depending on whether or not $3a + 5b$ exceeds $15D$. When it does, we consider the expected time minimization problem. When it doesn't, we consider the problem of maximizing the probability of a meeting.

### 17.3.1 Expected time minimization for $3a + 5b \geq 15D$

The expected time minimization problem is easy to state. Recalling that $\hat{T}$ denotes the expected meeting time, we define (for bounded distributions $F$ with maximum $D$

satisfying $3a + 5b \geq 15D$) the *bounded resources asymmetric rendezvous value* for the line as

$$R_{a,b}(D) = \min_{(f,g) \in \mathcal{S}_{a,b}} \hat{T}(f, g). \tag{17.7}$$

Recall from Section 16.5 that the optimal MWFM strategy pair for the unrestricted problem ($a = b = \infty$) with the atomic distribution had each player moving a distance $3D$. So for $b \geq 3D$ this strategy is still feasible, and consequently $R_{a,b}(D)$ should be the same as the unrestricted asymmetric rendezvous value $R^a(F)$, where the superscript simply stands for asymmetric. If $b = 0$, then Player II is simply a stationary "object," and the problem reduces to the bounded resources Linear Search Problem studied by Foley, Hill, and Spruill (1991). Finally, if $a = \infty$ and $b = 0$, then this is the original Linear Search Problem posed by Bellman (1963) and Beck (1964).

Since the special cases of this problem are not fully solved, we naturally cannot expect a complete solution to the more general problem (17.7). However, in the case of the atomic distribution $d = D$ we can give a complete solution which generalizes that given in Section 16.5 for infinite resources. For simplicity, we normalize the initial distance $D$ to 1.

**Theorem 17.4** *Suppose that two unit speed players are placed a unit distance apart on the line and faced in random directions. Players I and II can respectively travel a total distance of at most a and b, where $a \geq b$ and $5a + 3b \geq 15$. Then their least expected meeting time $R_{a,b}$ is given by*

$$R_{a,b} = \begin{cases} (31 - 5a - 3b)/8, & \text{if } 2a + b \leq 6, \ (I) \\ (19 - a - b)/8, & \text{if } 2a + b \geq 6, a \leq 3, \ (II) \\ (16 - b)/8, & \text{if } b \leq 3 \leq a, \ (III) \\ (13/8), & \text{if } 3 \leq b, \ (IV). \end{cases}$$

The detailed proof can be found in Alpern and Beck (1997). Here, we will give sample optimal strategies for each of the regions I to IV which define a particular class. These regions are drawn in Figure 17.2.

A typical example for Region I would be $a = 2, b = 1.8$. Here the solution is for Player I to move forward at unit speed from time 0 to time 0.6, rest until time 1.4, and then move backward at unit speed until time 2.8. Total distance traveled: $0.6 + 1.4 = 2$. For Player 2, the optimal strategy is to move forward at unit speed until time 0.4, rest until time 0.6, go backward at unit speed until time 1.4, rest until time 2.6, and finally forward at unit speed until time 3.2. Total distance: $0.4 + 0.8 + 0.6 = 1.8$. The four meeting times are 0.6, 1.4, 2.6, and 3.2, as illustrated below in Figure 17.3. The expected meeting time is 1.95.

If the resources are slightly increased to $a = 2.6, b = 2$, we enter Region II. The optimal strategy for this case is drawn below in Figure 17.4.

The meeting time has been reduced to 1.8 by employing a more efficient method of hitting all four sides of the square $\pm f \pm g = 1$. If we now increase $a$ past 3 but decrease $b$ to 0.4, we get the same meeting time as in the first case considered. The optimal strategy pair for this case is drawn below in Figure 17.5.

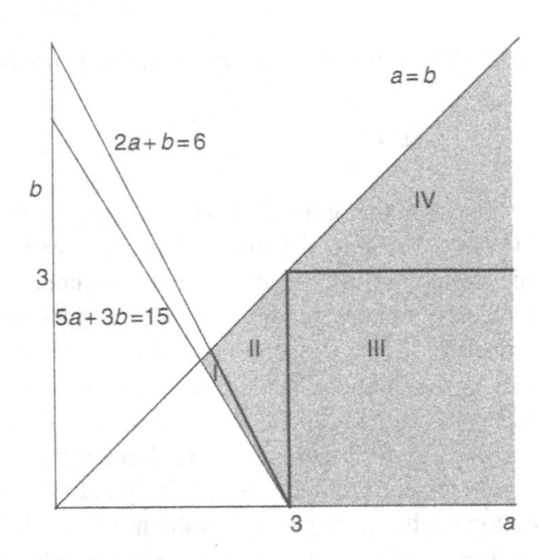


**Figure 17.2.** Regions I–IV in *ab* space

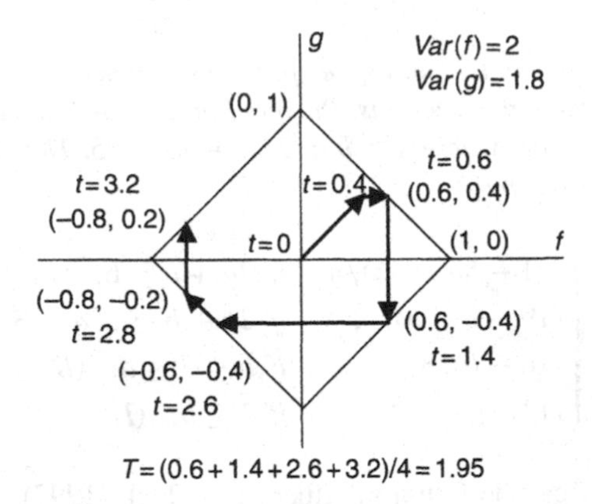


**Figure 17.3.** Optimal strategy, $a = 2$, $b = 1.8$

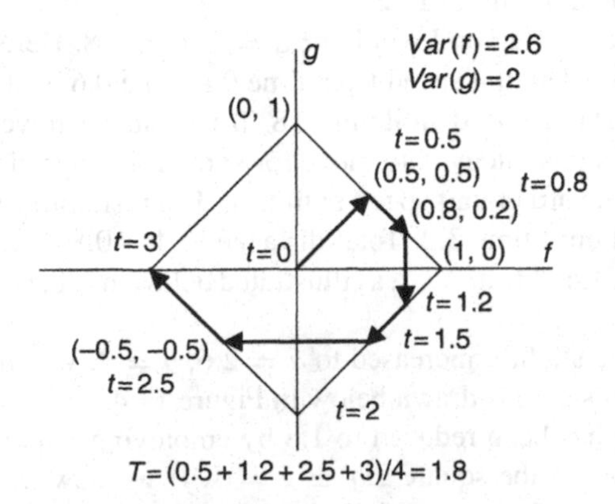


**Figure 17.4.** Optimal strategy for $a = 2.6$, $b = 2$

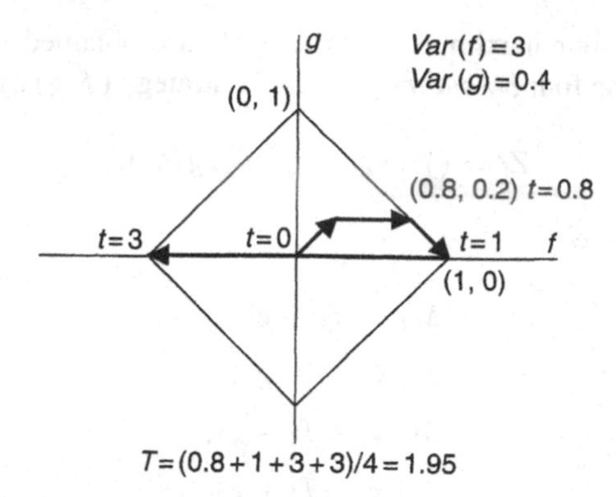


**Figure 17.5.** Optimal strategy for $a \geq 3, b = 0.4$

Of course if both $a$ and $b$ are at least 3, the optimal strategy for the unrestricted resource case, the MWFM strategy $\bar{f}, \bar{g}$, is optimal.

An important distinction worth observing between optimal strategies for the bounded resources and unrestricted cases is that, unlike Theorem 16.13, periods of resting can now be part of an optimal strategy. However as long as Player II has some resources (i.e., $b > 0$), he can always improve the expected meeting time by moving, so that WFM is still never optimal.

### 17.3.2 Probability maximization

According to Theorem 17.3, when the resource pair $a, b$ available to the players does not satisfy the required inequality $5a + 3b \geq 15$, the players cannot ensure they will meet. Hence in that case the expected meeting time corresponding to *any* feasible strategy in $\mathcal{S}_{a,b}$ is infinite. So for this situation another optimization criterion suggests itself. Namely maximizing the probability of a meeting. That is, we seek the *rendezvous probability value*

$$P(F; a, b) = \max_{(f,g) \in \mathcal{S}_{a,b}} P_{f,g}, \tag{17.8}$$

where $P_{f,g}$ is the probability that players adopting strategies $f$ and $g$ will meet (eventually). That is, $P_{f,g} = P_{f,g}(\infty)$ in the notation of the previous chapter. Strategies where the maximum is achieved will be called probability maximizing, or (when the context is clear) simply optimal. If we define the four maximum distances $M_i$ by $M_i(\infty)$ in (16.2), then $P_{f,g}$ is given by

$$P_{f,g} = \frac{1}{4} \sum_{i=1}^{4} F(M_i). \tag{17.9}$$

We note that for this problem the distribution $F$ need not be bounded. The results of this subsection are taken from Alpern and Beck (1999b). (However, the reader should be warned that the subscripts 2 and 4 (of $M$ and $Z$) are transposed from that paper to obtain consistency with previous chapters.)

Suppose that the four maximum distances $M_i$ are obtained respectively at first times $t_i$, and denote the four *boundary points* of a strategy $(f, g)$ by $Z_i$, where

$$Z_i = (f_i, g_i) = (f(t_i), g(t_i)).$$

In this notation, we have

$$\begin{aligned} M_1 &= f_1 + g_1, \\ M_2 &= f_2 - g_2, \\ M_3 &= -f_3 - g_3, \\ M_4 &= -f_4 + g_4. \end{aligned}$$

We may classify strategies by the temporal order in which these four maxima are achieved. That is, by the ordering of the four times $t_1, t_2, t_3$, and $t_4$. For example, if for some strategy we have $0 < t_1 < t_2 < t_3 < t_4$, then the parametric plot of this strategy hits the four sides of the bounding rectangle in cyclic clockwise order, and we say such a strategy has *order type* 1234. By suitably changing the signs of $f$ or $g$ (or both), we may assume that $t_1$ is the least of the four $t_i$, so for convenience we will assume that all strategies have this property, or equivalently, that all order types begin with a 1. Particular attention will be paid to the *cyclic* order types 1234 and 1432.

In the following subsections we will determine properties of optimal strategies for arbitrary distributions, certain classes of distributions, and some particular distributions. For this section we will say that a strategy is $F$*-dominated* by another if the latter has at least as large a value of $P$ for that $F$, and *dominated* if this holds for all $F$. We identify a strategy $(f, g)$ with its parametric plot in the plane.

**Arbitrary Distributions**

Even for arbitrary distributions $F$, it turns out that we may restrict our strategies to certain types without reducing the maximum probability of meeting. The following combines two observations of Alpern and Beck (1999b, Lemmas 2 and 3).

**Theorem 17.5** *For any distribution $F$ and any strategy $(f, g)$, there is an $F$-dominating strategy with either of the following properties: The path between any consecutive boundary points $Z$ and $Z'$ is a single straight line, or it consists of two straight lines, each parallel to a coordinate axis. Furthermore, if $Z$ and $Z'$ are located on adjacent sides of the bounding rectangle, then the path between them is a single straight line parallel to one of the coordinate axes.*

**Proof.** Let $Z = (x, y)$ and $Z' = (x + \Delta x, y + \Delta y)$ be the consecutive boundary points. The total variations of $f$ and $g$ over the intervening time interval $I$ are at least $|\Delta x|$ and $|\Delta y|$. So any modification of $f$ and $g$ which is restricted to $I$ and has variations $|\Delta x|$ and $|\Delta y|$ over $I$ will not have a smaller meeting probability $P_F$ or a larger total variation. This can always be achieved by a straight line between $Z$ and $Z'$ and can also always be achieved in two ways by successive lines parallel to the axes.

To establish the "furthermore" part, we give a brief outline of how to modify any given strategy to another one which has the specified properties, a bounding rectangle

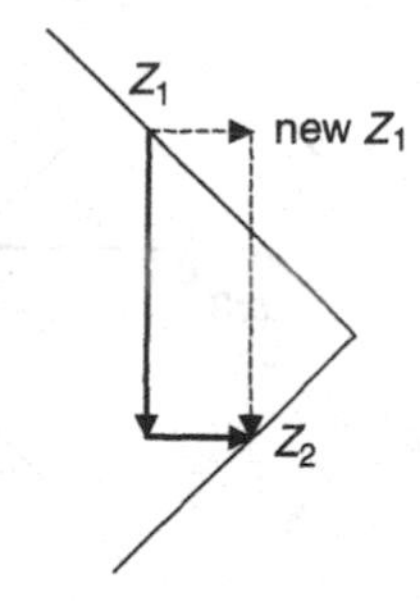


**Figure 17.6.** Modification for $f_2 < f_1$

which includes that of the original strategy and uses no more resources for either player. So we may first modify the original strategy by applying the last claim of the first paragraph (about lines parallel to the axes) four times: first to the origin and $Z_1$, then to $Z_1$ and the next boundary point, and so on. Since the two original endpoints are inside the new curve, the values of the relevant $M_i$ cannot decrease.

For the next modification suppose, for example, that $Z_2$ immediately follows $Z_1$ and that the path between these has a horizontal as well as a vertical line. This implies that $Z_1$ and $Z_2$ have distinct horizontal components ($f_1 \neq f_2$). If $f_2 > f_1$, the original path (shown in Figure 17.6 by a solid line) must have the vertical before the horizontal. The modification here is to change the order, putting the horizontal line first (shown in dashed lines in Figure 17.6). The new path has the same total variation in each coordinate, the same $Z_2$ and a new $Z_1$ with the same horizontal coordinate as $Z_2$ and an increased value of $M_1$. Consequently, the meeting probability $P$ of (17.9) cannot be smaller than before. An analogous modification can be made for the alternate case $f_2 < f_1$. ■

### The Atomic Distribution

We now consider the problem of maximizing the rendezvous probability for the atomic distribution $F_A$ with an atom at 1. That is, the players start at a known distance of 1 and have respective resources $a$ and $b$. Since in this situation $F(M_i)$ can only take the values 0 or 1, it is clear from the rendezvous probability formula (17.9) that $P_{f,g}$ and hence also its maximum value $P(F_A; a, b)$ can take on only the values 0, 1/4, 2/4, 3/4, and 1. Graphically, the strategy $(f, g)$ will satisfy $P_{f,g} = k/4$ if the parametric plot of $(f, g)$ meets $k$ sides of the square $S^1$ with corners at (0,1), (1,0), $(-1, 0)$, and $(0, -1)$, as pictured below in Figure 17.7. Since the atomic distribution $F_A$ is bounded with maximum distance $D = 1$, we already have a necessary and sufficient condition that $R(F_A; a, b) = 1$. Namely, the condition $3a + 5b \geq 15D = 15$ of Theorem 17.3. Necessary and sufficient conditions for the other possible values to be $R(F_A; a, b)$ are given in the first four parts of the following result of Alpern and Beck (1999b), which is based in part on the strategies in the following figure.

**Theorem 17.6** *Suppose two players are placed a unit distance apart (the atomic distribution A) on the line and faced in random directions so that neither knows the direction of the other. Suppose the players can move at maximum speed 1 and have resources*

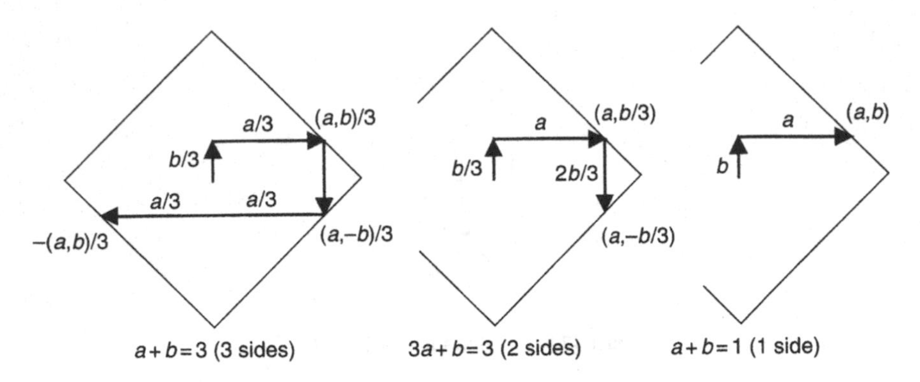


**Figure 17.7.** Strategies with rendezvous probability 3/4, 1/4, 1/4

*$a, b$ that bound the total distance each can travel, where $a \geq b$. Then the maximum probability with which they can meet is given by*

$$P(F_A; a, b) = \begin{cases} 0, & if\ a + b < 1, \\ 1/4, & if\ a + b/3 < 1 \leq a + b, \\ 1/2, & if\ a/3 + b/3 < 1 \leq a + b/3, \\ 3/4, & if\ a/3 + b/5 < 1 \leq a/3 + b/3, \\ 1 & if\ 1 \leq a/3 + b/5. \end{cases}$$

**Proof.** The fact that $P(F_A; a, b)$ is at least as large as the stated values is verified by checking the strategies in the above figure. To show that there are no better strategies, we adopt the "sum" metric $\rho$ in the plane in which the distance between two points is the sum of the absolute values of the differences between corresponding coordinates. (e.g., $\rho((5, 6), (4, 8)) = |5 - 4| + |6 - 8| = 3$.) This is useful because of the fact that the sum of the total variations of the two coordinates of any path cannot be less than the $\rho$ distance between its endpoints. Note that the unit sphere with respect to this metric is simply the square $S^1$, so that no side can be reached (from the origin) when $a + b < 1$. Similarly, in order that three sides of $S^1$ be reached, two opposite sides must be reached. Consequently, in this case combined resources (total variations) of 1 (to reach first side) plus 2 (to go between opposite sides), or $a + b \geq 3$, are required. It remains only to show that when $a + b/3 < 1$ two sides cannot be reached. We know from the previous argument that two opposite sides cannot be reached. If two adjacent sides are reached, we may assume by Theorem 17.5 that the boundary points are $(x, y)$ and $(x, -y)$ or $(x, y)$ and $(-x, y)$, in either case with $x + y = 1$. In the first case, the resources required are $a \geq x$ and $b \geq 3y$, so that $1 = x + y \leq a + b/3$. In the second case we would need $b + a/3 \geq 1$ and hence (because $a \geq b$)$a + b/3 \geq 1$. ■

**Concave distributions**

We now show that when the initial distribution $F$ is concave, we may further restrict our optimality search to strategies of a special type. First we show that we need only consider strategies with one of the two cyclic order types 1234 and 1432.

**Theorem 17.7** *Let $F$ be a concave distribution. Then every strategy is $F$-dominated by a strategy with a cyclic order type, which we may take to be 1234 or 1432.*

**Proof.** Let $(f, g)$ be a given strategy with a noncyclic order type $i_1, i_2, i_3, i_4$. This means that either $i_1$ is not adjacent to $i_2$ or $i_2$ is not adjacent to $i_3$, where 1 is considered to be adjacent to 4.

First assume the former case, that $i_1$ is not adjacent to $i_2$. Since we are assuming that $i_1 = 1$, this means that the type is either 1324 or 1342. Assume for the moment it is 1324. The parametric plot of this strategy from $Z_1$ may be assumed to be the three straight lines parallel to the axes joining of the points $Z_1$, $Z_3$, $Z_2$ and $Z_4$. The line joining $Z_1$ and $Z_3$ intersects the line joining $Z_2$ and $Z_4$ at some point $X$. So we may rewrite the original path from $Z_1$ as $Z_1, X, Z_3, Z_2, X, Z_4$. Now replace this path with the order type 1234 path $Z_1, X, Z_2, Z_3, X, Z_4$. The new path consists of the same line segments as the original, though they are traversed in different orders and directions. Nevertheless, the total variations in each coordinate remain the same. Consequently the 1324 path is dominated by the one of cyclic order 1234. Note that the assumption of concavity was not used in this part of the proof.

Now assume that $i_2$ is not adjacent to $i_3$, as in the case 1243. By the first part of Theorem 17.5 we may assume that $(f, g)$ is linear between the successive points 0, $Z_1$, $Z_2$, $Z_4$, $Z_3$, and by the "furthermore" part also that $f_1 = f_2$ and $f_4 = f_3$. Recall that $f_k$ and $g_k$ denote the coordinates of $Z_k$.

We will consider here only the more difficult case where $g_4 < 0$; the other case can be found in Alpern and Beck (1996). We will show that the original path is $F$-dominated by the path which is linear between the points 0, $Z_1$, $Z_2$, $(f_4, 2g_4 - g_3)$. Call the resulting strategy $(f^*, g^*)$. This is the same as the original path, except that the final segment from $Z_4$ to $Z_3$ has been reflected 180° about $Z_4$. Consequently, it uses the same pair of total variations (resources) as $(f, g)$. Since $Z_1^* = Z_1$ and $Z_2^* = Z_2$, we have the same maxima $M_1^* = M_1$ and $M_2^* = M_2$. Hence we only need to show that $F(M_3^*) + F(M_4^*) \leq F(M_3) + F(M_4)$. The individual calculations are as follows

$$
\begin{aligned}
M_4 &= -f_4 + g_4, \\
M_3 &= -f_4 - g_3, \\
M_4^* &\geq -f_4 + 2g_4 - g_3, \quad \text{(evaluated at } (f_4, 2g_4 - g_3)) \\
M_3^* &\geq -f_4 - g_4, \quad \text{(evaluated at } (Z_4 = (f_4, g_4))).
\end{aligned}
$$

Hence

$$
\begin{aligned}
&4(P_{f^*,g^*} - P_{f,g}) \\
&= F(-f_4 + 2g_4 - g_3) + F(-f_4 - g_4) - F(-f_4 + g_4) - F(-f_4 - g_3) \\
&= [F(-f_4 + 2g_4 - g_3) - F(-f_4 + g_4)] - [F(-f_4 - g_3) - F(-f_4 - g_4)] \\
&= [F(-f_4 + g_4 + [g_4 - g_3]) - F(-f_4 + g_4)] - [F(-f_4 - g_4 + [g_4 - g_3]) \\
&\quad - F(-f_4 - g_4)] \\
&< 0 \quad \text{(by the assumed concavity of} F).
\end{aligned}
$$

■

## 17.4 Unknown Initial Distribution

In the usual formulation of the rendezvous search problem, the initial placement of the players is assumed to have a known joint distribution in the search region $Q$. When $Q$ is the real line $R$ (the subject of this chapter) it is the initial distance $d$ whose distribution $F$ is assumed to be common knowledge. However, the initial placement may be considered in the context of uncertainty rather than risk, in that there may be no prior distribution accepted by the players. This type uncertainty is sometimes modeled by assuming that what is not known is chosen maliciously, which gives a worst case analysis, or what is known as a *game against Nature*. More generally, we may assume that some antagonistic agent (Nature or other) chooses the initial placement of the two rendezvousers. This problem was proposed by Alpern (1995) as follows (with the name of the search region changed here to $Q$):

> Adversary-rendezvous games: Suppose the two searchers' initial positions in $Q$ are chosen by another player who wishes to maximize their meeting time. ... This becomes a two-person zero-sum game where the rendezvousers are a single player.

In the context of rendezvous on the line, it is the distribution $F$ of the players' initial distance $d$ that is chosen by the adversary player. Of course, the adversary can make the meeting time arbitrarily large simply by choosing an $F$ for which small values of $d$ are excluded. To avoid this, we limit the choice of distributions $F$ to those with a given mean $\lambda$. This corresponds to the approach taken in Beck and Newman (1970) for the case of a single searcher seeking a stationary object hidden on the line according to a distribution chosen by an adversary, as discussed in Chapter 8.

### 17.4.1 Asymmetric strategies

Suppose the players can use distinct strategies against an unknown distribution of initial distances. One natural strategy pair to consider is the best strategy of Wait For Mommy (WFM) type. That is, one player stays still, while the other uses the best linear search strategy to find him, or equivalently an object hidden equiprobably at $\pm d$, with $d$ drawn from a distribution $F$ with mean $\lambda$. In Chapter 8 we showed that a best strategy of this type finds the object (or the stationary child-rendezvouser) in expected time at most $(4+2\sqrt{2})\lambda \doteq 6.83\lambda$. This result uses the minimax properties of geometric series derived in Chapter 7. Baston and Gal (1998) made the following interesting heuristic extrapolation argument to guess the best possible coefficient of $\lambda$: Recall that for the asymmetric rendezvous problem on the line with an atomic distribution ($D = 1$) WFM has expected meeting time $(1+3)/2 = 2$, while the optimal (MWFM) strategy has expected meeting time 13/8, an improvement of about 19%. A 19% improvement on $6.83\lambda$ would give a bound of about $5.5\lambda$ for the rendezvous problem. While they presented a strategy pair with expected meeting time $5.73\lambda$, the following result of Alpern and Beck (2000) shows that their extrapolation is indeed very accurate in this case.

**Theorem 17.8** *Let $(\breve{f}, \breve{g})$ denote the following asymmetric strategy pair. There are infinitely many time periods $[t_i, t_{i+1}]$, $0 < t_i < t_{i+1}$, $-\infty < i < \infty$, with each of length $\hat{a}$ times the previous one and $\lim_{i\to\infty} t_i = \infty$. In every period, each player* *moves in a fixed direction at maximum speed 1. At times $t_i$, $i$ odd, Player I reverses direction. At time $t_i$, $i$ even, Player II reverses direction. Take*

$$\hat{a} = \frac{1}{3} + \frac{2^{2/3}}{3(8+\sqrt{62})^{1/3}} + \frac{2^{1/3}(8+\sqrt{62})^{1/3}}{3} \doteq 1.5994.$$

*Then for any distribution $F$ with mean $\lambda$, the expected meeting time for these strategies satisfies*

$$\hat{T}(\check{f}, \breve{g}; F) \leq V\lambda,$$

*where*

$$V = \frac{1}{2} + (1/8)\left(\frac{8}{\hat{a}-1} + 7 + 5\hat{a} + 3\hat{a}^2 + \hat{a}^3\right) \doteq 5.514.$$

*Furthermore, this is the best possible bound of this type within the class of alternating strategies.*

**Proof.** The proof follows Alpern and Beck (2000), and analyzes this rendezvous problem on the line in terms of the equivalent double linear search problem (DLSP) on the line. The DLSP and its equivalence to the rendezvous problem ARPL are discussed in the previous chapter. We assume a stationary object is hidden at a distance $H$ from the origin of one of two given lines. One searcher starts at the origin of each line. The searchers alternate, each going at speed 2. A particularly significant class of DLSP strategies are the alternating strategies $A$. An alternating strategy $x \in A$ is an increasing positive sequence $x = \{x_i\}_{-\infty}^{\infty}$ interpreted as follows. In the $n$-th period one of the searchers goes at speed 2 from a distance $x_{n-2}$ on one side of his origin to $x_n$ on the other side. The two searchers move in alternate periods. In general (unless $x_i = 0$ for all $i$ less than some integer) such a strategy involves an infinite number of oscillations in arbitrarily small initial time segments. (However, the general strategy of this type may be uniformly approximated by one that starts with an initial period where the searcher moves from the origin at speed 2 on some initial time interval, and then the other does the same.) For any alternating strategy, the time $t_n$ when the $n$-th period ends (with a searcher reaching distance $x_n$) is given by

$$2t_n = x_n + x_{n-1} + 2s_{n-2}, \quad \text{where } s_i = \sum_{j=-\infty}^{i} x_j.$$

If the object is hidden at distance $H$, with

$$x_n < H \leq x_{n+1},$$

then it will be found by strategy $x$ equiprobably in one the periods $n+k$, $k = 1, 2, 3, 4$. If found in period $n+k$, it will be found at time

$$t_{n+k} - (x_{n+k} - H)/2.$$

Hence the expected time $E(x, H)$ required to find it is given by

$$4E(x, H) = t_{n+1} + t_{n+2} + t_{n+3} + t_{n+4} + (4H - x_{n+1} - x_{n+2} - x_{n+3} - x_{n+4})/2.$$

Consider the problem in which the searcher picks $x \in A$ so as to minimize the cost function given by

$$C(x, H) = E(x, H)/H,$$

in the worst case with respect to the unknown distance $H$. We can think of $E(x, H)$ as the expected time for the searchers to find an object at distance $H$ or for rendezvousers using the equivalent rendezvous strategy to find each other. Let $v(x)$ denote the maximum cost of the strategy $x$, so

$$v(x) = \sup_H C(x, H) = \sup_{n, x_n < H \le x_{n+1}} C(x, H).$$

From the above calculation of $E(x, H)$, we see that

$$\begin{aligned} v(x) &= \frac{1}{2} + \sup_n \frac{2t_{n+1} + 2t_{n+2} + 2t_{n+3} + 2t_{n+4} - x_{n+1} - x_{n+2} - x_{n+3} - x_{n+4}}{8x_n} \\ &= \frac{1}{2} + \sup_n \frac{x_n + x_{n+1} + x_{n+2} + x_{n+3}}{8x_n} + \frac{2(s_{n-1} + s_n + s_{n+1} + s_{n+2})}{8x_n} \\ &= \frac{1}{2} + \sup_n \frac{8\sum_{j=-\infty}^{1} x_{n+j} + 7x_n + 5x_{n+1} + 3x_{n+2} + x_{n+3}}{8x_n}. \end{aligned}$$

We wish to show that the above expression is minimized when the alternating strategy $x$ consists of a sequence of integer powers of some constant greater than 1. It follows from Corollary 7.11 on the optimality of geometric series that

$$\begin{aligned} v(x) - 1/2 &= \sup_n \frac{8\sum_{j=-\infty}^{1} x_{n+j} + 7x_n + 5x_{n+1} + 3x_{n+2} + x_{n+3}}{8x_n} \\ &= \min_{a>0} \sup_n \frac{8\sum_{j=-\infty}^{1} a^{n+j} + 7a^n + 5a^{n+1} + 3a^{n+2} + a^{n+3}}{8a^n} \\ &= \min_{a>1} (1/8)\left(\frac{8}{a-1} + 7 + 5a + 3a^2 + a^3\right). \end{aligned}$$

The above formula for $v$ as a function of $a$ has a minimum below 5.514, which is attained at the value stated in the theorem for $\hat{a}$. The rendezvous equivalent of the optimal DLSP strategy is described in the statement of the theorem. ■

It is well known that in zero-sum games the minimizer can sometimes lower his worst case expected payoff by choosing a mixed rather than a pure strategy, and this turns out to be the case here. In this game a pure strategy for the rendezvousers (minimizers) is a pair of trajectories. So the corresponding mixed strategies are probability distributions over the pure strategy pairs $\mathcal{S} \times \mathcal{S}$. Such strategies may also be considered "correlated strategies" if one interprets the two rendezvousers as distinct players. The best known strategy of this type is given in Baston and Gal (1998), which gives the following estimate.

**Theorem 17.9** *There is a probability measure $\omega$ on $\mathcal{S} \times \mathcal{S}$ with the following property: Let $F$ be any distribution the of the initial distance between two rendezvousers on the* *line which has a finite mean $\lambda$. Then the expected meeting time of two rendezvousers who jointly choose their strategies based on $\omega$ satisfies*

$$\int_{S\times S} \hat{T}((s_1, s_2); F)\, d\omega(s_1, s_2) \leq 4.42\lambda.$$

### 17.4.2 Symmetric strategies

We now consider the player-symmetric version of the unknown initial distance linear rendezvous problem. Recall that even for known initial distributions $F$, the players cannot achieve a finite expected meeting time with a common pure strategy because then if they start facing the same direction they will never meet. So the hardest reasonable problem for the rendezvousers occurs when they are forced to choose the same mixed strategy. The following is based on the strategy outlined in Baston and Gal (1998).

**Theorem 17.10** *There exists a mixed search strategy $\bar{\nu} \in S^*$ with the following property. Let $F$ be an initial distribution of distance between the players that has finite mean $\lambda$. If the two rendezvousers simultaneously adopt $\bar{\nu}$ with independent randomization, then their expected meeting time satisfies is no more than* 13.33$\lambda$. *That is,*

$$\int_S \int_S \hat{T}(s_1, s_2; F)\, d\nu(s_1) d\nu(s_2) \leq \left(7 + 2\sqrt{10}\right)\lambda \doteq 13.33\lambda.$$

A better estimate, also taken from Baston and Gal (1998), is available if the rendezvousers can randomize over their common mixed strategies.

**Theorem 17.11** *There exists a probability distribution over the mixed search strategies $S^*$ with the following property. Let $F$ be an initial distribution of distance between the players that has finite mean $\lambda$. If a mixed strategy is picked according to this distribution and then adopted simultaneously by both players, the expected meeting time of the total process is no more than* 11.4 $\lambda$.

## 17.5 Multiplayer Rendezvous

We will consider two versions of multiplayer rendezvous on the line: the players may wish either to minimize the expected time or the maximum time of some specified type of multiple rendezvous. In both cases the problem begins by placing $n$ agents randomly at $n$ consecutive integer locations on the line, which for simplicity we take to be $1, 2, \ldots, n$. They are faced randomly in either direction. They move at maximum speed 1 until the first time $T_{m,n}$ that a given number $m \leq n$ of them are at the same location. Depending on the version, they may wish to minimize either the expected or maximum value of $T_{m,n}$. The agents may or may not be required to use the same strategy (player symmetry or player asymmetry).

Multiple rendezvous has certain aspects not present in the two-player version. A strategy for a player must include instructions regarding what to do if he meets another player. What information should he exchange? Which of the two players who first meet should determine their future movements? Should players who meet stay together, or should they split up to increase the region they can search? Note that this last question of "stickiness" is not the same as the superficially similar notion discussed for the context of labeled networks in Chapter 13. (There it was theoretical *agents* of a single player who were required to stay together, while here we are talking about actual physical *players* who may meet.)

Multiplayer rendezvous in another context is considered in the important recent paper of Thomas and Pikounis (2001).

### 17.5.1 Expected time minimization

We first consider the player-symmetric version of the problem with $m = n$. That is, we seek the least expected time required for all $n$ of the agents to meet together at a single location, given that they all use the same strategy. Clearly, such a minimizing (optimal) strategy must be mixed, since if they all use the same strategy and are unlucky enough to all begin facing the same direction, there will never be any meetings at all. This problem may seem too difficult, as we saw earlier that even the two-player case of symmetric rendezvous on the line is an open problem. However, it turns out that the asymptotic value of the least expected time $R^s_{n,n}$ can be determined. In the following analysis we will call the part of a player's strategy that says what to do before he meets anyone his *Stage I strategy*. As with the previous rendezvous strategies we have seen (but not those to follow), such a strategy simply specifies a net motion function in a player's forward direction.

Unusually for rendezvous problems, in this version it is the *lower* bound on the rendezvous value that is easy to determine. Note that this lower bound holds even for the asymmetric version, where players may use distinct strategies.

**Lemma 17.12** *For any* $n$, $R^s_{n,n} \geq R^a_{n,n} \geq n/2$.

**Proof.** We prove this simple result for the rightmost inequality, as the symmetric value cannot be smaller. Observe that no meeting of players can occur before time 1/2. Let $f$ and $g$ be the strategies followed by the players who are placed at the two end locations 1 and $n$. Their expected distance at time 1/2 is their initial distance, $n-1$, plus the average of the four values $\pm f(1/2) \pm g(1/2)$, the second of which is clearly 0. The expected time for the end players to meet has a lower bound given by 1/2 plus half their expected distance at time 1/2, or

$$\frac{1}{2} + \frac{n-1}{2} = \frac{n}{2}.$$

Since a meeting of all $n$ players requires in particular a meeting of the end players, the result follows. ■

Better lower bounds on $R^s_{n,n}$ for small $n$ are clearly obtainable by similar methods. But as we shall see, this bound is exactly what will be needed to establish the asymptotic value as $n$ goes to infinity. The following result obtained by Lim, Alpern, and Beck (1997) established a suitable asymptotic upper bound.

**Theorem 17.13** *The symmetric and asymmetric multiple rendezvous values $R^s_{n,n}$ and $R^a_{n,n}$ are both asymptotic to $n/2$. In particular, they satisfy the inequality*

$$\frac{n}{2} \leq R^a_{n,n} \leq R^s_{n,n} \leq \frac{n}{2} + 5. \tag{17.10}$$

**Proof.** Given the previous lemma, it is enough to establish the rightmost inequality. To do this, we present a symmetric rendezvous strategy with the required expected time to achieve a total meeting of all $n$ players. All motion is, as usual, taken at maximum speed 1.

Before describing the strategy in full, we first state a rule that overrides anything that follows.

*Overriding rule*: If you meet someone who says "follow me," then follow him. That is, use the same future path.

Aside from this special rule, the strategy has three stages.

*Stage I strategy*: At each time $t = 2k$ for integer $k$, choose a random direction (equiprobably and independently of previous choices) to call forward. Go distance 1/2 in this direction, then distance 1 in the opposite direction, then forward again a distance 1/2. This brings you back to your starting point at time $2(k + 1)$. Repeat this until the first time when you are back at your start after having met another player. Then proceed to Stage II.

*Stage II strategy*: When beginning this Stage, you have recently met exactly one other player while you were in Stage 1, and you are at your original starting location. Go in the opposite direction from this meeting, a maximum distance of 1. If you meet someone (either after time 1/2 or time 1), then return immediately to your starting point and wait. If after time 1 in Stage II you have not met anyone new, go to Stage III.

*Stage III strategy*: Go back in the direction of your starting point (continuing past it), instructing anyone you meet to follow you.

We now seek to determine the expected total meeting time for this strategy. Let $\omega$ denote a particular way of all the possible chance moves $\Omega$ being picked (by Nature in the initial configuration and by the players in their independent randomizations). Let $t_i(\omega)$ denote the time when player $i$ (the one starting at position $i$) enters Stage II. Suppose that for some $\omega$ Player $i$ enters Stage II at time $2k$ (that is, $t_i(\omega) = 2k$) and that his neighbor, say $i - 1$, has not entered this Stage by time $2k$. Then Player $i - 1$ will meet Player $i$ either at time $2k + 1/2$ or $2k + 1$. So at time $2(k + 1)$ he will be back at his start after having met Player $i$. Consequently, $t_{i-1} = t_i + 1$. In general, this argument shows that for all $i = 1, \ldots, n - 1$ we have $|t_1 - t_{i+1}| \leq 1$, and by induction $|t_1 - t_n| \leq n - 1$. There is a bidirectional domino effect according to which adjacent players enter Stage II recursively after any player does.

We now compute the time $T = T(\omega)$ required for all the players to meet at a single location. Observe that at time $t_1 + 1$, Player 1 will be at location 0, having been in Stage II for time 1. Similarly Player $n$ will be at location $n + 1$ at time $t_n + 1$. At these respective times they will have satisfied the last sentence of the Stage II definition. Consequently, they will begin Stage III and move toward each other at these respective times. Denoting their respective positions at time $t$ by $1(t)$ and $n(t)$, we have

$$1(t) = 0 + (t - [t_1 + 1]) \quad \text{for } t \geq t_1 + 1,$$
$$n(t) = n + 1 - (t - [t_n + 1]) \quad \text{for } t \geq t_n + 1.$$

Consequently, the two end players 1 and $n$ meet (together with all intervening players) at time

$$T = \frac{n + 3 + t_1 + t_n}{2}.$$

Hence we have

$$R^s_{n,n} \leq E(t) = \frac{n+3}{2} + E(t_1), \tag{17.11}$$

(since by symmetry $E(t_1) = E(t_n)$) where $E$ denotes the expectation operator on $\Omega$. To estimate $E(t_1)$ we may assume that $n = 2$, since for larger values of $n$ we have earlier meetings since Player 2 may go left by the Stage II rule. For $n = 2$ the two players will meet after time 1/2 ($t_1 = 1$) if they both initially moved toward each other (probability 1/4) and after time 3/2 ($t_1 = 2$) if they initially moved away. So we have $E(t_1) = \frac{1}{4}(1) + \frac{1}{4}(2) + \frac{1}{2}(2 + E(t_1))$, or $E(t_1) = 7/2$. For general $n$ we have $E(t_1) \leq 7/2$. Substituting this value into (17.11) gives the required upper bound (17.10). ■

The upper bound in (17.10) can easily be improved (see Lim, Alpern, and Beck, (1997)). However, these improvements are not close to the best estimate that can be obtained for small $n$. For example, Baston's strategy for the symmetric rendezvous problem with two players (Theorem 17.2) gives $R^s_{2,2} \leq 2.2091$. It is also worth noting that the strategy given in the above proof is robust in that it does not require that the players know the number of players $n$.

For small values of $m$ and $n$, the asymmetric rendezvous value can be found by branch and bound techniques. In particular, the problem of minimizing the expected time for a pairwise meeting among three agents was solved in (Lim, Alpern, and Beck, (1997), Theorem 5), where it is established that $R^a_{3,2} = 47/48$.

### 17.5.2 Maximum time minimization

In certain applications it is important that the rendezvousers meet by a specified critical time; meeting earlier may be only moderately useful, and meeting later may not even be a possibility. For example, hikers may need to meet before nightfall to establish a camp. Or one of the rendezvousers may be a medic who must meet the other (a patient) early enough to apply the antidote to a certain condition (e.g., snakebite, heart attack). In such cases it is useful to determine for each strategy the maximum time it may require for a meeting to take place. The desired strategy is then the one which minimizes this maximum. (Or in a specific application, any strategy whose maximum time is lower than the critical time would be acceptable.)

The minimax multi-rendezvous problem for the line was introduced by Wei Shi Lim and Alpern (1996) based on the following definition.

**Definition 17.14** *Suppose that n distinguishable players (who may use distinct strategies) are arbitrarily placed at n consecutive integer locations on the line and faced* *in either direction. The minimax time $M_n$ is the least time required to ensure that all n players can meet together at a single point.*

It is fairly easy to see that this problem is equivalent to one where the players move in discrete time units (corresponding to a half unit of time) between adjacent points of the grid of integers and half integers. Consequently, each minimax time $M_n$ will be of the form $k/2$, for some integer $k$.

When $n = 2$, this is not a multiple rendezvous problem at all, but actually one already solved in Theorem 16.7. It was shown there that for any bounded distribution with maximum initial distance $D$, the WFM strategy pair is minimax, with maximum rendezvous time $3D$. Reinterpreting that result in the present context with $D = 1$ gives the following.

**Theorem 17.15** $M_2 = 3$.

The first true case of multiplayer rendezvous is $n = 3$. It was shown by Lim and Alpern (1996) that $M_3 \geq 3.5$. It was incorrectly claimed in that paper (Lemma 6 and its consequence, Theorem 3, are false in that paper) that the exact value of $M_3$ was 4. However, V. Baston (1999) subsequently discovered the following strategy (which we call the Baston strategy), and demonstrated that it has a maximum three-way meeting time of 3.5, which by the earlier result ($M_3 \geq 3.5$) is the best possible.

**Definition 17.16** *The Baston Strategy for three-person rendezvous on the line is given in terms of the following Stage I (before any meeting) and Stage II (after any meeting) rules.*

- *Stage I strategies: Player 1 reverses at time 0.5. Player 2 reverses at times 0.5, 1.5, 2.5. Player 3 reverses at times 0.5, 1.5, 2, and 2.5. (Between reversals players go in a fixed direction at maximum speed 1).*
- *Stage II strategies:*
  1. *If Players 2 and 3 meet at time 0.5, then in the next time unit they first go back to their starting points and then return to their meeting point. They wait there until Player 1 meets up with them.*
  2. *If Players 2 and 3 are the first to meet and this meeting occurs after time 0.5, they remain stationary at this meeting point until Player 1 meets up with them.*
  3. *If Players 1 and 2 meet at time 0.5, then they each go backward for time 1.5 (by which time one has met Player 3 and keeps him) and then forward for time 1.5.*
  4. *If Players 1 and 2 are the first to meet and this meeting occurs at time 1, then Player 2 moves forward for time 0.5 and then backward for time 2, while Player 1 moves backward for time 2 and then forward for time 0.5.*
  5. *If Players 1 and 2 are the first to meet and they meet at time 2.5, then they both move backward together (using Player 1's orientation) for time 1.*
  6. *The cases in which players 1 and 3 are the first to meet are similar to those of (3.), (4.) and (5.) except that players 2 and 3 must have the opposite orientation to Player 1 in the case corresponding to (5).*

Subsequent to the discovery of this strategy by Baston (1999), Alpern and Lim (2002) showed how to derive it as the *unique* optimal strategy. The combined work on this problem can be summarized as follows.

**Theorem 17.17** $M_3 = 3.5$. *Furthermore the only strategy triple that has a maximum meeting time of* 3.5 *is the Baston strategy.*

The Baston strategy has the property that sometimes when two players meet they must adopt different future paths. Another version of the problem outlaws such unsocial behavior. The minimax rendezvous time when players who meet must stick together is denoted by $\tilde{M}_n$. Consider the WFM strategy (1 Mommy, 2 Children) for this case. Mommy goes in some direction until she reaches an unoccupied location, then turns around until both children are collected. Children wait until Mommy finds them and then are carried by her. The worst time for triple rendezvous is 5. This occurs in the case where Mommy starts in the middle, reaches an unoccupied location at time 2, and finds the last child at time 5. No other sticky strategy does better than this (Lim and Alpern, 1996):

**Theorem 17.18** *If players who meet must stick together, then the minimax rendezvous time for three players is* $\tilde{M}_3 = 5$.

We now consider the problem when there are a large number of players. In this case the problem amounts to finding a way for the two end players to realize that they are end players, and which end they are on. Once this happens, they simply go toward each other. The following asymptotic result is due to Lim and Alpern (1996); the specific upper bound is from Gal (1999).

**Theorem 17.19** *The minimax rendezvous time* $M_n$, *required for n players randomly placed on adjacent integers to all meet at a common point, is asymptotic to* $n/2$. *That is,* $\lim_{n\to\infty} M_n/n = 1/2$. *Furthermore,*

$$M_n \leq \frac{n}{2} + 2\log_2 n + 3.5.$$

**Proof.** Since the end players cannot meet in time less than $(n-1)/2$, the asymptotic result will follow from the claimed upper bound, which we establish by analyzing the strategy given by S. Gal. In the following description $k$ denotes $\lceil \log_2 n \rceil$, $\alpha$ denotes going 1/2 forward and then 1/2 backward, and $\beta$ denotes going 1/2 backwards and then 1/2 forward. (Either $\alpha$ or $\beta$ takes total time 1.) Each player begins (for $t \leq 2k+2$) with a sequence of the following form

$$\alpha, \beta, \gamma_1, \gamma_2, \ldots, \gamma_k, \bar{\gamma}_1, \bar{\gamma}_2, \ldots, \bar{\gamma}_k,$$

where each $\gamma_i \in \{\alpha, \beta\}$ and $\bar{\gamma}_i = \beta$ if $\gamma_i = \alpha$ and $\bar{\gamma}_i = \alpha$ if $\gamma_i = \beta$. Since we have chosen $k$ so that $2^k \geq n$, there are sure to be enough sequences $\gamma_1, \gamma_2, \ldots, \gamma_k$ so that each of the $n$ players can adopt a different one. If two adjacent players are initially faced in opposite directions, they will have met by time 2 (after both adopt $\alpha, \beta$). Otherwise, they are initially facing in the same direction and will meet when one of them uses $\alpha$ while the other uses $\beta$. It follows that at time $2k+2$ the players who started at the end positions 1 and $n$ will be back at their initial positions and will each be aware of the direction to all the remaining players. So at this time they should move in that direction and tell anyone they meet to come with them. Consequently, there will be an $n$-way meeting at time $2k + 2 + (n-1)/2 < n/2 + 2\log_2 n + 3.5$ ■

It is worth noting that, since the maximum meeting time cannot be less than the expected meeting time, the above result gives us another way of obtaining the earlier result (the easy part of Theorem 17.12) that $R^a_{n,n}$ is asymptotic to $n/2$.

## 17.6 Asymmetric Information

In some real situations the two players may have different information about the initial location of the other. For example, if two parachutists drop at different times, the location of the first (to drop) may be known to the second but not the other way around. This type of problem was first considered in the plane by Anderson and Fekete (2001). Their analysis is given in the next chapter. The analysis given here is due to Alpern (2001).

We suppose that Player I starts at a position known to both, which we call 0. Player II can be assumed to know the location 0 or simply to know the direction to I. In any case it is easy to see that the following trajectory of II dominates any other motion:

$$g(t) = \begin{cases} g(0) - t, & \text{if } g(0) > 0, \\ g(0) + t, & \text{if } g(0) < 0. \end{cases}$$

In this version of rendezvous on the line it is not necessary to assume that the distribution of $g(0)$ (Player II's initial location) is distributed symmetrically around 0. However, for notational convenience (to agree with that of Appendix B on Alternating Search) we will assume that II's initial location $g(0)$ is equally likely to be positive or negative. In the event that $g(0) > 0$ we define

$$F_1(x) = \Pr(0 < g(0) \le x),$$

and in the event that $g(0) < 0$ we define

$$F_2(x) = \Pr(-x \le g(0) < 0).$$

Since we can assume that $g(0)$ is not 0, the two cumulative probability distributions $F_1$ and $F_2$ determine a distribution $F$ of the initial position $g(0)$ on the line.

Suppose that Player I follows a path $f(t)$, with $f(0) = 0$ and maximum speed 1. If $g(0) > 0$, then rendezvous will have occurred by time $t$ if and only if $f(t) \ge g(0) - t$, or equivalently if

$$0 \le g(0) \le f(t) + t.$$

Similarly if $g(0) < 0$ then the rendezvous time $T \le t$ if and only if

$$f(t) - t \le g(0) < 0.$$

Consequently, the rendezvous probability is given by

$$F_f(t) \equiv \Pr(T \le t) = \tfrac{1}{2}[F_1(f(t) + t) + F_2(-f(t) + t)], \qquad (17.12)$$

and the expected meeting time $\hat{T}_f(F)$ is given by

$$\hat{T}_f(F) = \int_0^\infty t \, dF_f(t).$$

It turns out that the problem of minimizing the above expected time for $f$ is equivalent to the problem of alternating search discussed in Appendix B on Alternating Search, and that the analysis of that problem given by the first author and John Howard can be effectively applied. The formula (17.12) shows that the meeting probability is the probability that a single stationary object placed at $g(0)$ according to $F$ is found *either* by a searcher going along the positive real axis with motion $f(t) + t$ or by a searcher going along the negative real axis with motion $-f(t) + t$ (describing its distance from the origin). If we write this in terms of the alternation rule $\alpha(t)$,with $0 \leq \alpha(t) \leq t$, as described in Appendix B on Alternating Search, we find the equivalence

$$\alpha(t) = \frac{f(t) + t}{2}.$$

In this equivalence the positive real axis is identified with ray 1 and the negative real axis with ray 2. Consequently, we have the following.

**Theorem 17.20** *Consider the asymmetric information rendezvous problem on the line, in which Player I is placed at 0 and Player II is placed equally likely on either the positive or negative real lines. II's initial distance from 0 has cumulative probability distribution $F_1$ or $F_2$, conditioned on the respective ray. Then the least expected meeting time (rendezvous value) is given by the value $v(F_1, F_2)$ of the associated alternating search problem as defined in Appendix B on Alternating Search.*

The analysis given in Appendix B can be used to give a qualitative description of the optimal Player I motion $f(t)$ in certain cases. For example Theorem B.5 (part 2) gives the following sufficient condition for waiting to be optimal for Player I.

**Theorem 17.21** *Suppose that Player II is symmetrically distributed in the asymmetric information rendezvous problem on the line ($F_1 = F_2$). Then a sufficient condition for "waiting" (that is, $f(t)$ identically 0) to be optimal for Player I is that $F_1$ is concave. If $F_1$ is strictly concave, then waiting is the unique optimal solution for Player I.*

In some cases it is optimal for Player I to first move in one direction to meet an oncoming Player II until he realizes he has gone in the wrong direction, and then to move in the other direction. The following is an immediate consequence of Theorem B.3.

**Theorem 17.22** *Suppose that both distributions $F_1$ and $F_2$ are convex on their supporting intervals. Then there is an optimal solution of the asymmetric information rendezvous problem on the line in which Player I goes in a single direction until the first moment he is sure that II was in the other direction and then turns and goes in that direction until he meets II.*

Situations in which $F_1$ and $F_2$ do not satisfy the conditions of the two previous results can be solved by the algorithms given in Appendix B on Alternating Search and more generally in Alpern and Howard (2000). For example, if the initial distance between the players is known to be 1, then Player I goes at speed 1 a distance 0.5 in one direction and then a distance 0.5 in the other. He meets Player II equally probably at times 0.5 and 1, so the rendezvous value is 0.75. If the initial distance is uniformly distributed on [0, 1] then Player I can either wait, with average waiting time 0.5; or he can follow the previous strategy, in which he meets in average time 0.25 if he guesses the direction right and 0.75 if he guesses wrong.

# Chapter 18

# Rendezvous in Higher Dimensions

Up to now, most of the search regions $Q$ on which the rendezvousers are assumed to move have been one-dimensional. This chapter considers various ways in which more difficult higher dimensional rendezvous problems may be analyzed. To make this problem more tractable we will generally assume that the players are confined to move on a grid that models $n$-dimensional space. Then we will consider versions of the problem where the players have a common notion of direction, or one player knows the starting location of the other, or they have differing speeds and detection distances.

Work on planar rendezvous for asymmetric players was initiated by Thomas and Hulme (1997) and Anderson and Fekete (2001), and has been extended to higher dimensions and player-symmetry by Alpern (2001). See also the very recent preprint of Chester and Tutuncu (2001).

## 18.1 Asymmetric Rendezvous on a Planar Lattice

We first consider the asymmetric rendezvous problem on a planar lattice, where the players have no common notion of locations or directions. As observers, we will adopt Player I's notion of North. As shown in Chapter 13, it is often useful to model a continuous rendezvous problem by a discrete one. To avoid the problem of "passing without meeting" (transposing positions on adjacent nodes), we begin with a placement of both players on "even" nodes, so that when players always move to a distinct adjacent node they will continue to be on nodes of the same parity. The general notion of even and odd nodes described in Chapter 13 will not be needed here, as we will restrict our attention to the specific network analyzed by Anderson and Fekete (2001). Anderson and Fekete assume that the players have a common notion of a clockwise direction or, equivalently, of how to get East when facing North (turn right). However, the strategy they propose does not in fact require this assumption. We begin by assuming this common notion of clockwise but then analyze the more restricted version of the game where the players cannot rely on this. In terms of the given group $G$ of symmetries described in Chapter 12, this means we first leave out the reflections from $G$, and then we put them in.

The network consists of the integer lattice points $(m, n)$ in the plane, with $(m, n)$ adjacent to the four other nodes $(m, n \pm 1)$ and $(m \pm 1, n)$ to which it is connected by a vertical or horizontal line. This is essentially the "graph paper" lattice. In this setting, even nodes ($m + n =$ even) are adjacent only to odd nodes ($m + n =$ odd). We will consider starting positions in which both players are on nodes of the same parity (say even). A move consists of one of the compass directions $N = (0, 1)$, $E = (1, 0)$, $S = (0, -1)$, and $W = (-1, 0)$. However, while we may choose to view the problem from Player's I's perspective, we cannot assume that Player II's compass directions will be the same. We may assume that Player II is randomly faced in one of the four compass directions and that he calls this direction North.

Before proceeding with the two-dimensional analysis, it is worth formulating the asymmetric rendezvous problem on the line (ASRL) in this discrete type of setting. If the initial distance between the two players was two units (taking the atomic distribution with $D = 2$) on a one-dimensional lattice, it was shown in Section 16.5 that the uniformly optimal strategy is given by the pair $[E, E, W, W, W, W]$, $[E, W, E, E, W, W]$.

Anderson and Fekete (2001) consider a specific initialization of the problem on the planar lattice described above, in which Player I starts at the origin (0,0) and Player II starts equiprobably at the four nodes $(\pm 1, \pm 1)$ and equiprobably calls any of the four directions $N$. The information available to both players is that the other player is one horizontal plus one vertical step away. We may consider that there are 16 equiprobable agents of Player II, and I wishes to minimize the expected time required to meet an agent. The initial locations and directions (say, the direction they call North) of these 16 agents are shown in Figure 18.1.

Anderson and Fekete analyzed a strategy pair that we call the A–F strategy, given by $\bar{f} = [N, W, S, S, E, E, N, N]$, $\bar{g} = [N, S, N, S, N, S, N, S]$. Note that this is a MWFM (Modified Wait for Mommy) strategy with Mommy's exhaustive search given

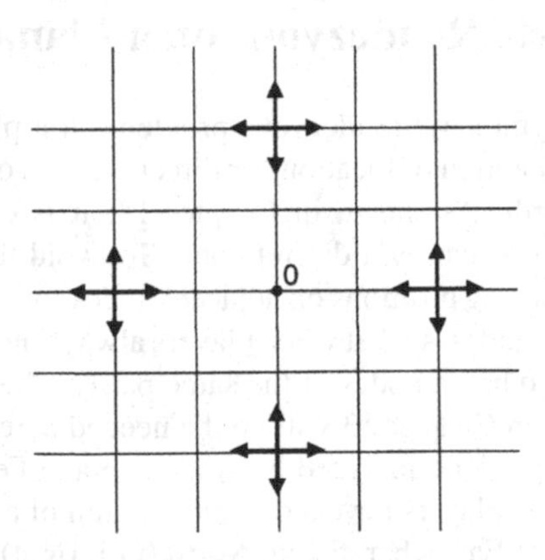


**Figure 18.1.** Initial placement of II's 16 agents

by $\bar{f}$, with the Child using $\bar{g}$ back at his start at all even times. The following table indicates the meeting times corresponding to $(\bar{f}, \bar{g})$ depending on the initial direction that II calls North (labeled in terms of what I calls it) and the initial location of Player II.

| Player II Initial direction | Starting point | | | |
|---|---|---|---|---|
| | (1,1) | (−1, 1) | (−1, −1) | (1, −1) |
| N | 8 | 2 | 3 | 6 |
| W | 1 | 2 | 4 | 5 |
| S | 7 | 2 | 4 | 6 |
| E | 8 | 1 | 4 | 6 |

For each time $t = 1, \ldots, 8$, the number of entries of the $4 \times 4$ matrix of meeting times that are equal to $t$ is denoted by $\bar{x}_t$ and the number that are less than or equal to $t$ is denoted by $\bar{y}_t$. Thus in the A–F strategy Player I meets $\bar{x}_t$ of the 16 Player II agents *at* time $t$ and $\bar{y}_t$ of these agents *by* time $t$. For a general strategy we will let $x_t$ and $y_t$ denote these numbers. For the A–F strategy we have

| $t$ | 1 | 2 | 3 | 4 | 5 | 6 | 7 | 8 |
|---|---|---|---|---|---|---|---|---|
| $\bar{x}_t$ | 2 | 3 | 1 | 3 | 1 | 3 | 1 | 2 |
| $\bar{y}_t$ | 2 | 5 | 6 | 9 | 10 | 13 | 14 | 16 |

(18.1)

The expected meeting time is $(1/16) \sum_{t=1}^{16} t\, y_t$, so for the A–F strategy this is

$$\frac{1}{16}(1 \cdot 2 + 2 \cdot 3 + 3 \cdot 1 + 4 \cdot 3 + 5 \cdot 1 + 6 \cdot 3 + 7 \cdot 1 + 8 \cdot 2) = \frac{69}{16}.$$

Anderson and Fekete establish the following result.

**Theorem 18.1** *The A–F strategy pair $(\bar{f}, \bar{g})$ is optimal for the least expected time problem, so that the asymmetric rendezvous value for this problem is 69/16. Furthermore, for any strategy pair, we have*

$$x_i \leq 3, \quad \textit{and} \tag{18.2}$$

$$x_i = 3 \quad \textit{implies} \quad x_{i+1} \leq 1.$$

**Proof.** We will not give a proof of the optimality of the A–F strategy here because we will later independently prove the stronger result that this strategy is uniformly optimal. However, we will prove (18.2), as we will need this later (Theorem 8.4) to establish the stronger property of uniform optimality. So suppose that $x_i \geq 3$, which means that Player I meets at least three agents of II at time $i$ at some location $A$. We first show that at time $i$ one of the players must be back at his start. Suppose not. Then agents of Player II starting at a common node must be at distinct locations. Hence all the agents that I meets at time $i$ must come from different starting points. Since all Player II agents are equally distant from their respective starting positions, the node $A$ must be equally distant (in the Manhattan or graph distance) from at least three of the start points of Player II. The only such location is the origin, which we take as Player I's initial location. So the meeting must be, as claimed, at a starting point. (Note that this implies that $i$ is even.) By symmetry of the players, we will assume that $A$ is one of the starting points of II. At time $i-1$ both Player I and one of the agents of II who started at $A$ must be at the same location. Hence $x_i \leq 3$ as claimed. Since all agents of II must be at their starting points at time $i$, and I is at one of these, he can meet at most one agent of II at time $i+1$. ■

We highlighted the constraints on the numbers $x_i$ because we will use these later to prove a stronger property of the A–F strategy, uniform optimality (see Definition 16.1). But first we will establish an extension of the class of optimal strategies. To do this, we define a mixed rendezvous strategy $(f^*, g^*)$ as follows. The strategy $f^*$ sends Player I cyclically around the square with corners $(\pm 1, \pm 1)$, equiprobably in one of the eight possible ways. These ways are determined by the first two moves (e.g., N, W for the A–F pure strategy), with the second direction resulting from a left or right turn. The strategy $g^*$ places Player II back at his starting point at all even integer times, moves in a random direction (independently of previous choices) at all odd times, except that the last odd move is the same direction as the first move.

To evaluate the expected meeting time for the mixed rendezvous strategy $(f^*, g^*)$ we can assume without loss of generality that Player I follows the pure strategy $\breve{f}$ determined by the first two choices N, E. The following table gives the location $\breve{f}$ of Player I at times $t$ and the probability $p_t$ that the first meeting time is at time $t$.

| $t$ | 1 | 2 | 3 | 4 | 5 | 6 | 7 | 8 |
|---|---|---|---|---|---|---|---|---|
| $f$ | (0,1) | (1,1) | (1,0) | $(1,-1)$ | $(0,-1)$ | $(-1,-1)$ | $(-1,0)$ | $(-1,1)$ |
| $p_t$ | $\frac{1}{4}\frac{1}{4}+\frac{1}{4}\frac{1}{4}$ | $\frac{1}{4}\frac{3}{4}$ | $\frac{1}{4}\frac{1}{4}$ | $\frac{1}{4}\frac{3}{4}$ | $\frac{1}{4}\frac{1}{4}$ | $\frac{1}{4}\frac{3}{4}$ | $\frac{1}{4}\frac{3}{4}\frac{1}{3}$ | $\frac{1}{4}\frac{3}{4}\frac{2}{3}$ |

This table is explained as follows. The probability $p_1$ arises from the possibility that II starts at $(-1, 1)$ and moves $E$ (in I's notation) or starts at (1,1) and moves $W$. For $t = 3, 5$ the probability $p_t$ is the probability that Player II starts at $\breve{f}(t+1)$ (just ahead of I's current position) and is lucky enough to move toward the oncoming Player I. If he was unlucky, this gives the probability $p_2$, $p_4$, or $p_6$. The probability $p_7$ is the probability that II started at $(-1, 1)$, did not initially go $E$, and did not go $S$ at move 7. The later probability is 1/3 for our strategy, as going $E$ is excluded. (Note that if we used an entirely random strategy for the second player, this probability would have been 3/64 rather than 4/64). It follows that the expected meeting time is $\Sigma_t t \cdot p_t = 69/16$. Since Theorem 18.1 says that this is the rendezvous value, we obtain the following.

**Theorem 18.2** *The mixed strategy $(f^*, g^*)$ has one player (say, I) equiprobably choose one of the eight cyclic search patterns, while the other player (say, II) chooses a strategy that is back at his start at all even times, picks a random direction at times 1, 3, 5 (independently of previous choices), and moves in the same direction at time 7 as that chosen at time 1. This strategy is optimal, giving an expected meeting time of 69/16, equal to the rendezvous value.*

Note that if a mixed strategy is optimal for the asymmetric rendezvous problem, it follows that every pair of pure strategies $(f, g)$ that occurs with a positive probability must be an optimal pair. Consequently, we have the following.

**Corollary 18.3** *A pure strategy pair $(f, g)$ is optimal if one of the players goes around the square $(\pm 1, \pm 1)$ in a cyclic fashion while the other moves in any direction at times 1, 3, 5, moves in the opposite direction from the previous at times 2, 4, 6, and moves in the same two directions at moves 7 and 8 as at moves 1 and 2. Such a strategy will be called a* ***generalized A–F strategy****.*

The description of a generalized A–F strategy given above shows that these strategies do not rely on a common notion of the clockwise direction. So they are feasible without this assumption. We will see below that for strategy pairs that do not rely on a common sense of clockwise, the A–F strategies are uniquely optimal and furthermore uniformly optimal. This will match the uniform optimality established for the similar one-dimensional problem in Section 16.5.

However, if the players *do* have a common sense of clockwise (i.e., if reflections are not included in the given symmetry group $G$ or if the players approach a vertically placed grid from the same direction), then there is another optimal strategy. This strategy was found by Vic Baston and the first author, so we call it the A–B strategy. It is an "Alternating Search For Mommy" (AWFM) strategy, in which Player I searches two of the possible initial location of Player II at times 2 and 6 (when II is back at his start), while Player II searches two of the possible initial locations of Player I at times 4 and 8 (when I is back at his start). Furthermore, Player I searches the two locations along upward sloping diagonals from his start, while Player II searches along the two downward sloping diagonals from his start. (Thus the players need a common notion of upward and downward sloping diagonals, which is equivalent.) One version of this strategy (that we call *the* A–B strategy) is $([N, E, S, W, S, W, N, E], [N, S, E, S, W, N, E, N])$. This strategy pair is plotted below in Figure 18.2, together with a version of II's strategy reflected in the NS axis.

Note that if the strategy for Player II is reflected about the NS axes (if he approaches the grid from behind when it is placed vertically), he will also be searching the upward-sloping diagonal. This corresponds to interchanging E and W in his strategy. So if the original placement (from I's point of view) has the diagonal between their initial

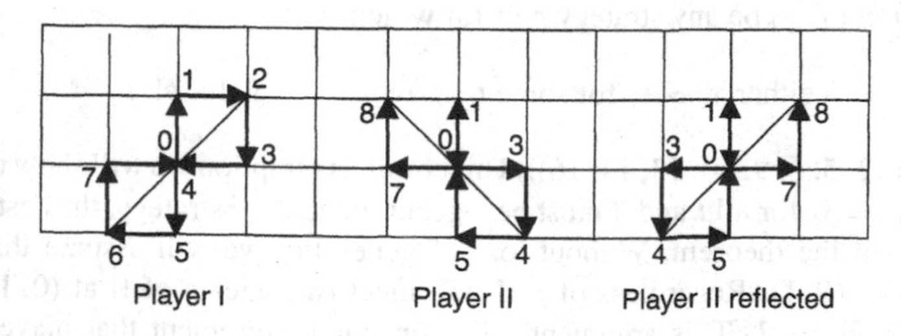


**Figure 18.2.** The A–B strategy

locations sloping upward, then they will not have met by time 8. The meeting times for the A–B strategy are given below.

| Player II, A–B Initial direction | Starting point | | | |
|---|---|---|---|---|
| | (1,1) | (−1, 1) | (−1, −1) | (1, −1) |
| N | 2 | 4 | 6 | 4 |
| W | 1 | 7 | 4 | 3 |
| S | 2 | 8 | 6 | 8 |
| E | 2 | 1 | 6 | 5 |

Observe that the number of meeting time $x_i$ at time $i$ for this strategy are the same as for the A–F strategy (18.1), so it follows that the A–F strategy is also optimal for the "common clockwise" form of the game. However, it does not even guarantee a meeting by time 8 if used in the "no common clockwise" version of the game.

Before dealing with the further optimality properties of the A–F strategy, we make the following observation regarding the $x_i$'s. It follows from the condition on the $x_i$ established in (18.2) that for any $i$ and $j$, we have

$$x_i + x_{i+1} + \cdots + x_{i+j} \leq 2j + 3.$$

In particular, we have that

$$\begin{aligned} x_i + x_{i+1} &\leq \bar{5}, \\ x_i + x_{i+1} + x_{i+2} &\leq \bar{7}, \\ x_i + x_{i+1} + x_{i+2} + x_{i+3} &\leq \bar{9}, \quad \text{and} \\ x_i + x_{i+1} + x_{i+2} + x_{i+3} + x_{i+4} &\leq \overline{11}. \end{aligned} \tag{18.3}$$

The bars over the numbers at the right will just be used to identify where these numbers (5, 7, 9, 11) come from in the analysis below (also $\bar{3}$, as stated earlier).

**Theorem 18.4** *A strategy pair $(f, g)$ is optimal for the "no common clockwise" version of the game if and only if it is a generalized A–F strategy. Furthermore, each of these strategies is uniformly optimal.*

**Proof.** Let $(f, g)$ be any strategy pair for which

$$\text{either } y_t > \bar{y}_t \text{ for some } t, \quad \text{or} \quad y_t \geq \bar{y}_t \text{ for all } t. \tag{18.4}$$

(Recall $\bar{y} = (2, 5, 6, 9, 10, 13, 14, 16)$.) Under this assumption we will show that $(f, g)$ must have $y_t = \bar{y}_t$ for all $t$ and it must be a generalized A–F strategy, thus establishing both claims of the theorem. Without loss of generality we will assume that I starts by going N to (0, 1). Regardless of $g$, I will meet two agents of II at (0, 1), so that $x_1 = y_1 = \bar{y}_1 = 2$. This argument relies on our requirement that players cannot stay still.

Unless I turns (E or W) and II returns to his start (or the other way around), the largest $x_2$ can be is 1 (corresponding to I continuing N and meeting an agent starting at $(-1, 1)$ or $(1, 1)$ at location $(0, 2)$). In this case $y_2 = 3$ and by (18.3) we have

$$y \leq (2, 3, 3 + \bar{3}, 3 + \bar{5}, 3 + \bar{7}, 3 + \bar{9}, 3 + \bar{11}, 16),$$

which violates our initial assumption (18.4). So we may assume without loss of generality that at time 2 Player I is at $(1, 1)$ and Player II is back at his start. (A symmetric case arises if I chooses NW and is at $(-1, 1)$.)

If I does not go to $(1, -1)$ in the next two steps, with II returning to his start at time 4, the largest values for $x_3$ and $x_4$ are, respectively, 1 and 2, obtained by Player I going to $(-1, 1)$. (Note that the A–B strategy would contradict the previous sentence, if it could be used in a "common clockwise" version.) However, any strategy using these two steps is strictly dominated by the symmetric strategy going to $(1, -1)$. Any other strategy gives at most $x_3 = 1$ and $x_4 = 2$, hence at most $y_3 \leq 6$ and $y_4 \leq 7$. Consequently, it has a cumulative distribution $y$ satisfying

$$y \leq (2, 3, 6, 7, 7 + \bar{3}, 7 + \bar{5}, 7 + \bar{7}, 16),$$

which again violates our assumption (18.4). So the strategy $(f, g)$ must be as claimed (and in particular a generalized A–F strategy) up to time 4.

If I does not move W to $(0, -1)$ at time 5, then $x_5 = 0$ and $y_5 = 9$. Hence

$$y \leq (2, 5, 6, 9, 9, 9 + \bar{3}, 9 + \bar{5}, 9 + \bar{7}),$$

which again violates our assumption (18.4). Hence I must go W to $(0, -1)$ at time 5.

If at time 6, II is not back at his start, and I at $(-1, -1)$, then $x_6 \leq 1$ and hence $y_6 \leq 11$. Consequently,

$$y \leq (2, 3, 6, 9, 9, 9 + \bar{3}, 9 + \bar{5}, 9 + \bar{7}),$$

again violating (18.4).

At time 6, I is at $(-1, -1)$ and there are three agents of II remaining at $(-1, 1)$. The only way that to ensure a meeting at time 7 at $(-1, 0)$ (that is, $x_7 = 1$) is for I to go to $(-1, 0)$ while one of the agents of II at $(-1, 1)$ also goes there. Player II must make sure that it is not the agent that I already met at time 1 that he meets there (not for the first time). The only way to ensure this is for II to go in the *same direction* at time 6 (move 7) as he went at time 0 (first move). So we may assume this, and consequently we have

$$y \leq \bar{y},$$

so we know that the first alternative in assumption (18.4) on $(f, g)$ is impossible. This proves that any strategy with cumulative distribution function $\bar{y}$ (and hence any generalized A–F strategy) is uniformly optimal. Finally, to show that only such strategies are optimal, we only have to observe that the only way to get $x_8 = 2$ and $y_8 = 16$ is for II to return to his start at time 8 while I continues to $(-1, 1)$. ■

## 18.2 The $n$-Dimensional Lattice $Z^n$

We now extend the model of Anderson and Fekete to higher dimensions. To this end, we have the players move on the $n$-dimensional integer lattice $Z^n$, consisting of all points in $R^n$ whose coordinates are all integers. We consider that two such points (called nodes) are adjacent if all of their coordinates are identical, except for one, where they differ by 1. A node will be called *even* if the sum of its coordinates is even; otherwise it will be called *odd*. In general, each player will be placed according to some distribution over even nodes and must move to an adjacent node in each time period. This will ensure that the players are always at nodes of the same parity and consequently avoids the possibility that they might pass each other on an edge without meeting at a node. In fact, the only starting position we will consider is where the initial difference vector between the players has length two and is parallel to one of the coordinate axes. Since the players are assumed to have no common labeling of nodes or directions (of the coordinate axes), this is equivalent to placing Player I at the origin and placing Player II equiprobably at one of the $2n$ nodes $\pm 2e_i$ (where $e_i$ are the coordinate vectors) and facing in one of the $2n$ possible directions in the lattice. (The strategies we propose do not require that the players even have a common ordering of the vectors $e_i$ , no "right hand rule," so this is analogous to the "no common clockwise" assumption of the previous section.) Hence there are $(2n)^2 = 4n^2$ possible agents of Player II, and the expected meeting time is the same as the expected time for Player I to meet the agents of Player II. This problem generalizes that of Section 16.5 with $D = 2$ and $n = 1$. For $n = 2$, the 16 possible initial placements of Player II at time zero are shown in Figure 18.3. We denote this rendezvous problem by $\Gamma(n)$ and denote the associated player asymmetric and symmetric rendezvous values by $R^a(n)$ and $R^s(n)$.

### 18.2.1 Asymmetric rendezvous

A useful observation regarding this configuration is that Player I can make a Traveling Salesman Tour of the four possible Player II starting locations, while still returning to his own starting position between each inspection. That is, he can pick any ordering of the

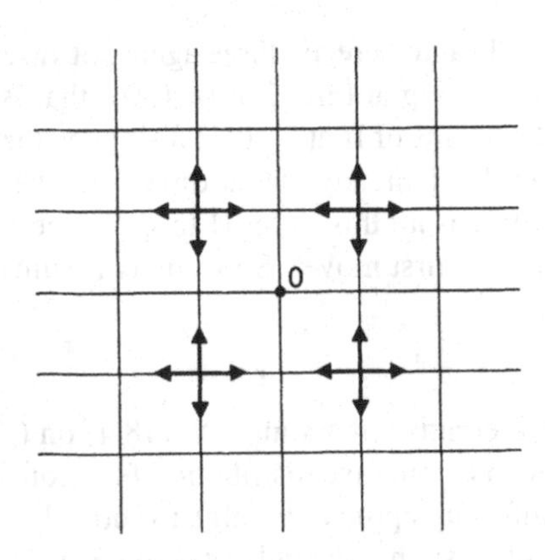


**Figure 18.3.** Starting states for Player II

four Player II starting locations ((0, 2), (2, 0), (0, −2), and (−2, 0)) and inspect them at times $t_i = 2 + 4i$, $i = 0, \ldots, 3$, while returning to his own starting location at time $q_i = 4i$, $i = 1, \ldots, 3$. So Player II can adopt a strategy so that he is back at his start at all times $t_i$ and similarly inspect Player I's possible starting locations at time $q_i$, in some arbitrary order. We call this strategy pair the Alternating Wait For Mommy (AWFM) Strategy, since the two players alternate playing the role of Mommy and that of the waiting Child. This strategy is well defined for all dimensions $n$, except that the upper bound of 3 on the index $i$ must be changed to $2n - 1$. By time $t_{2n-1} = 8n - 2$, Player I has searched all the initial locations of Player II, so the strategy has only to be defined up to this time. For $n = 1$, this strategy is identical with that shown in Theorem 16.9 to be uniformly optimal, the pair $[E, E, W, W, W, W]$ and $[E, W, E, E, W, W]$.

The AWFM strategy has a density of meeting times that is roughly linearly decreasing from $t = 0$ to $t = 8n - 2$. Let $\hat{x}_i$ denote the number of Player II agents that I meets at time $i$, assuming AWFM is adopted. At time $t = 1$, Player I meets exactly $\hat{x}_1 = 1$ agent of Player II. (This is true for any strategy pair.) At time $t = 2 = t_0$, Player I is at a starting point of II and will meet the remaining $\hat{x}_2 = 2n - 1$ agents who started there. At time $t = 4 = q_1$, Player I is back at his start and will meet the agents from the $2n - 1$ starting points of II not yet searched by I who go towards I's start. Thus $\hat{x}_2 = 2n - 1$. Similarly at time $t_1 = 6$, Player I will meet all the agents of II at this starting point of II except the one that found him at his start at the earlier time $q_1$, thus $\hat{x}_6 = 2n - 1$. In general, at times $q_i$ and $t_i$, $i = 1, \ldots, 2n - 1$, Player I will meet exactly $\hat{x}_{t_i} = \hat{x}_{q_i} = 2n - i$ agents of Player II.

For example, when $n = 2$, we have

| $i$ | 1 | 2 | 3 | 4 | 5 | 6 | 7 | 8 | 9 | 10 | 11 | 12 | 13 | 14 |
|---|---|---|---|---|---|---|---|---|---|---|---|---|---|---|
| $\hat{x}_i$ | 1 | 3 | 0 | 3 | 0 | 3 | 0 | 2 | 0 | 2 | 0 | 1 | 0 | 1 |
| $\hat{y}_i$ | 1 | 4 | 4 | 7 | 7 | 10 | 10 | 12 | 12 | 14 | 14 | 15 | 15 | 16 |

Techniques similar to those used to prove Theorem 18.4 can also be applied to show (Alpern, 2001) that these values of $\hat{y}_i$ are maximal, at least for $i \leq 7$. That is,

**Theorem 18.5** *For $n = 2$ and with the above starting situation, the AWFM strategy pair maximizes the probability of meeting by time i, for all $i \leq 7$.*

Since there are $4n^2$ agents in the $n$-dimensional problem $\Gamma(n)$, and $t_i + q_i = 8i + 2$, we can calculate the expected meeting time for the AWFM strategy as

$$\frac{1}{4n^2}\left(1 \cdot 1 + (2n-1) \cdot 2 + \sum_{i=1}^{2n-1} (2n-i)(8i+2)\right).$$

Simplifying this expected meeting time and setting it as an upper bound for the rendezvous value, we get the following estimate (which is exact for $n = 1$ as shown in Section 16.5).

**Theorem 18.6** *For the asymmetric rendezvous problem on the n-dimensional integer lattice, with no common directions or locations, and starting locations whose difference* *has length 2 and is parallel to a coordinate axis, we have*

$$R^a(n) \le \frac{32n^3 + 12n^2 - 2n - 3}{12n^2}.$$

*In particular, we have*

$$\lim_{n\to\infty} \frac{R^a(n)}{n} \le \frac{8}{3}.$$

### 18.2.2 Symmetric rendezvous

We now consider the player-symmetric version of the same problem. Recall that for the atomic distribution with a known initial distance $D = 2$ on the line, the best known estimate of Baston gives the estimate $R^s(1) \le 5.402$. To get an estimate for the symmetric rendezvous value $R^s(n)$ we will restrict ourselves to the case of large $n$, as even for $n = 1$ it is very hard to get exact values (see Section 17.2). The completely random strategy, of repeatedly going one step in a random direction and then back to your start, will meet on a given search with probability $1/4n^2$. Consequently, it takes an average of $4n^2$ such trials for a meeting, and therefore the expected meeting time will be $2(4n^2) - 1 = 8n^2 - 1$. A better strategy would be a randomized version of AWFM, with each player choosing equiprobably between the two roles (of I or II in AWFM), independently in each period of time $8n$. We need $8n$ rather than $8n - 2$ because we need to allow the players to return back to their starting locations to begin the next period. We may neglect the possibility that they meet while choosing the same strategy and observe that if they choose distinct strategies then by Theorem 18.6 they will meet in average time of $8n/3$, counting from the beginning of the period. Therefore they will meet in expected time $\hat{T}$, where $\hat{T}$ satisfies the equation

$$\hat{T} = \frac{1}{2}\left(\frac{8n}{3}\right) + \frac{1}{2}(8n + \hat{T}), \quad \text{with solution } \hat{T} = \frac{32n}{3}.$$

However the players can improve on this by randomizing between the two possible roles more often. Suppose that $n$ is large, $k$ is large, and $n/k$ is large, and the players divide the period of $8n+2k$ into $k$ subperiods of length $8n/k+2$. The extra two time units are needed simply to ensure that both players return to their respective starting positions for the beginning of the next subperiod. Since our aim is an asymptotic estimate of expected meeting time divided by $n$, and since $n/k$ goes to infinity we have $k/n$ going to zero, we can simplify the calculations by viewing the length of the period as $8n$ rather than $8n + 2k$. Before beginning each full period, the players each choose an order in which to search out the $2n$ possible starting locations of the other. In each subperiod, they search the next $2n/k$ of these before returning to their start. Call this the Randomized Alternating Wait For Mommy (RAWFM) strategy. If $k$ is large, then at any time within the full period they will have chosen opposite roles in about half the previous subperiods, so the density of meeting probability (or of the number of agents met) will be half that of the AWFM with distinct strategies.

Since we are only interested in the asymptotic behavior for large $n$ (and large $k$ and $n/k$), we will use a continuous model. In both the asymmetric case (with AWFM) and the symmetric case (with RAWFM) the probability density of meeting times decreases linearly with time. If it takes time $t = a$ for a meeting to be certain, the density and cumulative distributions are given by

$$f_a(t) = \frac{2}{a}\left(1 - \frac{t}{a}\right), \quad \text{and} \quad F_a(t) = \frac{2}{a}\left(t - \frac{t^2}{2a}\right).$$

For the asymmetric case, using AWFM, we have $a = 8n$ and so the expected meeting time is given by

$$\frac{2}{8n}\int_0^{8n} t\left(1 - \frac{t}{8n}\right) dt = \frac{8}{3}n.$$

This agrees with the value obtained by taking the limit in the discrete model covered by Theorem 18.6. For the symmetric case we have $a = 16n$ by the argument given above. So the probability $p$ that a meeting takes place using RAWFM within a period of length $8n$ is given by

$$p = F_{16n}(8n) = \frac{2}{16n}\left(8n - \frac{(8n)^2}{32n}\right) = \frac{3}{4}.$$

Given that such a meeting takes place within the period, the conditional density for $t \leq 8n$ is given by

$$\frac{4}{3}f_{16n}(t) = \frac{4}{3}\frac{2}{16n}\left(1 - \frac{t}{16n}\right).$$

and the expected meeting time is given by

$$L = \frac{4}{3}\frac{2}{16n}\int_0^{8n} t\left(1 - \frac{t}{16n}\right) dt = \frac{4}{9}(8n).$$

So if this process is repeated independently in each period of length $8n$, the expected meeting time $\hat{T}$ satisfies

$$\begin{aligned}
\hat{T} &= p(L) + (1-p)(8n + \hat{T}), \quad \text{or} \\
\hat{T} &= \frac{pL + (1-p)(8n)}{p} \\
&= \frac{4}{3}\left[\frac{3}{4}\left(\frac{32n}{9}\right) + \frac{1}{4}(8n)\right] \\
&= \frac{56}{9}n.
\end{aligned}$$

**Theorem 18.7** *For the symmetric rendezvous problem on the n-dimensional integer lattice, with no common directions or locations, and starting locations whose difference has length 2 and is parallel to a coordinate axis, we have*

$$\lim_{n\to\infty}\frac{R^s(n)}{n} \leq \frac{56}{9} \doteq 6.2222.$$

## 18.3 Continuous Rendezvous in the Plane

The previous sections of this chapter have considered higher dimensional rendezvous problems by restricting the players to a grid, or lattice, that in certain natural ways approximates $n$-dimensional Euclidean space. In this section we consider some attempts that have been made to directly attack the asymmetric rendezvous problem in the plane.

A natural formulation of the problem would be to place the players according to some known joint distribution, specify a common detection radius, and ask how to minimize the expected time for the players to come within this distance. In fact, all the attacks on the planar problem have added some additional interesting aspect to this problem. We shall discuss three of these. The first two are taken from the paper of Anderson and Fekete (2001) and respectively discuss the problems in which the players have a common sense of compass directions, and in which the initial location of one of the players is known to the other. The third, taken from the article of Thomas and Hulme (1997), considers the problem where a helicopter seeks a lost hiker: the helicopter goes faster, but the hiker can detect the helicopter from far away.

### 18.3.1 Common notion of direction

As in the case of the circle (Section 14.2) and the line (Section 16.2) we begin by looking at the rendezvous problem where the players have a common notion of direction. For the plane, we may assume that the rendezvousers are two hikers who each have accurate compasses and so have an identical idea of the compass directions. This problem, like the similar common-direction rendezvous problems on the circle and line, turns out to be equivalent to a one-person search problem for a stationary hidden object. (See Book I, Sections 3.7 and 6.2, for a worst case analysis.)

The formal problem begins by picking a vector describing the difference between the initial positions of Player II and Player I. If we take a coordinate system placing I at the origin, this vector simply describes the initial position of II. The distribution of Player II's initial position in the plane is known to both players and denoted by $F$. As described in Section 14.2, the optimal strategy pairs $(f, g)$ for this problem involve the players moving in opposite directions at their maximum speeds (that we take here to be 1). The problem is equivalent to a pure search problem faced by a single searcher who controls the speed 2 search pattern $z(t) = f(t) - g(t)$. Thus this type of rendezvous search problem is equivalent to that of a single searcher finding a stationary object hidden according to $F$ in the plane by following a speed 2 path from a given starting point 0. If we are considering a least expected time problem, then the infinite time path of the searcher must cover the support of $F$. Similarly, if we are considering a minimax time problem with minimum time $\bar{t}$, then the finite path $\bar{z}([0, \bar{t}])$ must contain the support of $F$. Once the optimal path $z$ for the search problem has been found, the rendezvousers should follow the unit speed motions $z/2$ and $-z/2$.

The problem of finding a stationary hidden object in the plane (in minimax or least expected time) is in general very difficult. We present two examples that illustrate certain classes of problems.

**Example 18.8** *Suppose the object is located at* $(0, 1)$ *or* $(0, -1)$*, each with probability* 0.4*; or it is located at* $(1, 0)$ *or* $(-1, 0)$*, each with probability* 0.1*. The minimax search* *strategies start by going directly to one of these points, and then follow three sides of the square determined by these points. The minimax time is $1 + 3\sqrt{2}$. This strategy (going first to a probability 0.4 location) finds the object in expected time $\frac{1}{2}[1.0 + 1.2\sqrt{2}] \doteq 1.3486$. However going first to the high probability locations $(0, \pm 1)$ and then to the low probability locations $(\pm 1, 0)$ gives the least expected time of $\frac{1}{2}[2.4 + 0.2\sqrt{2}] \doteq 1.3414$. In general, the minimax problem is similar to the well known Traveling Salesman Problem, while the least expected time problem is similar to the Minimum Latency Problem (see Blum et al., 1994).*

**Example 18.9** *Suppose the two rendezvousers know their distance (say 1) but not the direction to the other. Assuming they view all directions as equally likely, this results in a search problem of finding an object uniformly located on the unit circle (starting at the origin). The search must start by going to the circle along some ray, and then it is uniformly optimal to go around the circle (as in the "Columbus strategy" of Section 14.1). The minimax search (or rendezvous) time is $\frac{1}{2}[1 + 2\pi]$, while the least expected search (or rendezvous) time is $\frac{1}{2}[1 + \pi]$. For details and an elegant interpretation of the resulting rendezvous strategies in terms of "kissing circles," see the original article of Anderson and Fekete (2001).*

We believe that the assumption that the object is uniformly distributed on the circle can be relaxed by using techniques like those in Section 14.1, so that the distribution only needs to be *close to uniform*, in some sense. There are two differences with the analysis given there: first, the searcher's starting point on the circle is not specified, but rather a choice variable; second, the searcher may opt to travel along a chord to reach higher density parts of the circle quicker.

### 18.3.2 Asymmetric information

We now consider the scenario in which the initial position of one of the players (say, I) is known to the other (II). This asymmetric information version of rendezvous has been analyzed on the line in Section 17.6. Here we will make the additional assumption that the initial distance $D$ between the players is known (say, $D = 2$) and that the angle of the line connecting them is uniformly distributed. Taking I's initial location as our origin, this means that II is initially placed uniformly along the circle of radius 2 centered at the origin. Point capture will be required for a meeting to have occurred. We assume that the players have a common notion of clockwise direction.

Two strategies immediately present themselves. First is the WFM (Wait For Mommy), in which Player I waits at the origin until Player II reaches him at time 2, that is also the expected meeting time. Alternatively, Player I could move out to a randomly chosen point on the unit circle at $t = 1$, while Player II moves to the unique point on this circle he can reach by time $t = 1$. Thus, starting at time 1, they could choose to play the asymmetric rendezvous problem on the unit circle. Corollary 13.12 establishes that the optimal strategy pair for this problem is OP-DIR, in which the players go at unit speed around the circle in opposite directions. The expected meeting time for this strategy on the unit circle (not the circle of circumference 1) is $\pi/2$. Hence the expected meeting time for the combined strategy is $1 + \pi/2 \doteq 2.5708$, and this is worse than the expected time of 2 for WFM.

More generally, the players could arrive at the unit circle at time 1 (as above) but subsequently move along a circle of radius $r(t)$ for $t \geq 1$, where $r(1) = 1$ and $r'(t) < 0$ for $t > 1$. Depending on $r(t)$ and $r'(t)$ the players have some angular speed left to them that they can use to go in opposite angular directions (while keeping their common distance to 0 at $r(t)$). The determination of the optimal function $r(t)$ is an optimal control problem. By numerically approximating the solution of the differential equation corresponding to this control problem, Anderson and Fekete (2001) show that the rendezvous value for this problem is about 1.97.

### 18.3.3 Asymmetric speed and detection radius

Thomas and Hulme (1997) consider a rendezvous problem in which a fast helicopter and a slow hiker wish to rendezvous. The helicopter can see the hiker only if they are very close, but the hiker can see (or hear) the helicopter from a further distance. Unlike the other two-player rendezvous problems considered in this book, some new information may be received by a player during the course of play. In particular, the hiker may see the helicopter from a long distance. So a strategy for the hiker must take this into consideration.

This problem presents complexities not seen in other versions of rendezvous and needs to be attacked in a different manner, via simulation techniques. The search region was divided into a hexagonal grid. Three strategies for the hiker were considered:

1. (S) Stationary
2. (RW) Random Walk: At each move, go randomly to one of the six neighboring nodes, independently of the helicopter's motion or location.
3. (HWA) Head When Audible: Use (S) until helicopter is detected, then head toward helicopter along the estimated bearing.

Two strategies for the helicopter were considered:

1. (SCAN) Scanning Search: Move back and forth in parallel lines spaced according to the helicopter's detection radius.
2. (EXSQ) Expanding Square Search: This path starts at the center of the search area and spirals out.

The distribution of meeting times was determined through simulations. Against a stationary hiker (S), both SCAN and EXSQ perform identically, since the search is exhaustive and non-overlapping. Against a randomly walking (RW) hiker, EXSQ has a higher probability of finding the hike early, but eventually this reverses and SCAN does better. Against an intelligent hiker who uses HWA the simulations indicate that both SCAN and EXSQ are similar early on but that eventually SCAN does significantly better.

The simulations also reveal that a type of role reversal may be useful, where the helicopter tries to drag the hiker toward it. In this approach the helicopter adopts a decreasing spiral path. The reader should go to the original article (Thomas and Hulme, 1997) to see the detailed results of these simulations.

# Appendix A

# A Minimax Theorem for Zero-Sum Games

This appendix presents a general minimax theorem of the authors (Alpern and Gal, 1988), which establishes that every search game of the type considered in Book I has a value. Recalling equation (2.3), this means that if we consider all mixed search strategies $s$ and all mixed hider strategies $h$, we have

$$\inf_s \sup_h c\,(s, h) = \sup_h \inf_s c\,(s, h) = v \text{ (value of game)}, \tag{A.1}$$

where the cost function $c$ is the expected capture time. Furthermore the inf on the left is a min, which means there is an optimal mixed strategy for the searcher. (The hider may have only $\varepsilon$-optimal strategies.) The original and most direct approach to this result is given in Gal (1980, app. 1); the result given here is an extension Gal's result and applies to a wider class of zero-sum games. Although it is customary in the literature to have the first player be the maximizer, we follow the approach taken in the main text, where the first player (searcher) is the minimizer.

We will show that a result of the type (A.1) holds for a wide class of zero-sum two-person normal-form games which includes the search games considered in Book I. A zero-sum two person normal form game is characterized by a cost function $c$ (sometimes called a payoff function), which Player 1 (the searcher, in our games) wants to minimize and Player 2 (the hider, in our games) wants to maximize. The cost function $c$ is initially given in terms of the pure strategies of the two players, $c : \mathcal{S} \times \mathcal{H} \to R$. Here $\mathcal{S}$ is the set of pure strategies for the minimizing Player 1 and $\mathcal{H}$ is the set of pure strategies of the maximizing Player 2. We use the letters $\mathcal{S}$ and $\mathcal{H}$ only to make the identification with search games clear, although this model applies to any game, and the strategy sets and payoff may be very different for other games.

Keeping this framework in mind, we first consider games in which the number of pure strategies is finite for each player; say, $\mathcal{S} = S_1, \ldots, S_m$ and $\mathcal{H} = H_1, \ldots, H_n$. (This would be the case, for example, for the search game with immobile hider on a tree, where $H_j$ denote the leaves of the tree and the $S_i$ denote the ways of searching all the leaves in a given order.) For notational simplicity, we denote the payoffs corresponding to pairs of pure strategies in the matrix notation

$$C_{ij} = c(S_i, H_j).$$

We may view the $i, j$-th entry $C_{ij}$ of the $m \times n$ game matrix $C$ as the amount the first player (minimizer) pays the second player (maximizer). It may happen that the matrix $C$ will have a saddle point, i.e., an element $C_{\hat{i}\hat{j}}$ such that

$$\max_{1 \leq i \leq m} C_{i\hat{j}} = C_{\hat{i}\hat{j}} = \min_{1 \leq j \leq n} C_{\hat{i}j}.$$

In this case, the game would be in a state of equilibrium if the first player chooses his $\hat{i}$-th pure strategy and the second player chooses his $\hat{j}$-th pure strategy. The preceding strategies would be optimal, and thus this game could be solved using only pure strategies. Usually, however, such a saddle point does not exist because even in the simplest games (e.g., matching pennies, where $C = \begin{pmatrix} -1 & +1 \\ +1 & -1 \end{pmatrix}$), a player is at a disadvantage if he always uses the same pure strategy. The fact that usually there exists no optimal strategy in the set of pure strategies has led to the idea of using mixed strategies. Each player, instead of selecting a specific pure strategy, may choose an element from the set of his pure strategies according to a predetermined set of probabilities. Mixed strategies for the first and the second players will be denoted by $s = (x_1, x_2, \ldots, x_m)$ and $h = (y_1, y_2, \ldots, y_n)$, respectively, where $x_i, i = 1, \ldots, m$ is the probability that the first player will choose his $i$-th pure strategy $S_i$, and $y_j, j = 1 \ldots, n$ is the probability that the second player will choose his $j$-th pure strategy $H_j$.

When the first player plays a mixed strategy $s$ and the second player a mixed strategy $h$, the expected cost is given by the function

$$c(s, h) = \sum_i \sum_j C_{ij} x_i y_j$$

The fundamental theorem of two-person zero-sum finite games is due to Von Neumann (see Von Neumann and Morgenstern, 1953). It states that

$$\min_s \max_h c(s, h) = \max_h \min_s c(s, h).$$

This minimax value of $c$ is called the *value* of the game and denoted by $v$. An equivalent result is that there exists a pair of mixed strategies $\bar{s}$ and $\bar{h}$ such that

$$\min_s c(s, \bar{h}) = c(\bar{s}, \bar{h})(= v) = \max_h c(\bar{s}, h).$$

Thus, $c$ has a saddle point, or in other words, if the first player chooses the mixed strategy $\bar{s}$ and the second player uses the mixed strategy $\bar{h}$ then each of them can guarantee an expected payoff of $v$. Thus, $(\bar{s}, \bar{h})$ is a pair of optimal strategies and the game has a solution in mixed strategies. (e.g., in the matching pennies game, $\bar{s} = (\frac{1}{2}, \frac{1}{2})$, $\bar{h} = (\frac{1}{2}, \frac{1}{2})$, and $v = 0$).

The situation is more complicated if the game has an infinite number of pure strategies. In this case, the mixed strategies are probability measures on the set of pure strategies. Such a game is said to have a value $v$ if, for any positive $\varepsilon$, the first player has a (mixed) strategy $s_\varepsilon$ that limits him to an expected loss of at most $(1+\varepsilon)v$ and the second player has a (mixed) strategy $h_\varepsilon$ that guarantees him an expected payoff of at least $(1-\varepsilon)v$. If one of the players has an infinite number of pure strategies while the other player has only a finite number of pure strategies, then the game has a value. However, if both players have an infinite number of pure strategies, then the existence of a value is not assured. (For details see Luce and Raiffa, 1957, app. 7.)

In the search games considered in the main text, both players have an infinite number of pure strategies. Nevertheless, Gal (1980, app. 1) has proved that any search game has a value. Using the more general formulation of Alpern and Gal (1988) the minimax theorem can be stated as follows.

**Theorem A.1** *Let X be a compact Hausdorff space and (Y, A) a measurable space. Let $f : X \times Y$ be a measurable function that is bounded below and lower semicontinuous on X for all fixed y in Y. Let M be any convex set of probability measures (mixed strategies) on (Y, A) and B(X) the regular probability measures on X. Then*

$$\min_{\beta\in B(X)} \sup_{\gamma\in M} \iint f(x, y)\, d\beta\, d\gamma = \sup_{\gamma\in M} \min_{\beta\in B(X)} \iint f(x, y)\, d\beta\, d\gamma.$$

For our search trajectories we use the topology of uniform convergence for any finite interval. Since any $S \in \mathcal{S}$ is Lipshitz (with constant 1) it follows from the Ascoli theorem that $\mathcal{S}$ is compact. Under that topology $\mathcal{S}$ is also Hausdorff (two distinct trajectories always have disjoint neighborhoods). Since the capture time $C(S, H)$ can only "jump" down, it easily follows that it is *lower semicontinuous* ($C(\lim) \le \lim C(\cdot)$) in each of its arguments (see Gal, 1980). Thus, we can use the general minimax theorem A.1 and obtain

$$\min_s \sup_h \int C(S, H)\, d(s \times h) = \sup_h \min_s \int C(S, H)\, d(s \times h)$$

so that the search games considered in our book always have a value and an optimal search strategy. Note that the lower semicontinuity implies that $C(S, H)$ is Borel measurable, in both arguments. Thus, the above integral

$$\int C(S, H)\, d(s \times h)(\equiv c(s, h))$$

is well defined.

# Appendix B

# Theory of Alternating Search

In some cases, particularly for the undirected circle and line, the asymmetric rendezvous problem can be reduced to a problem in which two searchers act as a team to locate a stationary hidden object. The object is placed in one of two disjoint regions, with a searcher in each. The searchers can only move one at a time (hence the term "alternating"), and each has a maximum speed of 2. To make the space of search strategies closed, we also allow the limiting case in which the searchers can move simultaneously with a combined speed of 2.

In this appendix we review the relevant results of Alpern and Howard (2000) on the theory of *alternating search at two locations*. We will need only the special case of this theory in which each search region is a ray (a copy of $[0, \infty)$) and each searcher starts at the end (labeled 0). We will also assume that the object is equally likely to be on either ray.

Let $F_i(t), i = 1, 2$, denote the probability that the object is placed in the interval $[0, t]$ on ray $i$, given that it is somewhere on ray $i$. Thus $F_1$ and $F_2$ are probability distributions. The alternation of the two searchers may be described by a rule $\alpha$ that determines when each of the searchers is moving (at speed 2). In this interpretation, we may let $\alpha(t)$ denote twice the total time up to $t$ that Searcher 1 has been moving. In this case, Searcher 1 will have covered the interval $[0, \alpha(t)]$ on ray 1, while Searcher 2 will have covered the interval $[0, 2t - \alpha(t)]$ on ray 2. Hence the object will have been found by time $t$ with probability

$$F_\alpha(t) = \tfrac{1}{2}[F_1(\alpha(t)) + F_2(2t - \alpha(t))].$$

The expected time to find the object will be $\int_0^\infty t \, dF_\alpha(t)$, and the least expected time is denoted by

$$v(F_1, F_2) = \min_\alpha \int_0^\infty t \, dF_\alpha(t).$$

An $\alpha$ for which the minimum is attained is called an *optimal alternation rule*. In order to justify the *existence* of the minimum, we need to consider a wider class of alternation rules $\alpha$. The rule described above, with intervals of alternating motion of the searchers, could be described by a continuous piecewise linear function $\alpha : [0, \infty) \to [0, \infty)$, which has intervals of slope 2 (when Searcher 1 moves) and intervals of slope 0 (when Searcher 2 moves). As these intervals get smaller and smaller, the motions of the two searchers become, in the limit, simultaneous motions. This is easily formalized as follows. We define an alternation rule $\alpha$ to be any increasing function with maximum slope (Lipshitz constant) 2, with $\alpha(0) = 0$. The set of all such alternation rules $\mathcal{A}$ is compact with respect to the topology of uniform convergence on compact intervals. Since the integral $\int_0^\infty t \, dF_\alpha(t)$ is lower semicontinuous in $\alpha$ with respect to this topology, the existence of the minimum is established. An alternation rule $\alpha$ has a derivative almost everywhere. We interpret the position of searcher 1 to be $\alpha(t)$ and that of searcher 2 to be $2t - \alpha(t)$. Their speeds are $\alpha'(t)$ and $2 - \alpha'(t)$, which sum to 2.

In this presentation of the problem, the distributions $F_1$ and $F_2$ are given. This will be the case when the two regions are each rays, and the starting points are the ends of the rays. However, in general (e.g., alternating search on two circles) there may be *many* ways of searching a given region (say region 1), and each way will determining a *different* distribution $F_1$. We note that in the general case it is not necessarily true that $v(F_1, F_2)$ is minimized by taking two distributions $F_1^*$ and $F_2^*$ that individually minimize the expected time to find the object if it is certain to be in that region. However, we can say the following.

**Lemma B.1** *Suppose that $\hat{F}_1$ dominates $F_1$ in the sense that it is at least as large for all $t$. Then for any distribution $F_2$, we have*

$$v(\hat{F}_1, F_2) \le v(F_1, F_2).$$

This inequality holds simply because we can write the moment $\int_0^\infty t \, dF_\alpha(t)$ as $\int_0^\infty (1 - F_\alpha(t)) \, dt$, and the latter will not increase when $F_1$ is replaced by $\hat{F}_1$. The importance of this observation is that if there is a strategy (like the Columbus strategy in Chapter 14) or a family of search strategies for a given region that dominates any strategy, we may assume that a strategy of this type is used.

We now list some results obtained in Alpern and Howard (2000) on optimal alternation rules $\alpha$ corresponding to various assumptions on the distributions $F_1$ and $F_2$.

**Theorem B.2** *Suppose an optimal alternation rule searches in the two rays alternately in consecutive time intervals. Then the interval on line $i$ for which $F_i$ has the higher average density is searched first.*

**Theorem B.3** *If one of the distributions $F_i$ is convex on an interval, then there is an optimal alternation rule for which this interval (on line $i$) is traversed at maximum speed without interruption.*

**Theorem B.4** *If one of the distributions $F_i$ is constant on an interval (which has density zero and cannot contain the object), then **any** optimal alternation rule traverses this interval (on line $i$) at maximum speed without interruption.*

For the next result we assume that $F_1 = F_2$ and we denote by $\bar{F}$ the *concavification* of $F$, that is, the smallest concave function satisfying $\bar{F}(x) \ge F(x)$ for all $x$. The following presents a complete characterization of the optimal solution in terms of the $F_i$ and their concavifications.

**Theorem B.5** *Suppose the object has a common conditional distribution $F$ on both rays.*

1. *If $\bar{F}(x) > F(x)$ for all $a < x < b$, $\bar{F}(a) = F(a)$, $\bar{F}(b) = F(b)$, then there is an optimal alternation rule $\alpha$ which satisfies either $\alpha' = 2$ on $(a, (a+b)/2)$ and $\alpha' = 0$ on $((a+b)/2, b)$, or the reverse. That is, the intervals $[a, b]$ on the two rays are searched consecutively at speed 2.*
2. *If $F$ is concave then the constant alternation rule $\alpha' = 1$, which searches both rays in parallel, is optimal. If it is strictly concave, then this strategy is uniquely optimal.*

This work may be thought of as an extension of the Gittens Index to continuous time and general distribution. See Gittens (1989).

## B.1 Arbitrary Regions

The analysis given above for alternating search on two rays can be useful in solving the more general problem of searching two arbitrary regions $Q_1$ and $Q_2$. In the general problem, there is a searcher starting at a point $q_i \in Q_i, i = 1, 2$. The two searchers move subject to a maximum combined speed of 2. If searcher $i$ moves with maximum speed all the time, there are many possible cumulative distribution functions $F_i$ of capture time given that the hidden object lies in $Q_i$. The minimum time required for the two searchers to find the object is given by

$$\min_{F_1, F_2} v(F_1, F_2).$$

# Appendix C

# Rendezvous-Evasion Problems

The situations described in Books I and II are either search-evasion or rendezvous search. The novel element that was introduced in the article (Alpern and Gal, 2002) on which this appendix is based, is an uncertainty regarding the motives of the lost agent: he may be a mobile or immobile hider (evader) as in Book I, or he may be a cooperating rendezvouser who shares the same aim as the searcher, as in Book II. We assume that the probability $p$ of cooperation is known to the searcher and to the agent. In any given search context (search space and player motions) $G$, we obtain a continuous family $\Gamma_G(p)$ of search problems, $0 \leq p \leq 1$, where $\Gamma_G(0)$ is a search game with mobile or immobile hider and $\Gamma_G(1)$ is an asymmetric rendezvous problem. We obtain a unique continuous value function $V_G(p)$ that gives the least expected value of $T$ for the given (cooperation) probability $p$ that the target wants to be found. As we shall show, this uncertainty regarding *a priori* agent motives affects both the paths (strategies) chosen by the searcher and the paths chosen by the (cooperating or evading) agent. One might say that here we introduce the game theoretic notion of incomplete information into the theory of search. Our formalization of the cooperative portion of the problem will be that of asymmetric rendezvous search, where the searcher and agent may agree on the roles (paths) they will take in the event that the agent gets lost and wants to be found. For example a mother (the searcher) may tell her child (the agent) what to do in this event, knowing, however, that this instruction may be disregarded if the child does not want to be found. The child will follow these (rendezvous) instructions if he wants to be found and may use the knowledge of these instructions in deciding on an evasion strategy if he does not.

This uncertainty as to the agent's aims is common, for example, when a teenager is reported missing to the police by the parents. In such cases the police usually ascribe a probability ($1 - p$ in our notation) that the teenage is not lost or abducted but rather a runaway who does not want to be found. Parents usually complain that the police overestimate this probability, and indeed this is a subject of some controversy, as in the following passage from Clancy (1999):

> Bannister had gone to a local police station to make a report in person... From a police detective... he'd heard "Look, its only been a few

weeks . . . She's probably alive and healthy somewhere, and ninety nine out of a hundred of these cases turn out to be a girl who just wanted to spread her wings [an evader]." Not his Mary, Bannister had replied.

Another example of such incomplete information search is found in the novel *Hunt for Red October* (Clancy, 1995) where a Russian submarine of that name becomes lost to Soviet command. The difficulties faced by the Russian search effort are exactly those formalized in this article, as they are uncertain whether the sub is indeed lost (their first assumption) or is defecting to the West (as they gradually come to believe). The current SETI (Search for Extraterrestrial Intelligence) Project is based on an implicit assumption regarding not only the existence of EI but of a sufficiently large value of $p$. In the SETI context, $T$ would be the first time when transmitter and listener are on the same frequency and the message is understood to be nonrandom. Search and Rescue operations (seeking lost hikers, for example) also make judgements about $p$ in determining where to look first.

It would be natural for the searcher to behave in the following way: If $p$ is relatively high, then he assumes a cooperative agent and hence he first goes to the agreed meeting point, switching into a search-evasion mode if the agent is not found at this location. If $p$ is small, then the searcher assumes an evading agent right from the start and acts accordingly. This situation is characterized by a threshold value for $p$ that separates between the above two strategy types. For some cases this natural strategy is indeed optimal. For example:

**Immobile agent on a tree** This can represent the following search problem. A mother drives a teenager son to a roadside drop-off point $O$. From there he will hike in a large park. If it rains, he will go to one of a group of huts (none of which are at $O$), where he will wait, while the mother drives back to $O$ and, covered with proper raingear, begins a search. The Mother tells the child which hut he should go to if it rains, knowing full well that if he is enjoying himself at that time he will disregard her instructions. The huts and the drop-off point $O$ are all connected by a network of paths that forms a tree.

In this model the search domain will be a tree $Q$ with a distinguished node called $O$ where the searcher must start his search. (In keeping with the search and rendezvous literature, we will refer to the agents as "he" even when they adopt the "wait for mummy" strategy, or when the motivating example includes females.) Each edge of $Q$ has a certain length, and the sum of these is called the length of $Q$, denoted $\mu(Q)$. For simplicity we will normalize the length of the given tree so that $\mu = 1$. The agent in this problem is immobile; he simply picks a node of the tree other than $O$ and stays there. The searcher moves along the tree at unit speed until he reaches the node chosen by the agent, aiming to minimize the expected time $\hat{T}$ to reach this node. With probability $p$, the agent is a cooperator (or rendezvouser) who also wishes to minimize $\hat{T}$; and with probability $1 - p$ the agent is an evader who wishes to maximize $\hat{T}$. Simple domination arguments are sufficient to show that a cooperator will always chose a node adjacent to the starting node $O$, while the evader will always chose a terminal node of $Q$. We will denote by $v(Q, O, p)$ the minimal expected time for this specific problem.

Denote by $h^*(q, o)$ an optimal hiding distribution for a tree $q$ with root $o$ (see Section 3.3). We have the following theorem.

**Theorem C.1** *Let $Q$ be a unit-length tree with searcher starting node $O$. Let $X$ be any node at minimum distance $d_X$ to $O$ and let $Z$ be any node that determines a subtree $Q_Z$ of maximum total length $\mu(Q_Z) = w_Z$. Denote $Q \backslash Q_Z$ by $Q_{-Z}$. Then the value $v(Q, O, p)$ of the rendezvous-evasion problem on $(Q, O)$ with cooperation probability $p$ is given by*

$$v(Q, O, p) = \begin{cases} 1 - pw_Z, & \text{if } p \le \dfrac{d_X}{1 - w_Z}, \\ 1 - p + d_X, & \text{if } p \ge \dfrac{d_X}{1 - w_Z}. \end{cases}$$

*For $p \le d_X/(1 - w_Z)$, $Z$ is an optimal rendezvous strategy for the cooperator, a traversal of a minimal tour of $Q$ from $O$ (equiprobably in either direction) is optimal for the searcher, and an optimal strategy for the evader is to adopt $h^*(Q_{-Z}, O)$ with probability $\mu(Q_{-Z})/(1 - p)$ and $h^*(Q_Z, O)$ with the complementary probability. For $p \ge d_X/(1 - w_Z)$, $X$ is an optimal rendezvous strategy for the cooperator, a move to $X$ followed by a minimal tour of $Q$ from $X$ (equiprobably in either direction) is optimal of the searcher, and the strategy $h^*(Q, X)$ is optimal for the evader.*

However, this situation, of the above ($p$) threshold type strategy being optimal, does not always hold.

**Rendezvous-evasion in two cells** The searcher and agent are initially placed at time $t = 0$ into distinct cells, which we (and the players), respectively, call 1 and 2. At each time $t = 1, 2, 3, \ldots$, searcher and agent locate in cell $x_t$ and $y_t$, and meet at the first time when they are in the same cell. We may assume (by relabeling in each period) that the cooperator strategy is $(2, 2, \ldots)$. For $p$ close to 1, the optimal searcher strategy is $(2, b, b, \ldots)$ and the optimal evader strategy is $(1, b, b, \ldots)$, where $b$ denote the Bernoulli strategy of picking either location equiprobably and independently of previous choices. For $p$ near 0 the optimal strategy for both searcher and evader is $(b, b, b, \ldots)$. These strategy pairs give respective expected meeting times of $p(1) + (1-p)(1+2) = 3 - 2p$ and 2, which give the same time 2 when $p = 1/2$. However in this intermediate case ($p = 1/2$) both searcher strategies are dominated by the mixed strategy $0.9(2, b, b, \ldots) + 0.1(1, 2, b, b, \ldots)$. Since a best response of the evader is $(1, b, b, \ldots)$, this mixed searcher strategy ensures that the expected meeting time is no more than $(0.9)(3-2p) + (0.1)(2p + (1-p)) = 1.95 < 2$, for $p = 1/2$. This analysis shows that for intermediate values of $p$ the optimal searcher strategy is not optimal for either extreme case ($p = 0$ or 1). The determination of optimal strategies for all $p$, and the corresponding optimal expected meeting time $v(p)$, seems an interesting problem.

Alpern and Gal (2002) also analyze mobile agent on the line (the agent is assumed to have a smaller speed than the searcher) and suggest additional other interesting problems for further research:

**Rendezvous-evasion in $n$ cells** The problem is the same as for two cells, except that for $n > 2$ the two players can never achieve a common labeling of the cells. For $p = 1$ the cooperator stays still while the searcher picks a random permutation of the remaining cells. For $p = 0$ both searcher and evader should move randomly. What is the solution for general $p$? See Anderson and Weber (1990) for a discussion of a related problem.

**Mobile rendezvous-evasion on the circle** The searcher and agent are placed randomly (uniform probability density) on a circle of unit circumference and can both move with unit speed. For the search game $p = 0$ it has been shown Alpern (1974), and Zelikin (1972), that the optimal search strategy is "cohatu." This is short for "coin half tour": at times $i/2$, $i = 0, 1, 2, \ldots$, go half way around the circle equiprobably in either direction. All evader responses give the same expected meeting time of $v(0) = 3/4$. For the rendezvous problem $p = 1$, posed in Alpern (1995), the optimal strategy is for the searcher to move clockwise while the cooperator moves counterclockwise (for a proof see Section 14.2), with expected meeting time $v(1) = 1/4$. For $p$ near to 1 the searcher should use "clockwise-cohatu": this means clockwise for $t \leq 1/2$ (by which time he will have met the cooperator) and then use cohatu. Any evader strategy that goes clockwise for $t \leq 1/2$ is an optimal response. Thus this strategy has an expected meeting time of

$$p\left(\frac{1}{4}\right) + (1-p)\left(\frac{1}{2} + \frac{3}{4}\right) = \frac{5-4p}{4}.$$

The two strategies, cohatu and clockwise-cohatu, give the same expected time of 3/4 when $p = 1/2$. However, neither is optimal in this case ($p = 1/2$), since both are dominated by a mixture of 0.9 (clockwise-cohatu) and 0.1 (counterclockwise–clockwise-cohatu). The latter strategy goes in the indicated directions for times (0, 1/2) and (1/2, 1) and then in random directions. Against this mixture, an optimal evader must go clockwise for the first half time unit. Thus, if clockwise-cohatu is used, then the expected meeting time is the same (3/4) as for either of the two searcher strategies already calculated. However, in the case that counterclockwise–clockwise-cohatu is used, the searcher does better: The cooperator will be found in the same expected time $1/2 + v(1) = 3/4$ (no improvement), but the evader will be found in the smaller expected time $1/4$. This analysis is similar to that for two cells (there are two directions on the circle), and as in that problem the solution for intermediate values of $p$ would be interesting. This analysis assumes that the searcher and agent have a common notion of direction around the circle. The problem is also well defined (but distinct) without this assumption.

**Immobile rendezvous-evasion on networks** To what extent does the dichotomy of search strategies found for trees apply to other networks?

**Two-sided ambiguity of aims** In the analysis given in the article, only the aims of the agent are uncertain. How should the problem be analyzed if the agent has a probability $p_A$ of wanting to minimize $T$ (otherwise he wants to maximize) and the searcher has a similar probability $p_S$?

We believe that the introduction into search theory of models with uncertain target motives can present many interesting new problems into the area that may stimulate further research.

# Bibliography

[1] Agin, N. I. (1967). The application of game theory to ASW detection problems, Math. Rep., Princeton, New Jersey (September).

[2] Ahlswede, R. and Wegener, I. (1987). *Search Problems*. Wiley.

[3] Alpern, S. (1974). The search game with mobile hider on the circle. In *Differential Games and Control Theory* (E. O Roxin, P. T. Liu, and R. L. Sternberg, eds), pp. 181–200, Dekker, New York.

[4] Alpern, S. (1976). *Hide and Seek Games*. Seminar, Institut fur Hohere Studien, Wien, 26 July.

[5] Alpern, S. (1985). Search for point in interval, with high–low feedback. *Math. Proc. Cambridge Philos. Soc.* 98, no. 3, 569–578.

[6] Alpern, S. (1992). Infiltration games on arbitrary graphs. *J. Math. Anal. Appl.* 163, no. 1, 286–288.

[7] Alpern, S. (1995). The rendezvous search problem. LSE CDAM Research Report, 53, (1993). *SIAM J. Control Optim.* 33 (1995), 673–683.

[8] Alpern, S. (1998). Aisle Miles (letter). In *The Last Word* (New Scientist). Oxford Univ. Press, 37–38.

[9] Alpern, S. (2000). Asymmetric rendezvous search on the circle. *Dynam. Control* 10, 33–45.

[10] Alpern, S. (2001). Rendezvous search in one and more dimensions. The London School of Economics Mathematics Preprint Series, CDAM–2001–05.

[11] Alpern, S. (2002a). Rendezvous search: A personal perspective. *Oper. Res.* 50, no. 5 (scheduled).

[12] Alpern, S. (2002b). Rendezvous Search on Labelled Networks. *Naval Res. Logist.* 49, 256–274.

[13] Alpern, S. and Asic, M. (1985). The search value of a network. *Networks* 15, no. 2, 229–238.

[14] Alpern, S. and Asic, M. (1986). Ambush strategies in search games on graphs. *SIAM J. Control Optim.* 24, no. 1, 66–75.

[15] Alpern, S., Baston, V., and Essegaier, S. (1999). Rendezvous search on a graph. *J. Appl. Probab.* 36, no. 1, 223–231.

[16] Alpern, S. and Beck, A. (1997). Rendezvous search on the line with bounded resources: expected time minimization. *Eur. J. Oper. Res.* 101, 588–597.

[17] Alpern, S. and Beck, A. (1999a). Asymmetric rendezvous on the line is a double linear search problem. *Math. Oper. Res.* 24, no. 3, 604–618.

[18] Alpern, S. and Beck, A. (1999b). Rendezvous search on the line with bounded resources: Maximizing the probability of meeting. *Oper. Res.* 47, no. 6, 849–861.

[19] Alpern, S. and Beck, A. (2000). Pure strategy asymmetric rendezvous on the line with an unknown initial distance. *Oper. Res.* 48, no. 3, 1–4.

[20] Alpern, S. and Gal, S. (1988). A mixed strategy minimax theorem without compactness. *SIAM J. Control Optim.* 26, 1357–1361.

[21] Alpern, S. and Gal, S. (1995). Rendezvous search on the line with distinguishable players. *SIAM J. Control Optim.* 33, 1270–1276.

[22] Alpern, S. and Gal, S. (2002). Search for an agent who may or may not want to be found. *Oper. Res.* 50, no. 2, 311–323.

[23] Alpern, S. and Howard, J. V. (2000). Alternating search at two locations. *Dynam. Control* 10, 319–339.

[24] Alpern, S. and Lim, W. S. (1998). The symmetric rendezvous-evasion game. *SIAM J. Control Optim.* 36, no. 3, 948–959.

[25] Alpern, S. and Lim, W. S. (2002). Rendezvous of Three Agents on the Line. *Naval Res. Logist.* 49, 244–255.

[26] Alpern, S. and Pikounis, M. (2000). The telephone coordination game. *Game Theory Appl.* 5, 1–10.

[27] Alpern, S. and Reyniers, D. J. (1994). The rendezvous and coordinated search problems. *Proc. IEEE*, 33rd CD, no. 1, 513–517.

[28] Alpern, S. and Snower, D. (1987). Inventories as an information gathering device. World Bank, Development Research Department, Rep. no. DRD267.

[29] Alpern, S. and Snower, D. (1988a). Production decisions under uncertain demand: the high–low search approach. Disc. paper no. 223, Centre for Economic Policy Research, London.

[30] Alpern, S. and Snower, D. (1988b). A search model of optimal pricing and production. *J. Eng. Costs Prod. Econom.* 15, 279–284.

[31] Alpern, S. and Snower, D. (1988c). High–Low search in product and labour markets. *American Economic Review* (AEA Papers and Proceedings) 78, no. 2, 356–362.

[32] Alpern, S. and Snower, D. (1989). Geometric search theory and demand uncertainty. *J. Eng. Costs Prod. Econom.* 17, 245–251.

[33] Alpern, S. and Snower, D. (1991). Unemployment through "learning from experience." In *Issues in Contemporary Economics*, Vol. 2, (M. Nerlove, ed.), MacMillan, 42–74.

[34] Alspach, B. (1981). The search for long paths and cycles in vertex-transitive graphs and digraphs. In *Combinatorial Mathematics VIII*, Lecture Notes #884, (K. L. McAveney, ed.), Springer, Berlin, 14–22.

[35] Anderson, E. J. and Aramendia, M. A. (1990). The search game on a network with immobile hider. *Networks* 20, no. 7, 817–844.

[36] Anderson, E. J. and Aramendia, M. (1992). A linear programming approach to the search game on a network with mobile hider. *SIAM J. Control Optim.* 30, no. 3, 675–694.

[37] Anderson, E. J. and Essegaier, S. (1995). Rendezvous search on the line with indistinguishable players. *SIAM J. Control Optim.* 33, 1637–1642.

[38] Anderson, E. J. and Fekete, S. (2001). Two dimensional rendezvous search. *Oper. Res.* 49, 107–188.

[39] Anderson, E. J. and Weber, R. R. (1990). The rendezvous problem on discrete locations. *J. Appl. Probab.* 28, 839–851.

[40] Arnold, R. D. (1962). Avoidance in one dimension: A continuous-matrix game. Operations Evaluation Group, Office of Chief of Naval Operations, Washington, D.C.-OEG IRM-10 (AD 277 843), 14 pp.

[41] Aubin, T. and Jouventin, P. (1998). Cocktail-party effect in king penguin colonies. *Proc. R. Soc. London* B 265, 1665–1673.

[42] Auger, J. M. (1991a). On discrete games of infiltration. Differential games – developments in modelling and computation (Espoo, 1990), 144–150, *Lect. Notes Control Inf. Sci.* 156, Springer.

[43] Auger, J. M. (1991b). An infiltration game on *k* arcs. *Naval Res. Logist.* 38, no. 4, 511–529.

[44] Avetisyan, V. V. and Melikyan, T. T. (1999a). Guaranteed control of the search for a moving object in a rectangular domain. *Izv. Akad. Nauk Teor. Sist. Upr.*, no. 1, 58–66.

[45] Avetisyan, V. V. and Melikyan, T. T. (1999b). On the problem of guaranteed search for a moving object in a rectangular domain. *Izv. Akad. Nauk Teor. Sist. Upr.*, no. 2, 31–39.

[46] Baeza-Yates, R. A., Culberson, J. C., and Rawlins, G. J. E. (1993). Searching in the plane. *Inf. Comput.* 106, no. 2, 234–252.

[47] Baston, V. J. (1999). Two rendezvous search problems on the line. *Naval Res. Logist.* 46, 335–340.

[48] Baston, V. J. and Beck, A. (1995). Generalizations in the linear search problem. *Isr. J. Math.* 90, no. 1–3, 301–323.

[49] Baston, V. J. and Bostock, F. A. (1985). A high–low search game on the unit interval. *Math. Proc. Cambridge Philos. Soc.* 97, no. 2, 345–348.

[50] Baston, V. J. and Bostock, F. A. (1991). A basic searchlight game. *J. Optim. Theory Appl.* 71, 47–66.

[51] Baston, V. J. and Gal, S. (1998). Rendezvous on the line when the players' initial distance is given by an unknown probability distribution. *SIAM J. Control Optim.* 36, no. 6, 1880–1889.

[52] Baston, V. J. and Gal, S. (2001). Rendezvous search when marks are left at the starting point. *Naval Res. Logist.* 48, 722–731.

[53] Baston, V. J. and Garnaev, A. Y. (1996). A fast infiltration game on *n* arcs. *Naval Res. Logist.* 43, no. 4, 481–489.

[54] Baston, V. J. and Garnaev, A. Y. (2000). A search game with a Protector. *Naval Res. Logist.* 47, no. 2, 85–96.

[55] Beck, A. (1964). On the linear search Problem. *Naval Res. Logist.* 2, 221–228.

[56] Beck, A. (1965). More on the linear search problem. *Isr. J. Math.* 3, 61–70.

[57] Beck, A. and Beck, M. (1984). Son of the linear search problem. *Naval Res. Logist.* 48, no. 2–3, 109–122.

[58] Beck, A. and Beck, M. (1986). The linear search problem rides again. *Naval Res. Logist.* 53, no. 3, 365–372.

[59] Beck, A. and Beck, M. (1992). The revenge of the linear search problem. *Naval Res. Logist.* 30, no. 1, 112–122.

[60] Beck, A. and Newman, D. J. (1970). Yet more on the linear search problem. *Naval Res. Logist.* 8, 419–429.

[61] Beck, A. and Warren. P. (1973). The return of the linear search problem. *Isr. J. Math.* 14, 169–183.

[62] Bellman, R. (1956). Minimization problem. *Bull. Am. Math. Soc.* 62, 270.

[63] Bellman, R. (1963). An optimal search problem. *SIAM Rev.* 5, 274.

[64] Beltrami, E. J. (1963). The density of coverage by random patrols. NATO Conf. Appl. Oper. Res. to the Search and Detection of Submarines (J. M. Dobbie and G. R. Lindsey, eds) 1, 131–148.

[65] Ben-Asher, Y., Farchi, E., and Newman, I. (1999). *SIAM J. Comput.* 28, no. 6, 2090–2102.

[66] Benkoski, S., Monticino, M., and Weisinger, J. (1991). A survey of the search theory literature. *Naval Res. Logist.* 38, 469–494.

[67] Berman, P., Charikar, M., and Karpinski, M. (2000). On-line load balancing for related machines. *J. Algorithms* 35, no. 1, 108–121.

[68] Berge, C. (1973). *Graphs and Hypergraphs.* North-Holland, Publ Amsterdam.

[69] Blum, A., Chalasani, P., Coppersmith, D., Pulleyblank, W. R., Raghavan, P., and Sudan, M. (1994). The minimum latency problem. *Proc. Ann. Sympos. Theory Comput.* (STOC), 163–171.

[70] Blum, A., Raghavan, P., and Shieber, B. (1997). Navigating in unfamiliar geometric terrain. *SIAM J. Comput.* 26, no. 1, 110–137.

[71] Borodin, A and El-Yaniv, R. (1998). *Online Computation and Competitive Analysis.* Cambridge University Press, New York.

[72] Bostock, F. A. (1984). On a discrete search problem on three arcs. *SIAM J. Algebraic Discrete Methods* 5, no. 1, 94–100.

[73] Bram, J. (1963). A 2-player N-region search game. Operations Evaluation Group, Office of Chief of Naval Operations, Washington, D. C., OEG IRM–31 (AD 402 914).

[74] Breiman, L. (1968). *Probability*. Addison-Wesley, Reading, Massachusetts.

[75] Brocker, C. A. and Lopez-Ortiz, A. (1999). Position-independent street searching. Algorithms and data structures (Vancouver, BC, 1999), 241–252, *Lect. Notes Comput. Sci.* 1663, Springer.

[76] Brocker, C. A. and Schuierer, S. (1999). Searching rectilinear streets completely. Algorithms and data structures (Vancouver, BC, 1999), 98–109, *Lect. Notes Comput. Sci.* 1663, Springer.

[77] Bruss, T. F. and Robertson, J. B. (1988). A survey of the linear-search problem. *Math. Sci.* 13, no. 2, 75–89.

[78] Burley, W. R. (1996). Traversing layered graphs using the work function algorithm. *J. Algorithms* 20, no. 3, 479–511.

[79] Burningham, N. (1999). Personal communication (email).

[80] Chester, C. and Tutuncu, R. H. (2001). Rendezvous search on finite domains. Preprint, Carnegie Mellon University.

[81] Christofides, N. (1975). *Graph Theory: An Algorithmic Approach*. Academic Press, New York.

[82] Chudnovsky, D. V. and Cudnovsky, G. V. (eds) (1989). *Search Theory: Some Recent Developments*. Marcel Dekker, New York.

[83] Clancy, T. (1995). *The Hunt For Red October*. Harper Collins, New York.

[84] Clancy, T. (1999). *Rainbow Six*. Berkley Books, New York.

[85] Crawford, V. P. and Haller, H. (1990). Learning how to cooperate: Optimal play in repeated coordination games. *Econometrica* 58, no. 3, 571–596.

[86] Danskin, J. M. (1968). A helicopter versus submarine game. *Oper. Res.* 16, 509–517.

[87] DiCairano, C. (1999). Asymmetric rendezvous search on the line. Ph.D. Dissertation, Faculty of Mathematical Studies, University of Southampton.

[88] Djemai, I., Meyhofer, R., and Casas, J. (2000). Geometric games between a host and a parasitoid. *Am. Naturalist* 156, no. 3, 257–265.

[89] Dobbie, J. M. (1968). A survey of search theory. *Oper. Res.* 16, 525–537.

[90] Dudek, J. and Jenkin, M. (2000). *Computational Principles of Mobile Robots*. Cambridge University Press, 294 pp.

[91] Edmonds, J. (1965). The Chinese postman problem. *Bull. Oper. Res. Soc. Am.* 13, Suppl. 1, B-73.

[92] Edmonds, J. and Johnson, E. L. (1973). Matching Euler tours and the Chinese postman problem. *Math. Program.* 5, 88–124.

[93] Eiselt, H. A., Gendreau, M., and Laporte, G. (1995). Arc routing problems. I. The Chinese postman problem. *Oper. Res.* 43, 231–242.

[94] Essegaier, S. (1993). The rendezvous problem: A survey on a pure coordination game of imperfect information, *LSE CDAM Research Report*, 61.

[95] Even, S. (1979). *Graph Algorithms*, Computer Science Press, Rockville, MD.

[96] Fan, K. (1953). Minimax theorems. *Proc. Natl. Acad. Sci. U.S.A.* 39, 42–47.

[97] Feller, W. (1971). *An Introduction to Probability Theory and its Applications*, Vol. II, 2nd ed. Wiley, New York.

[98] Fershtman, C. and Rubinstein, A. (1997). A simple model of equilibrium in search procedures. *J. Econom. Theory* 72, no. 2, 432–441.

[99] Finch, S. (1999). Lost in a forest, http://www.mathsoft.com/asolve/forest/forest.html.

[100] Foley, R. D., Hill, T. P. and Spruill, M. C. (1991). Linear search with bounded resources. *Naval Res. Logist.* 38, 555–565.

[101] Fomin, F. V. (1999). Note on a helicopter search problem on graphs. *Discrete Appl. Math. 95, 241–249.*

[102] Foreman, J. G. (1974). The princess and the monster on the circle. In *Differential Games and Control Theory* (E. O. Roxin, P. T. Liu, and R. L. Sternberg, eds), pp. 231–240. Dekker, New York.

[103] Foreman, J. G. (1977). Differential search game with mobile hider. *SIAM J. Control Optim.* 15, 841–856.

[104] Foster, K. (1999). Personal communication (email).

[105] Franck, W. (1965). On an optimal search problem. *SIAM Rev.* 7, 503–512.

[106] Fraenkel, A. S. (1970). Economic traversal of labyrinths. *Math. Mag.* 43, 125–130.

[107] Fraenkel, A. S. (1971). Economic traversal of labyrinths. *Math. Mag.* 44, 12.

[108] Fristedt, B. and Heath, D. (1974). Searching for a particle on the real line. *Adv. Appl. Probab.* 6 , 79–102.

[109] Gal, S. (1972). A general search Game. *Isr. J. Math.* 12, 32–45.

[110] Gal, S. (1974a). Minimax Solutions for linear search problems. *SIAM J. Appl. Math.* 27, 17–30.

[111] Gal, S. (1974b). A discrete search game. *SIAM J. Appl. Math.* 27, 641–648.

[112] Gal, S. (1978). A stochastic search game. *SIAM J. Appl. Math.* 31, 205–210.

[113] Gal, S. (1979). Search games with mobile and immobile hider. *SIAM J. Control Optim.* 17, 99–122.

[114] Gal, S. (1980). *Search Games*. Academic Press, New York.

[115] Gal, S. (1989). Continuous search games, Chapter 3 of *Search Theory: Some Recent Developments* D. V. Chudnovsky and G. V. Chudnovsky (eds), Marcel Dekker, New York.

[116] Gal, S. (1999). Rendezvous search on the line. *Oper. Res.* 47, 974–976.

[117] Gal, S. (2000). On the optimality of a simple strategy for searching graphs. *Int. J. Game Theory* 29, 533–542.

[118] Gal, S. and Anderson, E. J. (1990). Search in a maze. *Probab. Eng. Inf. Sci.* 4, 311–318.

[119] Gal, S. and Chazan, D. (1976). On the optimality of the exponential functions for some minimax problems. *SIAM J. Appl. Math.* 30, 324–348.

[120] Garnaev, A. Y. (1991). Search game in a rectangle. *J. Optim. Theory Appl.* 69, no. 3, 531–542.

[121] Garnaev, A. Y. (1992). A remark on the princess and monster search game. *Int. J. Game Theory* 20, no. 3, 269–276.

[122] Garnaev, A. Y. (2000). *Search Games and Other Applications of Game Theory*. Springer-Verlag, Berlin.

[123] Garnaev, A. Y. and Garnaeva, G. Y. (1996). An infiltration game on a circumference. *Nova J. Math. Game Theory Algebra* 5, no. 3, 235–240.

[124] Garnaev, A. Y., Garnaeva, G., and Goutal, P. (1997). On the infiltration game. *Int. J. Game Theory* 26, no. 2, 215–221.

[125] Garnaev, A. Y. and Sedykh, L. G. (1990). A search game of the searcher–searcher type. *Soviet J. Automat. Inform. Sci.* 23, no. 3, 70–73.

[126] Gilbert, E. (1962). Games of identification and convergence. *SIAM Rev.* 4, 16–24.

[127] Gittins, J. C. (1979). Bandit processes and allocation indices, *JR Stat. Soc. Ser. B*, 41, 148–177.

[128] Gittins, J. C. (1989). *Multi-armed Bandit Allocation Indices*. Wiley.

[129] Gittins, J. C. and Roberts, D. M. (1979). The search for an intelligent evader concealed in one of an arbitrary number of regions. *Naval Res. Logist. Quart.* 26, no. 4, 651–666.

[130] Glicksberg, I. L. (1950). Minimax theorem for upper and lower semi-continuous payoffs. Rand Corp., Res. Memo. RM-478.

[131] Gluss, B. (1961a). An alternative solution to the lost at sea problem. *Naval Res. Logist. Quart.* 8, 117–121.

[132] Gluss, B. (1961b). The minimax path in a search for a circle in the plane. *Naval Res. Logist. Quart.* 8, 357–360.

[133] Goemans, M. and Kleinberg, J. (1998). An improved approximation ratio for the minimum latency problem. Networks and matroids, Sequencing and scheduling. *Math. Program.* 82, Ser. B, 111–124.

[134] Golumbic, M. C. (1987). A general method for avoiding cycling in networks. *Inf. Process. Lett. 24, 251–253.*

[135] Griffiths, R. and Tiwari, B. (1995). Sex of the last wild Spix's macaw. *Nature* 275, 454.

[136] Gross, O. (1955). A search problem due to Bellman. Rand Corp., RM-1603 (AD 87 962).

[137] Gross, O. (1964). The rendezvous value of a metric space. In *Advances in Game Theory* (M. Dresher, L. S. Shapley, and A. Tucker, eds), pp. 49–53. Princeton Univ. Press, Princeton, New Jersey.

[138] Haley, B. K. and Stone, L. D. (eds) (1980). *Search Theory and Applications*, Plenum Press, New York.

[139] Halmos, P. (1950). *Measure Theory*. D. Van Nostrand Company Inc., New York.

[140] Hammar, M., Nilsson, B. J., and Schuierer, S. (1999). Parallel searching on *m* rays. STACS 99 (Trier), 132–142, *Lect. Notes Comput. Sci.*, 1563, Springer.

[141] Harary, F. (1972). *Graph Theory*. Adison-Wesley, London.

[142] Hardy, G. H., Littlewood, J. E., and Po1ya, G. (1952). *Inequalities*. Cambridge Univ. Press, London and New York.

[143] Hassin, R. and Tamir, A. (1992). Minimal length curves that are not embeddable in an open planar set: The problem of a lost swimmer with a compass. *SIAM J. Control Optim.* 30, 695–703.

[144] Hipke, C., Icking, C., Klein, R., and Langetepe, E. (1999). How to find a point on a line within a fixed distance. *Discrete Appl. Math.* 93, 67–73.

[145] Houdebine, A. J. (1963). Etude sur l''efficacité d''un barrage anti-sous-marin constitue'. NATO Conf. Appl. Oper. Res. to the Search and Detection of Submarines (J. M. Dobbie and G. R. Lindsey, eds) 1, 103–120.

[146] Howard, J. V. (1999). Rendezvous search on the interval and circle. *Oper. Res.* 47, no. 4, 550–558.

[147] Iida, K. (1992). Studies in the Optimal Search Plan. *Lect. Notes Statist.* (J. Berger et al., eds) 70, Springer-Verlag.

[148] Isaacs, R. (1965). *Differential Games*. Wiley, New York.

[149] Isbell, J. R. (1957). An optimal search pattern. *Naval Res. Logist. Quart.* 4, 357–359.

[150] Jaillet, P. and Stafford, M. (2001). Online searching. *Oper. Res.* 49, no. 4, 501–515.

[151] Johnson, S. M. (1964). A search game. In *Advances in Game Theory* (M. Dresher, L. S. Shaply, and A. W. Tucker, eds), pp. 39–48. Princeton Univ. Press, Princeton, New Jersey.

[152] Juniper, T. and Yamashita, C. (1990). The conservation of Spix's macaw. *Oryx* 24, 224–228.

[153] Kao, M., Reif, J. H., Tate, and Stephen, R. (1996). Searching in an unknown environment: an optimal randomized algorithm for the cow-path problem. *Inf. Comput.* 131, no. 1, 63–79.

[154] Kella, O. (1993). Star search – a different show. *Isr. J. Math.* 81, no. 1–2, 145–159.

[155] Kelley, J. L. (1955). *General Topology*. Van Nostrand Reinhold, Princeton, New Jersey.

[156] Kikuta, K. (1990). A hide and seek game with traveling cost. *J. Oper. Res. Soc. Japan 33*, no. 2, 168–187.

[157] Kikuta, K. (1991). A search game with traveling cost. *J. Oper. Res. Soc. Japan* 34, no. 4, 365–382.

[158] Kikuta, K. (1995). A search game with traveling cost on a tree. *J. Oper. Res. Soc. Japan* 38, no. 1, 70–88.

[159] Kikuta, K. and Ruckle, W. H. (1994). Initial point search on weighted trees. *Naval Res. Logist.* 41, no. 6, 821–831.

[160] Kikuta, K. and Ruckle, W. H. (1997). Accumulation games. I. Noisy Search. *J. Optim. Theory Appl.* 94, no. 2, 395–408.

[161] Kikuta, K. and Ruckle, W. H. (2000). Accumulation games. II. Preprint.

[162] Koopman, B. O. (1946). Search and screening. Operations Evaluation Group Rep. no. 56, Center for Naval Analysis, Rosslyn, Virginia.

[163] Koopman, B. O. (1956). The theory of search Part II. Target detection. *Oper. Res.* 4, 503–531.

[164] Koopman, B. O. (1980). *Search and Screening: General Principles with Historical Applications*, Pergamon Press, New York.

[165] Koutsoupias, E., Papadimitriou, C., and Yannakakis, M. (1996). Searching a fixed graph. Automata, languages and programming (*Lect. Notes Comput. Sci.* 1099) 280–289, Springer.

[166] Kramarz, F. (1996). Dynamic focal points in N-person coordination games. *Theory Decis.* 40, 277–313.

[167] Lalley, S. P. (1988). A one-dimensional infiltration game. *Naval Res. Logist.* 35, no. 5, 441–446.

[168] Lalley, S. P. and Robbins, H. E. (1988a). Stochastic search in a square and on a torus. *Statistical Decision Theory and Related Topics*, IV, Vol. 2 (West Lafayette, Ind., 1986), 145–161, Springer.

[169] Lalley, S. P. and Robbins, H. E. (1988b). Stochastic search in a convex region. *Probab. Theory Relat. Fields* 77, no. 1, 99–116.

[170] Langford, E. S. (1973). A continuous submarine versus submarine game. *Naval Res. Logist. Quart.* 20, 405-417.

[171] Lawler, E. L. (1976). *Combinatorial Optimization: Networks and Matroids*. Holt, New York.

[172] Lim, W. S. (1997). A rendezvous-evasion game on discrete locations with joint randomization. *Adv. Appl. Probab.* 29, 1004–1017.

[173] Lim, W. S. (1999). Rendezvous-evasion as a multi-stage game with observed actions. In *Advances in Dynamic Games and Applications* (Annals of the International Society of Dynamic Games, Vol. 5), (Filar et al., ed.), Birkhauser, 137–150.

[174] Lim, W. S. and Alpern, S. (1996). Minimax rendezvous search on the line. *SIAM J. Control Optim.* 34, 1650–1665.

[175] Lim, W. S., Alpern, S., and Beck, A. (1997). Rendezvous search on the line with more than two players. *Oper. Res.* 45, no. 3, 357–364.

[176] Lindsey, G. R. (1968). Interception strategy based on intermittent information. *Oper. Res.* 16, 489–508.

[177] Lopez-Ortiz, A. and Schuierer, S. (1997). Position-independent near optimal searching and on-line recognition in star polygon. Algorithms and Data Structures. *Lect. Notes Comput. Sci.* 1272, 284–296.

[178] Lopez-Ortiz, A. and Schuierer, S. (1998). The ultimate strategy to search on $m$ rays? In (W. L. Hsu and M. Y. Kao, eds), Proc. 4th Int. Conf. on Computing and Combinatorics, *Lecture Notes in Computer Science*, Vol. 1449, 75–84.

[179] Lovasz, L. (1970). Unsolved problem II. In *Combinatorial Structures and their Applications* (Proc. Calgary Int. Conf. on Combinatorial Structures and their Applications, 1969) (R. Guy et al., ed.), Gordon and Breach, New York.

[180] Lucas, E. (1882) *Recreations Mathematique*. Paris.

[181] Luce, R. D. and Raiffa, H. (1957). *Games and Decisions*. Wiley, New York.

[182] McCabe, B. J. (1974). Searching for a one-dimensional random walker. *J. Appl. Probab.* 11, 86–93.

[183] McNab, A. (1994). *Bravo Two Zero*. Corgi Books, London.

[184] Matthews, R. (1995). How to find your children when they are lost. *Sunday Telegraph* (London, U.K.). Science page, 26 March.

[185] Matula, D. (1964) A peridic optimal search. *Am. Math. Mon.* 71, 15–21.

[186] Megiddo, N. and Hakimi, S. L. (1978). Pursuing mobile hider in a graph. The Center for Mathematical Studies in Economics and Management Sciences, Northwestern University, Evanston, Illinois, Disc. paper no. 360, 25 pp.

[187] Megiddo, N., Hakimi, S. L., Garey, M. R., Johnson, D. S., and Papadimitriou, C. H. (1988). The complexity of searching a graph. *J. Assoc. Comput. Mach.* 35, no. 1, 18–44.

[188] Meyhofer, R. , Casas, J., and Dorn, S. (1997). Vibration-Mediated interactions in a host-parasitoid system. *Proc. R. Soc. London, Ser. B-Biol. Sci.* 264: (1379) 161–266.

[189] Murakami, S. (1976). A dichotomous search with travel cost. *J. Oper. Res. Soc. Japan* 19, 245–254.

[190] Nakai, T. (1986). A search game with one object and two searchers. *J. Appl. Probab.* 23, no. 3, 696–707.

[191] Nakai, T. (1990). A preemptive detection game. *J. Inf. Optim. Sci.* 11, no. 1, 1–15.

[192] Neuts, M. F. (1963). A multistage search game. *J. SIAM* 11, 502–507.

[193] Norris, R. C. (1962). Studies in search for a conscious evader. Lincoln Lab., MIT Tech. Rep. no. 279 (AD 294 832), 134 pp.

[194] Papadimitriou, C. H. and Yannakakis, M. (1991). Shortest paths without a map. *Theoret. Comput. Sci.* 84, no. 1, Algorithms Automat. Complexity Games, 127–150.

[195] Parsons, I. D. (1978a). Pursuit-evasion in a graph. In *Theory and Application of Graphs* (Y. Alavi and P. R. Lick, eds). Springer-Verlag, Berlin.

[196] Parsons, T. D. (1978b). The search number of a connected graph. Proc. 9th Southwestern Conf. on Combinatorics, Graph Theory, and Computing, Boca Raton, Florida.

[197] Pavillon, C. (1963). Un probleme de recherche sur zone. NATO Conf. Appl. Oper. Res. to the Search and Detection of Submarines (J. M. Dobbic and G. R. Lindsey, eds), 1, 90–102.

[198] Pavlovic, L. (1993a). Search game on an odd number of arcs with immobile hider. *Yugosl. J. Oper. Res.* 3, no. 1, 11–19.

[199] Pavlovic, L. (1993b). Search game on the union of $n$ identical graphs joined by one or two points. *Yugosl. J. Oper. Res.* 3, no. 1, 3–10.

[200] Pavlovic, L. (1995a). Search for an infiltrator. *Naval Res. Logist.* 42, no. 1, 15–26.

[201] Pavlovic, L. (1995b). A search game on the union of graphs with immobile hider. *Naval Res. Logist.* 42, no. 8, 1177–1189.

[202] Pavlovic, L. (2002). More on the search for an infiltrator. *Naval Res. Logist.* 49, no. 1, 1–14.

[203] Ponssard, J.-P. (1994). Formalisation des connaissances, apprentisage organisationnel et rationalité interactive (Chapitre 7). In *Analyse Economiques des Conventions*, (A. Orléan, ed.), P.U.F., Paris.

[204] Reijnierse J. H. (1995). Games, graphs and algorithms. Ph. D Thesis, University of Nijmegen, The Netherlands.

[205] Reijnierse, J. H. and Potters, J. A. M. (1993a). Search games with immobile hider. *Int. J. Game Theory* 21, 385–394.

[206] Reijnierse, J. H. and Potters, J. A. M. (1993b). Private communication.

[207] Reyniers, D. (1988). A high–low search model of inventories with time delay. *J. Eng. Costs Prod. Econom.* 15, 417–422.

[208] Reyniers, D. (1989). Interactive high–low search: The case of lost sales. *JORS* 40, no. 8, 769–780.

[209] Reyniers, D. (1990). A high–low search algorithm for a newsboy problem with delayed information feedback. *Oper. Res.* 38, 838–846.

[210] Reyniers, D. (1992). Information and rationality asymmetries in a simple high–low search wage model. *Econom. Lett.* 38, 479–486.

[211] Roberts, D. M. and Gittins, J. C. (1978). The search for an intelligent evader: Strategies for searcher and evader in the two region problem. *Naval Res. Logist. Quart.* 25, 95–106.

[212] Ross, S. M. (1969) A problem in optimal search and stop. *Oper. Res.* 984–992.

[213] Ross, S. M. (1983). *An Introduction to Stochastic Dynamic Programming*. Academic Press, New York.

[214] Roy, N. and Dudek, G. (2001). Collaborative robot exploration and rendezvous: Algorithms, performance bounds and observations. *Autonomous Robots* 11, 117–136.

[215] Ruckle, W. H. (1983a). *Geometric Games and Their Applications*. Pitman, Boston.

[216] Ruckle, W. H. (1983b). Pursuit on a cyclic graph. *Int. J. Game Theory* 10, 91–99.

[217] Ruckle, W. H. and Kikuta, K. (2000). Continuous accumulation games in continuous regions. *J. Optim. Theory Appl.* 106, no. 3, 581–601.

[218] Rudin, W. (1973). *Functional Analysis*. McGraw-Hill, New York.

[219] Schelling, T. (1960). *The Strategy of Conflict*. Harvard Univ. Press, Cambridge.

[220] Schuierer, S. (1999). On-line searching in simple polygons. Sensor Based Intelligent Robots. *Lect. Notes Artif. Intell.* 1724, 220–239, Springer.

[221] Schuierer, S. (2001). Lower bounds in on-line geometric searching. *Comp. Geom. Theor. Appl.* 18, no. 1, 37–53.

[222] Stewart, T. J. (1981). A two-cell model of search for an evading target. *Eur. J. Oper. Res.* 8, no. 4, 369–378.

[223] Stone, L. D. (1989). *Theory of Optimal Search*, 2nd ed. Operations Research Society of America, Arlington, VA.

[224] Tarry, G. (1895). La problem des labyrinths. *Nouvelles Annales de Mathematiques* 14, 187.

[225] Tetali, P. and Winkler, P. (1993). Simultaneous reversible Markov chains. *Bolyai Society Mathematical Studies, Combinatorics, Paul Erdos is Eighty*, Vol. 1, 433–451.

[226] Thomas, L. C. and Hulme, P. B. (1997). Searching for targets who want to be found. *J. OR Soc.* 48, Issue 1, 44–50.

[227] Thomas, L. C. and Pikounis, M. (2001). Many player rendezvous search: stick together or split and meet? *Naval Res. Logist.* 48, no. 8, 710–721.

[228] Thomas, L. C. and Washburn, A. R. (1991). Dynamic search games. *Oper. Res.* 39, no. 3, 415–422.

[229] *The Times*, Issue 65084, 13 October 1994, London.

[230] Von Neumann, J. and Morgenstern, O. (1953). *Theory of Games and Economic Behavior*. Princeton Univ. Press, Princeton, New Jersey.

[231] Von Stengel, B. and Werchner, R. (1997). Complexity of searching an immobile hider in a graph. *Discrete Appl. Math.* 78, 235–249.

[232] Wagner, I. A., Lindenbaum, M., and Bruckstein, A. M. (1996). Smell as a computational resource – a lesson we can learn from the ant. Israel Symposium on Theory of Computing and Systems (Jerusalem, 1996), 219–230, IEEE Comput. Soc. Press, Los Alamitos, CA.

[233] Wagner, I. A., Lindenbaum, M., and Bruckstein, A. M. (1998). Efficiently searching a graph by a smell-oriented vertex process. Artificial intelligence and mathematics, VIII (Fort Lauderdale, FL, 1998). *Ann. Math. Artif. Intell.* 24, no. 1–4, 211–223.

[234] Wagner, I. A., Lindenbaum, M., and Bruckstein, A. M. (2000). Ants: Agents on networks, trees, and subgraphs, *Future Generation Computer Systems* 16, 915–926.

[235] Washburn, A. R. (1981). An upper bound useful in optimizing search for a moving target. *Oper. Res.* 29, 1227–1230.

[236] Washburn, A. R. (1995). Dynamic programming and the backpacker's linear search problem. *J. Comput. Appl. Math.* 60, no. 3, 357–365.

[237] Wilson, D. J. (1972). Isaacs' princess and monster game on the circle. *J. Optim. Theory Appl.* 9, 265–288.

[238] Wilson, D. J. (1977). Differential games with no information, *SIAM J. Control Optim.* 15, 233–246.

[239] Worsham, R. H. (1974). A discrete game with a mobile hide. In *Differential Games and Control Theory* (E. O. Resin, P. T. Liu, and R. L. Sternberg, eds), pp. 201–230. Dekker, New York.

[240] Zelikin, M. 1. (1972). On a differential game with incomplete information. *Soviet Math. Dokl.* 13, 228–231.

# Index